The RFF Reader

IN ENVIRONMENTAL AND RESOURCE MANAGEMENT

EDITED BY WALLACE E. OATES

University of Maryland
and
Resources for the Future

D1190050

RESOURCES FOR THE FUTURE • WASHINGTON, DC

Printed in the United States of America

Published by Resources for the Future
1616 P Street, NW, Washington, DC 20036–1400

Library of Congress Cataloging-in-Publication Data

The RFF reader in environmental and resource management : edited by Wallace E. Oates
     p.   cm.
    Includes bibliographical references and index.
    ISBN 0–915707–96–9 (pbk.)
    1. Environmental policy. I. Oates, Wallace E.
GE170.R447 1999
363.7—dc21   98-50115
                                   CIP

The paper in this book meets the guidelines for permanence and durability of the Committee on Production Guidelines for Book Longevity of the Council on Library Resources.

This book was copyedited and coordinated by Betsy Kulamer and designed by Diane Kelly, Kelly Design.

Resources for the Future

About Resources for the Future

Resources for the Future is an independent nonprofit organization engaged in research and public education with issues concerning natural resources and the environment. Established in 1952, RFF provides knowledge that will help people to make better decisions about the conservation and use of such resources and the preservation of environmental quality.

RFF has pioneered the extension and sharpening of methods of economic analysis to meet the special needs of the fields of natural resources and the environment. Its scholars analyze issues involving forests, water, energy, minerals, transportation, sustainable development, and air pollution. They also examine, from the perspectives of economics and other disciplines, such topics as government regulation, risk, ecosystems and biodiversity, climate, Superfund, technology, and outer space.

Through the work of its scholars, RFF provides independent analysis to decisionmakers and the public. It publishes the findings of their research as books and in other formats, and communicates their work through conferences, seminars, workshops, and briefings. In serving as a source of new ideas and as an honest broker on matters of policy and governance, RFF is committed to elevating the public debate about natural resources and the environment.

Contents

Part 3. Environmental Regulation

Part 4. Environmental Federalism

Part 5. Resource Management

Part 6. Biodiversity

Part 7. Environmental Justice

Part 8. Global Climate Change

Part 9. Sustainable Development

Part 10. Environmental Problems in Developing and Transitional Countries

Foreword

In May 1959, Resources for the Future began publication of its quarterly periodical *Resources* to address an urgent need. The new organization had assembled an impressive roster of researchers and laid the foundations for an innovative long-term project to assess the adequacy of America's resource base to meet the expected needs of the future. But policymakers, professors, journalists, and the many people involved in or dependent upon the natural resource industries were all making clear their need for near-term, well-focused, easy-to-understand reports on RFF's progress. What were our projects beginning to indicate? Were they discerning any apparent trends or surprises, whether to date or yet to come?

Resources, then as now, attempts to meet this demand. Think of it as RFF's principal means of communicating the results of its research in a timely and understandable way to all interested parties. This is so even while its outreach work is both supplemented and complemented by RFF's books and reports and, most recently, its Web sites, where *Resources* also has been available since the summer of 1996. The pioneering research that has won RFF wide acclaim over the decades has resulted largely from the free and open-ended fashion in which it has been conducted. Yet in a very real way, our contribution to the public's understanding of complex issues can be viewed equally through the success of this modest periodical. In publishing *The RFF Reader in Environmental and Resource Management,* we are demonstrating as well as facilitating its impact in one vitally important arena—academia.

Resources is widely used in colleges and universities. We know this to be true from periodic reader surveys, from the number of reprint requests we receive from schools, from the publication's numerous citations in scholarly literature, from the comments we receive from younger researchers joining the RFF staff, and from the direct testimony of faculty members—such as Wallace E. Oates—who populate our research family. The short, relatively jargon-free, and timely research articles appearing in *Resources* have proved to be an ideal classroom tool in the hands of a skilled instructor.

The long-standing success and usefulness of this tradition of writing to our educational mission has driven it in recent years beyond the pages of *Resources:* one can now find such work throughout RFF's web sites. The lead essay in this volume's section on climate change is just one example, having appeared initially in *Weathervane,* RFF's online forum of analysis and commentary on U.S. and global policy initiatives related to climate change.

We try to fulfill our educational mission in other ways as well. We run a weekly seminar series on Wednesdays where the environmental policy community can hear about new technical developments in applied and analytical fields of research, as well hear RFF researchers and other presenters talk more directly about policy design and implementation. We also sponsor conferences and lectures to

similar ends; Robert Solow's contribution to this volume was written and delivered as an RFF anniversary lecture.

The articles collected here span more than a dozen years. They address issues fundamental to environmental and resource management—such as the roles of science, regulation, and benefit-cost analysis—as well as cutting-edge topics such as global climate change, environmental justice, and sustainable development. RFF as an institution has been at the forefront in addressing such issues, and our publications and communications efforts have sought to elevate and enrich public understanding of key environmental and resource topics. We hope that our decision to publish this collection of writings will advance that effort.

Paul R. Portney
President, Resources for the Future

Preface

In teaching a variety of courses over the years that address environmental and natural resource issues, I have found myself making increasing and good use of various sets of readings from *Resources,* a quarterly publication of Resources for the Future (RFF). *Resources* provides admirably succinct, yet quite incisive, nontechnical treatments of analytical techniques for the evaluation of environmental programs, useful findings from specific pieces of research, and balanced assessments of important policy issues. I have found that these short pieces can supplement the more lengthy (and sometimes ponderous) discussion in standard textbooks or, in certain instances, to substitute for them altogether.

The articles in *Resources* address a broad audience. Many of them can serve the needs of students very effectively. But they are also of real interest to the larger community concerned with environmental and resource management. In particular, the papers make clear the important contribution that an economic perspective can bring both to our understanding of environmental issues and the design of measures for protection of the environment. For these reasons, it seemed to me useful to pull together a substantial selection of the best of these articles in a single place to provide ready access both for courses in environmental studies and for a more general audience on environmental matters.

Nearly all the articles in this collection (in fact, all but two) come directly from various issues of *Resources.* They are concise—typically about five or six pages in length. Some are methodological in orientation: they examine, for example, specific issues in benefit-cost analysis (such as discounting the future) or the role of science in environmental policy determination. Others address particular policy issues, such as Superfund or global climate change. But in all cases, the papers make their points in a nontechnical and readily accessible way. I have found some of them especially helpful in clarifying certain subtle but important matters, such as the valuation of life for purposes of evaluating environmental programs.

In two instances, I have included RFF pieces from outside *Resources*. The first is the seminal and widely cited essay by Robert Solow on sustainable growth. The second is a specially commissioned background piece by John Anderson on global climate change; this essay was prompted by the major importance of the issue and the attention that it is receiving in the global community.

For organizational purposes, I have grouped the papers by general topics in a sequence that roughly parallels the structure of some university courses. The papers appear in their original form from *Resources*. However, we offered the authors an opportunity to provide a short update, where appropriate, and/or to append to their selections some suggested further reading. In some instances where there have been important events, analytical advances, or new empirical findings, the reader will thus find an "update box" that notes briefly important material available subsequent to the writing of the article. Finally, I want to call attention to the rich source of current information and analysis of environmental and resource issues that is available on the RFF web site (http://www.rff.org).

For those unfamiliar with the organization, RFF is an independent, nonprofit institution, located in Washington, DC, that undertakes research and public education on natural resource and environmental issues. Through the years, RFF has been the source of much fundamental and important research, both basic and applied, into environmental management. Indeed, the first recipients of the prestigious Volvo Prize in Environmental Science were Allen Kneese and John Krutilla of RFF for their path-breaking work in the analysis of environmental problems. I have been most grateful through the years to have a continuing association with RFF as a University Fellow and am delighted to have this opportunity to assemble a collection of RFF readings that I hope many will find as useful as I have.

Wallace E. Oates
University of Maryland and
Resources for the Future

An Economic Perspective on Environmental and Resource Management
An Introduction

Wallace E. Oates

The central concern of economics is the allocation of scarce resources. The basic problem is one of using our limited means to provide an array of goods and services that satisfies peoples' preferences in an efficient and equitable manner. It doesn't require much reflection to realize that our environmental resources are scarce. Clean air and water, the diversity of species, and perhaps even a stable global climate are clearly not available in unlimited supply, irrespective of human activities. Perhaps economics has something useful to say about the management of our environment.

This is indeed the case. I shall suggest here that economics has three basic and important messages for environmental protection. First, economic analysis makes a compelling case for the proposition that an unfettered market system will generate excessive pollution. A market system, in a sense, "overuses" many of the services provided by the environment, resulting in excessive environmental degradation. Thus, economics makes a basic and persuasive case for the need for public intervention in the form of environmental regulation.

Second, economics provides some guidance for the setting of standards for environmental quality. It provides one approach to answering the question: How clean should the environment be? In fact, this approach is simply a straightforward application of the general economic principle that any activity should be extended to the point where the marginal benefits equal the marginal costs.

And third, once we have determined the standards or targets for environmental quality (and even if—incidentally—this determination is made irrespective of marginal analysis), economics has some important things to say about the design of the policy instruments to achieve these standards. In particular, economic analysis suggests how we can structure policy measures so as to realize our environmental goals in the most effective and least-cost ways.

In this introduction, I want to explain and explore these three ideas, for nearly all the papers in this volume draw on this conceptual framework in one way or another. In fact, it is the purpose of this volume to show how basic economic analysis can help us to understand the causes of environmental degradation and to design policies to protect and improve the environment.

Free Markets and the Environment

Economists have a deep appreciation of the market system. Guiding the individualized choices of both consumers and producers, a system of markets has the capacity to channel our limited resources into their most highly valued uses. In pursuing their own gain, individuals (as Adam Smith put it) "are led by an invisible hand" to promote the social good.

Markets generate and make use of a set of prices that serve as signals to indicate the value (or cost) of resources to potential users. Any activity that imposes a cost on society by using up some of its scarce resources must come with a price, where that price equals the social cost. For most goods and services ("private goods" as economists call them), the market forces of supply and demand generate a market price that directs the use of resources into their most highly valued employment.

There are, however, circumstances where a market price may not emerge to guide individual decisions. This is often the case for various forms of environmentally damaging activities. In the first half of this century at Cambridge University, A.C. Pigou set forth the basic economic perspective on unpriced goods (encompassing pollution) in his famous book, *The Economics of Welfare*. Since Pigou, many later economists have developed Pigou's

insights with greater care and rigor. But the basic idea is straightforward and compelling: the absence of an appropriate price for certain scarce resources (such as clean air and water) leads to their excessive use and results in what is called "market failure."

The source of this failure is what economists term an *externality*. A good example is the classic case of the producer whose factory spreads smoke over an adjacent neighborhood. The producer imposes a real cost in the form of dirty air, but this cost is "external" to the firm. The producer does not bear the cost of the pollution it creates as it does for the labor, capital, and raw materials that it employs. The price of labor and such materials induces the firm to economize on their use, but there is no such incentive to control smoke emissions and thereby conserve clean air. The point is simply that whenever a scarce resource comes free of charge (as is typically the case with our limited stocks of clean air and water), it is virtually certain to be used to excess.

Many of our environmental resources are unprotected by the appropriate prices that would constrain their use. From this perspective, it is hardly surprising to find that the environment is overused and abused. A market system simply doesn't allocate the use of these resources properly. In sum, economics makes a clear and powerful argument for public intervention to correct market failure with respect to many kinds of environmental resources. Markets may work well in guiding the production of private goods, but they cannot be relied upon to provide the proper levels of "social goods" (like environmental services).*

But if we can't rely on markets to "manage" our environmental resources, what principles should we employ to regulate their use? To this I turn next.

---

*Two qualifications are worthy of note here. First, there are some cases where voluntary negotiations among a small group of affected parties can effectively resolve an externality. Such cases are the subject of the famous paper by Ronald Coase, "The Problem of Social Cost," *Journal of Law and Economics* (October 1960), pp. 1–44. Although the Coasian treatment has gotten considerable attention

The Setting of Standards for Environmental Quality

There is a basic economic principle that indicates the efficient level of any economic activity: extend that activity to the level at which the benefits from an additional "unit" of the activity equal the costs. Economists sometimes refer to these extra units as *incremental* or *marginal*. Thus, the condition for the economically correct level of any activity can be stated simply as the equality of marginal benefits with marginal cost.

The intuition here is straightforward. So long as higher levels of a particular service yield additional (marginal) benefits that exceed the additional (marginal) costs, we are obviously better off providing the additional units of the service than not providing them. But it clearly would not be a good idea to go past the point where marginal benefits equal marginal cost, for any units past this point would cost more than they are worth (i.e., marginal cost would exceed marginal benefits for such units).

The moral of this exercise for environmental policy, from the standpoint of economic efficieny at least, is that we should set standards for environmental quality such that the benefits at the margin from tightening the standards further exactly equal the marginal cost of pollution abatement (often called *marginal abatement cost*). Note that this implies that, in general, the economically efficient

---

in the literature, its applicability remains limited. The major environmental problems, including, for example, urban air pollution and water pollution, cannot be addressed through voluntary market mechanisms; they require public regulatory intervention. Second (and closely related), one might envision a system where markets were supplemented by a perfect tort system such that polluters were fully liable for the costs of any damages that they impose on society. Such liability could, in principle, provide the needed incentives for efficient levels of pollution abatement. Liability rules, in fact, have an important role to play in environmental protection, but the various imperfections inherent in any practicable legal system for environmental protection leave a large role for other forms of regulatory measures.

level of pollution is not zero. The cost of a perfectly pure environment would simply be too much to make it worthwhile. Economics is, in a sense, rather pragmatic when it comes to setting standards for things. It recognizes that tradeoffs and compromises are needed in order to make the best use of our limited resources.

While this guidance for the setting of environmental standards seems straightforward and sensible in principle, it is not so easy to implement. Consider, for example, the case of improved air quality. In considering the benefits from a proposal to introduce a more stringent standard for clean air, we must somehow quantify the improvement in well being that comes with the associated reduced levels of illness and increased longevity. And this, along with any other benefits (such as reduced damage to materials and wildlife), must be compared to the additional abatement costs that the measure would entail. Such quantitative analyses are not easy, but neither are they impossible. Part II of this volume presents a series of short essays that take up some of the difficult problems that arise in *benefit-cost analysis*.

It is interesting in this regard that the early major pieces of environmental legislation in the United States almost completely ignored the economic approach to the setting of environmental standards. The Clean Air Act Amendments of 1970, which still embody the basic principles for air-quality management in the United States, literally directed the U.S. Environmental Protection Agency to set standards for air quality so stringent that *no one anywhere in the United States* would suffer any adverse health effects from air pollution. The courts have consistently held that, since this law was silent on the role of costs in setting air quality standards, they may *not* be taken into account. Two years later, the U.S. Congress declared in the Clean Water Act Amendments of 1972 that our goal was the complete elimination of "all discharges into the navigable waters by 1985."

Some of these extreme strictures have been relaxed in later legislation and their implementation modified by presidential executive orders. For example, in Executive Order 12291, President Ronald Reagan required benefit-cost studies for all major new regulatory measures (as President Jimmy Carter had done under an earlier executive order). Such systematic studies of the benefits and costs of proposed programs continue in the executive branch. Moreover, there have been overtures in Congress to override provisions in laws that prohibit costs from being considered, but they have not come to a vote. In fact, we find ourselves currently subject to a somewhat puzzling and conflicting set of requirements. First, the legislation for some programs explicitly rejects benefit-cost studies while others require them. Second, and more maddeningly, even where a law prohibits regulators from considering costs, they must still conduct benefit-cost analyses for major rules!

This is not to say that the findings from a benefit-cost study should constitute the sole criterion for deciding whether or not to undertake a new environmental program. The uncertainties inherent in such studies and the importance of other objectives suggest that it would probably be unwise to institute a rigid rule requiring that any proposed program pass a benefit-cost test. At the same time, such analyses surely provide important information that should be an integral part of the decisionmaking process.

The Choice of Policy Instruments

Once we have set specific targets for environmental programs, there remains the critical and challenging task of designing a set of regulatory measures to attain the targets. Here again, economics provides some valuable insights. In particular, it is important that a regulatory regime achieve its targets effectively and in the least costly way. A good system of regulatory instruments will both minimize abatement costs in the short run and provide incentives over the longer term for polluters to discover and introduce yet better techniques for controlling polluting waste emissions.

It is here that a set of incentive-based policy instruments has real appeal. Our earlier discussion

suggested that excessive pollution results from the absence of an appropriate price to induce controls on waste emissions. The implication is that we can correct the resulting market failure through the introduction of the missing price. Economic analysis thus points directly to a concrete policy proposal: the introduction of a surrogate price in the form of a unit tax on polluting waste emissions. Such a tax can play the role of the missing price by providing the needed incentive to polluters to economize on their use of the environment (see chapter 10).

To take an example, suppose that we have set a standard for air quality that requires that sulfur emissions in a particular region be cut by 50 percent. One way of achieving this goal would be to introduce a tax per pound of sulfur emissions and simply raise the tax to a sufficiently high level to induce a 50-percent reduction in sulfur discharges. Moreover, such a regulatory strategy has some important properties. It is straightforward to show that such a system of effluent taxes can attain the target at the minimum total cost to society. In addition, it provides an incentive over the longer term to seek out new and cheaper ways to control waste emissions, for such R&D efforts by polluting firms can reduce their tax bills and increase their profits. Systems of environmental taxes (or "green taxes," as some call them) effectively redirect the powerful profit motive of the market to the protection of the environment.

The environmental-tax approach is not the only way to mobilize economic incentives on behalf of the environment. An interesting alternative, one actually being used in the United States to reduce sulfur emissions on a national scale, is a system of tradable emissions permits (see chapters 11 and 12). Under this approach, the environmental authority issues a limited number of permits, each of which allows a certain number of pounds of pollutants per year to be emitted into the environment. The total number of such permits is limited to ensure that the predetermined standard for environmental quality is attained. But these permits have the important property that they can be traded: they are bought and sold in a market. Like the tax approach, such a system of tradable permits generates a price for polluting waste emissions that promotes a least-cost pattern of pollution-control efforts among sources and likewise provides the needed longer-run incentive for the search and introduction of improved control technologies. Some form of such a system, incidentally, is under serious consideration for use on a global scale to address the problem of global warming (see chapter 36).

As with the setting of environmental standards, the economic approach to the choice of policy instruments was essentially ignored in early environmental legislation. Instead, environmental authorities employed *command-and-control* (CAC) techniques for pollution control. Such CAC regimes often consisted of directives from the environmental authority to individual polluters that specified, at times in considerable detail, the precise forms of control measures that were to be adopted. Many studies have documented the unnecessarily high costs that these programs imposed on polluters and the economy. Not only this, but such measures typically provided little in the way of incentives for efforts to develop more effective control technologies.

Over time, we have come to appreciate the need for attaining our environmental objectives efficiently. For one thing, if we can keep control costs down, we will be in a position to do more in the way of environmental cleanup. There has, in consequence, been a growing interest in the use of incentive-based policy instruments, including not only taxes and systems of transferable permits, but such things as deposit-refund systems and various forms of legal liability that can, in certain instances, give polluters appropriate inducements for adopting control measures. Even where the CAC approach is still used, it is recognized that it is important not to specify control technologies, but to allow the source some flexibility in determining the most effective and least-cost way to comply with the limit that the regulatory authority imposes on its emissions.

Some Concluding Observations

In this introduction, I have focused attention on three basic ideas or lessons that economic analysis provides for environmental and resource management. Economics does, of course, have some interesting and important things to say on other matters. In Part 4 of this volume, for example, there are two essays addressing the issue of "environmental federalism," the question of how to assign regulatory responsibility for environmental management among the different levels of government. These essays suggest some principles for making this assignment. Part 7 takes up the matter of "environmental justice." Even here, where ethical considerations are so prominent, we find that an understanding of the economics of the problem is essential to an appreciation of the complexity of the policy issue.

More generally, the reader will find that the three ideas discussed in this essay manifest themselves in a wide variety of forms in the essays that make up this book. Even the issue of biodiversity in Part 6, again one with important ethical content, involves choices in the use of our scarce resources and thus inescapably has economic dimensions. Perhaps the most challenging of all—because of its enormous potential consequences, scientific uncertainties, and distant time horizon—is the issue of global climate change. Part 8 offers five essays on this critical issue. In view of its importance and inherent difficulty, researchers at RFF have, and are, devoting a major effort to the study of climate change and the range of available policy responses on both a national and global scale.

The final section of the volume presents a few papers on the problems of addressing pressing environmental issues in the developing countries and those nations making the transition from formerly communist regimes to more democratic and market-oriented systems. I want to acknowledge that this topic probably deserves more attention than it receives here. The course of environmental management in the developing world is clearly going to have a profound impact on the future of the global environment. And here the issue of scarce resources presses especially hard on the capacity to introduce ameliorative measures for the environment. Sensible goals and efficient policy measures may be even more important in this setting than in the industrialized world.

During the past thirty years, the sense of the importance of the economic perspective on environmental and resource management has been steadily growing. I surely don't mean to suggest that this is the whole story; obviously it's not. But our experience with environmental legislation and policy has made it clear that ignoring the lessons of economics takes a heavy toll on our efforts to clean up the environment and to do so in a relatively inexpensive way. This last point is becoming increasingly important as we try to raise environmental quality yet further. We have come a long way in cleaning up the air in our cities and our polluted lakes and rivers. This has been relatively easy in the sense that there existed straightforward measures for improving the environment when initiating cleanup programs. In the current lingo, we have picked the "low-hanging fruit," and we now must invest in more difficult and higher-cost methods for further environmental improvements. This means that there will be an even larger premium both on the selection of sensible environmental targets and the design of cost-effective regulatory measures to attain these targets.

Part 1

Science and Environmental Policy

What the Science Says
How We Use It and Abuse It To Make Health and Environmental Policy

James Wilson and J.W. Anderson

Under an ideal scenario, only the best science, pure and undefiled, would flow directly into policy as it is made to protect human health and the environment. But that wish isn't realistic. The best assurance of good public policy seems to lie not only in scientific knowledge per se but in open debate, caution, and a regulatory system capable of self correction.

Environmental policy is always based on science—up to a point. But defining that point is often a matter of fierce dispute and political combat. Then the quality of the science involved becomes an issue.

Decisions are easiest when threats and benefits are immediately visible to the naked eye. No one questioned, for instance, the proposition that burning soft coal in fireplaces and furnaces meant smoky skies over St. Louis. When people got sufficiently tired of the smoke, as they finally did in 1937, this source of home heating was outlawed with no argument over causation. But much of the modern environmental protection movement has been a response to menaces that are invisible, indirect, and detectable only through advanced technology. The effect has been to draw subtle and complex scientific issues into the arenas of politics.

The debates burn hottest where scientific uncertainty is the greatest and economic stakes are the highest. Scientific uncertainty comes in many forms.

About-Face on Thresholds

When science changes, environmental regulation has great difficulty adapting. One dramatic example is the issue of carcinogens' thresholds—whether there are doses below which carcinogens have no adverse effect on health. On that one, the consensus among scientists has reversed twice in less than fifty years.

Originally published in *Resources,* No. 128, Summer 1997.

Until the 1950s, it was a settled principle of toxicology that every poison had a threshold below which the dose was too slight to do harm. But with rising anxiety about the environmental causes of cancer, especially in the context of the debates about nuclear radiation and weapons testing, it began to seem more prudent to assume that carcinogens generally had no thresholds. One result was the famous Delaney Clause that Congress wrote into the 1958 Food, Drug, and Cosmetics Act.

The Delaney Clause banned all carcinogens from any processed food. At the time Congress, like the experts advising it, was under the impression that carcinogens were few and readily identifiable. But over time research found more and more substances that, if fed to rats in sufficiently massive amounts, could cause cancer. Some were naturally present in common foods—including orange juice. At the same time the increasing sophistication of measuring techniques identified traces of widely used pesticides and fungicides in many foods.

The regulatory system generally responded to these unwelcome findings by ignoring them. But at the same time the science was changing. Improved understanding of the processes by which cancers originate and develop made it seem increasingly likely that thresholds exist after all. The regulators themselves became convinced of that, although the Delaney Clause remained the law. The Food and Drug Administration quietly whittled away at the clause until the courts told them to go no further.

There's a high cost to society when government must enforce laws that make no sense to the people charged with enforcing them. It engenders cynicism among the regulators, and among the public it erodes confidence in both the law and its enforcement. But while Congress increasingly understood that the law was unenforceable, it refused to consider any reform that might be attacked as lowering the standard of health protection.

Lawsuits Force the Issue

A lesson for science policy lies in the way this paralysis was ended. It wasn't the advance of science that

did it, although the science was certainly advancing. Instead, as often happens in environmental affairs, the issue was forced by litigation—in this case, litigation brought by people who wanted the Delaney Clause enforced more literally. In 1992 a federal appellate court decision raised the prospect that the Environmental Protection Agency would be required to ban many widely used pesticides, with drastic implications for farmers' crops and retail food prices. That got the attention of Congress, and in 1996 it replaced Delaney's flat ban with a more realistic standard of "reasonable certainty" of no harm. According to its authors, the phrase was intended to mean a lifetime risk of cancer of no more than one in a million. With this change, the law is now back in conformity with scientific opinion and the regulators' actual practice.

Opinion Masked as Science

If it is possible to draw up a list of the circumstances that generate strife over the application of science to policy, along with changing science, disputes among scientists must also be near the top. To many laymen, certainty and precision is the essence of science: as they understand it, a scientific question can have only one right answer. But especially in matters of public health, it is often essential to make policy decisions long before the science is entirely clear. When people's lives and welfare are at stake, it is not possible to wait until every technical doubt has been resolved.

The situation is frequently aggravated by scientists who underestimate the uncertainties in their own work, leading them to blur the line between science and policy. Endless examples have turned up in the congressional hearings in 1997 on the EPA's proposals to revise the air quality standards for ozone and particulate matter. The EPA's Clean Air Science Advisory Committee (CASAC) set up a special panel of experts on ozone, and the panel came to general agreement that, within the range of standards under discussion, there was no "bright line" to distinguish any of them as being "significantly more protective of public health" than the others. Setting the standard, they said, was purely a policy choice. But the law specifically authorizes CASAC panels to offer policy advice, and more than half of the panel went on to offer EPA their various and conflicting personal opinions as to where the standard should be set. CASAC is deliberately organized to represent a wide range of views and interests.

The policymakers, most of them trained as lawyers, seized whichever of these personal opinions agreed with their own and cited them as the voice of science itself. In congressional hearing after hearing, EPA's Administrator, Carol Browner, defended her proposed standards as merely reflecting "the science." Her adversaries then quoted back to her the opinions of scientists who disagreed, some of them members of CASAC and others officials of the Clinton administration.

A more productive way to approach policy choices is to acknowledge uncertainty and take it explicitly into account. Do you go on a picnic if the weather report forecasts a 60 percent chance of rain? Do you commit society to a complex new air quality regulation if there's a 40 percent chance that it will not provide health benefits as intended? Attempting to quantify risk is an important step in making policy decisions. Unfortunately, it violates the current style of politics, in which it is safer to minimize responsibility and discretion by suggesting that decisions are determined solely by the science.

But which science? Toxicology looks for the mechanisms of damage to health at the molecular level, in terms that can be demonstrated in the laboratory, and tends to dismiss anything less specific as mere speculation. Epidemiology, on the other hand, sees reality in the statistical associations between the presence of a pollutant and the evidence of damage. As Mark Powell has pointed out in his RFF discussion paper on EPA's use of science in setting ozone policy, the tension within the agency between the toxicologists and the epidemiologists is as old as EPA itself. On clean air, CASAC is similarly divided.

In the current round of debate over clean air rules, the policymakers who support tighter standards cite the epidemiologists. Those who resist tighter standards cite the toxicologists. At present the differences between the two specialties' positions on particulate matter is substantial, and there is no one view that represents settled and accepted scientific truth.

Science as Proxy for Other Issues

In the vehement debates over science, scientific uncertainty often becomes the proxy for other issues—in the case of the Clean Air Act, for the forbidden subject of economic costs. The act prohibits EPA from taking costs into account in setting standards. Opponents of proposed regulations, unable to pursue their argument that the costs will outweigh any prospective benefits to health, go after the scientific basis of the regulations instead.

Confusion also arises when science asks the wrong question—sometimes because the law requires it. Here again the Clean Air Act provides examples. To take a prominent one, the act wants science to tell the regulators what effects each of six common pollutants has on human health. Since the pollutants are regulated separately, the health effects have to be studied separately. Scientists have been trying to tell the regulators for some years that it would be far more useful to investigate these pollutants mixed together, in the "soup" that people actually breathe, because the presence of one compound can affect the impact of another. But Congress has never responded to that advice because the concept of mixtures doesn't fit easily into the existing statutory framework for regulation. When environmental reality collides with statutory tradition, it's not always the statute that gives way.

Sometimes the Wrong Battle

Science, or what seems to be science, can sometimes be flatly wrong. The process of scientific inquiry is self-correcting over time. That is its greatest strength. But policy doesn't always wait for the corrections.

The Superfund program originated, notoriously, in response to mistaken and exaggerated scientific judgment. The Love Canal, in Niagara Falls, NY, had been well-known locally as a toxic chemical dump that was leaking insecticide into Lake Erie. But it suddenly became a national news story and a symbol of a new range of hidden environmental dangers, when in the summer of 1978 the state's health commissioner declared it a threat to the health of people living there. It was an election year in New York, and suddenly politicians at all levels, including President Carter, were competing to show concern and protect the residents. The following year a scientific consultant to the local homeowners association reported findings that indicated a wide range of threats to health. Then another consultant engaged by EPA reported evidence of high rates of chromosome damage among residents. Those claims established the atmosphere in which Congress began to draft the Superfund legislation.

Subsequently, review panels within EPA severely criticized the contractor's chromosome report, and a special committee of scientists set up by the governor of New York dismissed all of the health findings as inconclusive. But by the time that happened, the Superfund bill was approaching final passage.

It would be pleasant to think that some mechanism might be invented to allow the best science to flow, pure and undefiled, directly into policy. But that's hardly realistic, amidst the turbulence of rapidly developing science and especially in a field that, like environmental and health protection, has emerged as one of the leading battlegrounds of national politics. The best assurance of good public policy seems to lie in open debate, caution, and a regulatory system capable of self-correction.

Research Needs Funding

One point on which improvement is both possible and badly needed is the funding of scientific research relevant to regulatory decisions. Private and public spending in this country to meet the federal requirements for pollution control and abatement during the mid-1990s was in the range of $140 billion a year. Congress gives EPA less than half of one percent as much to spend on all its scientific and technological work for all purposes, a sadly disproportionate effort to ensure that environmental rules have the best possible scientific base.

It's not only the general pressure to cut the budget that inhibits adequate spending on science to support environmental regulation. Concerns about global warming have led to substantial outlays of federal science money on other purposes, and on other agencies than EPA. Currently, the EPA science budget is only about 10 percent of total federal spending on environmental scientific research and development.

The purpose of balancing the budget is to enhance the economy's efficiency and promote future growth. But budget cuts won't help the economy if they lead to the waste of resources on misguided policy.

2

Using Science Soundly
The Yucca Mountain Standard

Robert W. Fri

Nuclear waste is piling up at electric power plants and weapons production facilities. When the federal government decided to build a national repository for radioactive waste, it was hoped that "sound science" could resolve concerns about the safety of such a facility. In Fri's view, science can surely help, but some of the most difficult problems are bigger than science alone can solve.

Using "sound science" to shape government regulation is one of the most hotly argued topics in the ongoing debate about regulatory reform. Of course, no one is arguing that the government should rely on *unsound* science for its decisions. But supposing, as some reform advocates apparently do, that even the best science will sweep away regulatory controversy is equally foolish.

My experience as the chair of a National Research Council (NRC) committee that studied the scientific basis for regulating high-level nuclear waste disposal drove home this conclusion for me. I learned that science alone could resolve few of the key regulatory questions. More often, science could only offer a useful framework and starting point for policy debates. And sometimes, science's most helpful contribution was to admit that it had nothing to say.

A Short History of Nuclear Waste Regulation

Both commercial generation of electric power and government production of nuclear weapons result in high-level (long-lasting and highly radioactive) nuclear waste. At present, these wastes are stored at nearly a hundred sites around the United States, but federal policy mandates that the wastes ultimately be placed in a mined underground geologic repository. In 1987, Congress decreed that the first such repository be located at Yucca Mountain, which is near Las Vegas, Nevada.

The basic idea of geologic disposal is to use permanent natural barriers as a principal means of isolating nuclear waste

Originally published in *Resources,* No. 120, Summer 1995.

from the environment. Over time, however, some of the radioactive material will escape from even the best repository. At Yucca Mountain, for example, the casks in which nuclear waste will be initially stored will eventually break down, allowing the waste to migrate to the water table, which is located several hundred feet below the repository, and contaminate the flow of groundwater away from the repository site.

This process may take many thousands of years, but the nuclear waste will retain some of its radioactivity for more than a million years. Once the groundwater is contaminated, then the people who use it for drinking and irrigation will be exposed to radionuclides. Given this inevitability, the goal at Yucca Mountain is to design a repository that will limit to an acceptable level over very long periods of time the human health effects associated with nuclear waste releases.

Developing a standard that defines this acceptable level is one of Washington's longest running regulatory dramas. After ten years of work, the U.S. Environmental Protection Agency (EPA) first promulgated a standard in 1985. But following a successful court challenge in 1987, the standard was remanded to the agency for revision. Before EPA could issue the new standard, however, Congress enacted the Energy Policy Act of 1992, which mandated a new and different process for setting the standard for the proposed repository at Yucca Mountain.

Congress clearly wanted to curtail the debate over the standard. To do this, it reposed considerable faith in sound science. It required the National Academy of Sciences (through the National Research Council) to evaluate the scientific basis for a Yucca Mountain standard and directed EPA to promulgate a new standard "based on and consistent with" the findings of the academy. At the time, the idea of constraining regulators with the findings of a scientific panel was unfamiliar to the agency and the academy. Since a similar idea is afoot in regulatory reform, the Yucca Mountain experience may be instructive for that debate.

The Yucca Mountain Standard

Developing a standard that specifies a socially acceptable limit on the human health effects of nuclear waste releases involves many decisions. As the NRC committee learned in evaluating the scientific basis for the Yucca Mountain standard, a scientifically best decision rarely exists. The trick is to make the best use of the science that is available.

The first decision that EPA faces is how to measure safety. This decision entails setting a socially acceptable limit on some aspect of the repository's performance. As a technical matter, for example, the limit could be stated in terms of how much radioactivity the repository releases per year, how much radiation people will be exposed to as a result of releases, or people's risk of dying from this exposure. The committee recommended to EPA a standard stated in terms of risk of death.

The evolving scientific understanding of the relationship between radiation doses and the health effects that they cause certainly influenced this recommendation. Over the years, successive scientific reviews typically have concluded that a given dose of radiation may cause more deaths than scientists had previously believed. As a result of this trend in science, it makes sense to state the standard as a limit on the number of additional deaths attributable to releases from the repository. Doing so would mean that the standard would not have to change as the science continues to evolve. This observation also weighed heavily in the committee's preference for a risk-based standard.

Although a scientific fact lies behind it, this recommendation is clearly not dictated by science. Changing a standard to incorporate new information is technically not a problem. The preference for a stable, risk-based standard rests on the belief that changing so controversial a standard as one that specifies the acceptable level of human health effects associated with nuclear releases is socially, politically, and administratively undesirable.

This intersection of science and policy permeates the other decisions that have to be made in setting the standard for determining whether the Yucca Mountain repository would adequately protect human health. In particular, EPA has to specify what level of protection is to be afforded, to whom, and over what time period. For only one of these decisions does science provide reasonably conclusive guidance.

Establishing the level of risk that the standard will allow is a question of policy, not science. In other contexts, however, EPA and other organizations have set limits on a variety of nuclear risks that range from one additional death per hundred thousand persons to one in a million. At best, this information provides a scientifically defensible starting point for debating the acceptable level of risk at Yucca Mountain. It certainly does not predestine the outcome. Acknowledging this reality, the NRC committee could only recommend a reasonable range of risks for EPA to consider in crafting its regulatory proposal.

To determine whether a repository provides the acceptable level of protection, the risk that repository releases could impose on a specific individual or group must be calculated. How this person or group is defined can determine whether the standard is met. It has a particularly significant effect on whether the standard is met at Yucca Mountain, because the geology of the site lends itself to the creation of spots—for example, places in a groundwater plume—at which radiation tends to concentrate. A clever opponent of the repository could define the person to be protected as someone drawing water for drinking and irrigation only from one of these hot spots. An advocate for the repository would naturally assume that the affected parties were located at a safe distance from these areas.

As a matter of policy, the NRC committee preferred to avoid these extreme assumptions. Given this policy, it looked to science (or at least to careful scientific thinking) to contribute a methodology for calculating compliance with the standard that resists extreme cases. The methodology that the committee chose was the "critical group method,"

which calculates the average risk to a member of the group at greatest risk.

Guidance for the time period over which the standard should provide protection is provided by the fact that radioactivity associated with high-level nuclear waste will not dissipate for more than a million years. Ideally, then, compliance with the standard would be tested over the full duration of this period in order to determine the time at which the greatest effect on human health occurs. Whether this determination is possible depends on the ability of scientists to evaluate the behavior of the repository over very long periods of time.

Here, for a change, is a question of science rather than policy. The committee answered it by saying that compliance assessment is feasible for most physical and geological aspects of repository performance on the order of a million years at Yucca Mountain. Still, this answer is based on the expert scientific judgment that the fundamental geologic structure will be relatively stable for this long, not on the testable hypotheses of scientific method. Thus, other experts might reach a different conclusion.

Running Out of Science

The NRC committee was able to recommend the foregoing elements of the standard with at least one foot in the realm of science. Unfortunately, however, science can contribute little to answering three of the most controversial questions that bothered Congress about the standard in the first place. For three of these questions, the scientific basis for decisionmaking essentially does not exist.

What Is a Negligible Risk?

The main concern of a standard for a nuclear waste repository is to protect populations living near the repository. In principle, however, a very large and dispersed population could be affected by releases of nuclear waste. In the case of Yucca Mountain, radioactive carbon dioxide gas could escape from nuclear waste canisters and be inhaled by people living far away from the repository. The carbon-14 problem, named after the radioactive isotope present in the waste, is one of the most vexing problems with which EPA must deal. Because carbon-14 releases from Yucca Mountain would be mixed with the global atmosphere, the health risk to any one individual is exceedingly small. On the other hand, the number of people exposed worldwide over the life of the repository is astronomical. If we multiply the very small risk by this very large number of people, we can calculate that many additional deaths could occur over a very long time period.

But how do we interpret a number computed in this way? No adverse health effects may occur at the very low doses of carbon-14 to which people would be exposed; but lacking data to show that this would be the case, experts in the field say that the prudent course is to assume that health effects will occur. Making this assumption could produce a scenario that leads either to abandoning the Yucca Mountain site or to spending a great deal of money to contain carbon dioxide gas.

To the dismay of policymakers, science cannot make this problem go away. Faced with this dilemma, the committee could only observe that the risk to any one individual in the global population would be very small—perhaps ten thousand times lower that the one-in-a-million level at which the basic standard might be set. A responsible decisionmaker could conclude that such risks are so negligible that they should not affect the design of the repository, but he or she would have to do so without much definitive guidance from the scientific community.

Can We Guard against Future Human Intrusion at a Repository?

One way to project significant human exposure to radiation releases from repositories is to assume that someone intrudes after they close. For example, a future oil explorer could drill into a waste canister and bring radioactive material directly to the surface. In crafting its charge to the NRC, Congress specifically asked whether any scientific

basis exists for evaluating this risk or for assuming that it can be prevented.

The answer to both questions is no. The committee found no scientific basis for predicting the behavior of humans thousands of years into the future. Since neither the probability of human intrusion nor the effectiveness of preventive measures is predictable, the committee concluded that these issues should not be considered in the assessment of compliance with a risk-based standard. (We did, however, offer an alternative analysis to test the resilience of the repository to an assumed intrusion.)

In this case, the absence of a scientific basis is probably a help to decisionmaking. Admitting the limits of science should greatly reduce the considerable analysis and controversy lavished on speculation about the likelihood of human intrusion. I should note, however, that if regulators were deciding whether to dispose of waste at scattered surface sites instead of in a geologic repository, as at Yucca Mountain, analyzing the risks of human intrusion might be crucial.

What Assumptions Do We Make about Exposure Scenarios?

In all of the above issues, the committee walked the line between science and policy without dissent. But consensus failed when it came to specifying the exposure scenario to use in calculating compliance with the standard.

The exposure scenario describes how radiation that is released from the repository passes through the biosphere to expose humans. The scenario thus must specify whether and how water wells are drilled into the groundwater underlying Yucca Mountain, whether the water is used for drinking or irrigation, how much of a person's food intake is contaminated by this irrigation, and so on. Science can put bounds on many of these assumptions; for example, people can drink only so much water, and plants retain radionuclides at predictable rates. Developing exposure scenarios, even for the distant future, is therefore not entirely a blue-sky exercise.

Update

Three years after our report was issued, the decision about how to regulate health risks remains unresolved. At present, the issue is whether to protect groundwater as a resource, in addition to its contribution to health risks. The prospect is for perpetual regulatory hairsplitting.

In retrospect, I realize that my 1995 article failed to mention what is perhaps the most important role of "sound science"—to guide policymakers toward the most useful questions. Asking whether long-term radioactive waste disposal creates a risk that exceeds some threshold turns out to be not a very good question. It leads to a compliance calculation that is not empirically testable and so largely unsupported scientifically. Perhaps a better regulatory question would be comparative: given that the nuclear waste has to go somewhere, where is the least risky place to put it?

Still, science cannot predict human behavior. This consideration is important in the Yucca Mountain case, because the area is sparsely settled—one good reason for locating a repository there. Given this, what should an exposure scenario assume about whether someone is present to be exposed to any release that might occur?

Remember that the committee recommended a standard that would protect the people at greatest risk, while avoiding the trap of extreme assumptions. It would be inconsistent with this principle to base the exposure scenario on, say, the expectation that millions of people will move into the Yucca Mountain neighborhood. A more reasonable assumption is that farmers scattered about the area will comprise the population at greatest risk. Insisting on such a cautious but reasonable approach to narrow the range of assumptions about the distribution of population in the distant future is no small accomplishment. Indeed, doing so would considerably circumscribe the current debate about Yucca Mountain.

Even within this narrowed range of options, however, members of the committee disagreed on the exact population-distribution assumption that should be used. One member felt strongly that the exposure scenario should assume that a subsistence farmer will always be living at the place where exposure to radiation will be highest over the life of the repository. The other members believed that the physical features of the site naturally lead to a dispersed population and that the exposure scenario should take account of this fact.

These alternative views can excite considerable passion on the part of their proponents. In my view, however, such controversy obscures two crucial points. One is that the population-distribution assumption cannot be resolved on the basis of science. No one can predict where people will live in the future; therefore, regulators must make a judgment call in choosing an assumption about population distribution in the exposure scenario. The other point, noted above, is that the debate is over a fairly narrow range of assumptions. Despite the passion attendant on it, this debate is far more manageable than the open-ended debate to which EPA might be exposed if the committee had not narrowed the range of assumptions.

The Role of Science in Regulatory Decisions

The lessons that the NRC committee learned in studying the scientific basis for the Yucca Mountain standard may be important to those involved in the regulatory reform debate. The chief lesson is that the soundest science rarely provides black-and-white answers for regulatory decisionmaking; it only brightens a bit the familiar gray space in which decisions are made.

To be sure, science can sometimes have a conclusive effect on a regulatory decision. In the Yucca Mountain case, the conclusion that the standard should be applied without time limit rests almost entirely on expert scientific judgment. By contrast, the current EPA standard applies only over a 10,000-year duration. Accepting the scientific judgment of the Yucca Mountain study would thus have a profound effect on the design of the standard.

Admitting that science has nothing to say also can powerfully affect decisionmaking. For example, the committee found no scientific basis for evaluating the probability of human intrusion. Therefore, it concluded that the issue should not be considered in assessing compliance with a risk-based standard. If EPA accepts this conclusion, a significant line of argument that could distract the regulatory debate will be closed off.

Mostly, however, the Yucca Mountain study shows that science is helpful, but not conclusive, in arriving at reasonable decisions—such as setting the acceptable level of protection, defining the people to be protected, and specifying the exposure scenarios to be used for compliance analysis. In these instances, the committee avoided asserting that sound science provided a complete answer, but did try to use scientific judgment to define a reasonable starting point and a bounded range of options for EPA to consider. In this way, science can be quite helpful in fostering constructive debate.

Finally, the Yucca Mountain study indicates that science cannot protect public officials from hard decisions. Advocates of the Yucca Mountain repository would like nothing better than for science to make the carbon-14 problem go away. But science cannot do that; it can only note that the risk from carbon-14 emissions to an average individual in the global population is exceedingly small. Whether these risks are so small as to be negligible is a tough political call that science cannot—and should not—make.

In short, the Yucca Mountain study clearly illustrates that excessive faith in the power of sound science is more likely to produce messy frustration than crisp decisions. A better goal for regulatory reform is the sound use of science to clarify and contain the inevitable policy controversy.

Suggested Reading

National Research Council. 1990. *Rethinking High-Level Radioactive Waste Disposal.* Washington, DC: National Academy Press.

National Research Council. 1994. *Science and Judgment in Risk Assessment.* Washington, DC: National Academy Press.

National Research Council. 1995. *Technical Bases for Yucca Mountain Standards.* Washington, DC: National Academy Press.

Part 2

Benefit-Cost Analysis

3

Economics Clarifies Choices about Managing Risk

A. Myrick Freeman III and Paul R. Portney

Like it or not, environmental risks cannot be completely eradicated. Difficult choices must be made about how best to control particular risks using the limited resources available. These choices invariably involve tradeoffs. Economics can help with these decisions by providing information on the pros and cons of particular courses of action.

Government officials making decisions about such issues as allowable pesticide residues in foods, nuclear reactor safety standards, and air quality standards face a difficult problem. On the one hand, there is the evident desire of the public to reduce the risks inherent in modern life. On the other hand, reducing these risks is costly. So choices about risk policy involve trade-offs. Risk management refers to the process through which a variety of considerations—scientific, legal, political, economic (benefits and costs), and even philosophical—are taken into account and a decision is reached concerning an environmental regulatory problem.

Economics can contribute in a number of ways to managing risks to health and the environment. At a basic level, economics can help to inform decisionmakers about how much various regulatory approaches or pollution control options will cost society. Upon first blush, this might seem pedestrian and straightforward. In fact, it might even appear that engineers rather than economists are better able to make such determinations, especially when the options under consideration involve primarily structures and equipment.

But appearances are deceiving. One of the real, albeit subtle, virtues of economics is its focus on what are called *opportunity costs*—that is, what society must give up in the form of other desirable things in order to pursue a desired goal such as reduced environmental risk. Under some circumstances, expenditures for pollution control equipment, cleaner fuels, or the like will closely approximate true opportunity costs. However,

Originally published in *Resources*, No. 95, Spring 1989.

often this correspondence between money expenditures and opportunity cost is lacking. For example, rules on private behavior such as mandatory recycling of household wastes or limits on eating fish caught by sports fishermen involve no direct money outlays, but they impose costs in the form of time or reduced satisfaction. An economic perspective on costs provides valuable insights about the nature and magnitude of these forgone opportunities.

An important criterion for the rational management of risk is that any reduction in risk be accomplished at the lowest possible economic cost. Economic analysis can help to identify the least costly way to accomplish a particular reduction in environmental risk. Used in this way—how we can accomplish X for as little as possible—the application of economics goes by the name of cost-effectiveness analysis.

Besides helping to identify and properly measure costs, economics can help us to understand how these costs (as well as benefits) are distributed among the population. For instance, we might be interested in knowing whether residents of rural areas would bear a disproportionate share of the costs of an acid rain control program. Or, we might wish to know whether financing Superfund cleanups via direct budgetary outlays is more or less regressive in its impacts than financing those same cleanups through taxes on manufacturing firms. Again, we might want to determine whether the favorable effects of a policy are distributed equally among current and future generations. Economics can help us answer these questions.

Normative Guidance

Using economics merely to supply information about the costs of different options is one of its less controversial applications in risk management. The challenge comes when economics is used to answer questions like: What should we do about the problem of pesticide residues in foodstuffs? Which cleanup strategy is best at the XYZ site? Here economics is being asked to go beyond the purely informational—beyond describing what would

happen here or there—and instead is being asked to provide normative guidance to decisionmaking—that is, to help us answer the question, What *ought* we do?

To answer normative questions like these, economists generally rely on a branch of economics known as benefit-cost analysis. (See the box, "Benefit-Cost Analysis and Risk.") Economists view benefit-cost analysis as akin to common sense. This is because after peeling away the analytical veneer, formal benefit-cost analysis essentially asks: If we pursue a particular policy option, what good will come of it and what will we have to sacrifice to get it? It is a simple extension to ask whether the former is worth the latter.

Although benefit-cost analysis can clarify the pros and cons of taking particular actions, its application to the problems of environmental risk management has not gone smoothly. It is not embraced in any major environmental statutes except the Toxic Substances Control Act of 1976 and the Federal Insecticide, Fungicide, and Rodenticide Act

Benefit-Cost Analysis and Risk

Benefit-cost analysis is based on the twin premises that (1) the purpose of economic activity is to increase the well-being of the individuals who make up the society, and (2) each individual is the best judge of how well off he or she is in any given situation. These premises provide the foundation for the widespread application of benefit-cost analysis in such areas as investments in water resources, transportation projects, and human resource development. They are applicable, as well, to the analysis of public policies toward risk.

If society is to make the most of its endowment of scarce resources, it should compare what it gains from a risk management policy (that is, the benefits) with what it gives up by taking resources from other uses (the costs). The benefits and costs should be valued in terms of their effects on individuals' well-being. Society should undertake risk management activities only if the results are worth more in this sense than what society would forgo by diverting resources from other uses.

of 1972. Nor is it the rule in other regulatory statutes protective of public health (for example, those having to do with occupational safety and health or with consumer products). In fact, the balancing of benefits and costs appears to be *prohibited* when the Environmental Protection Agency sets most standards for air and water pollution and the regulation of active or abandoned hazardous waste disposal sites. Similarly, the well-known Delaney clause in the Federal Food, Drug, and Cosmetic Act of 1938 explicitly prohibits the head of the Food and Drug Administration from considering the health benefits associated with certain food additives if these additives are known or suspected of causing cancers in humans. Moreover, although the last three presidents have issued executive orders mandating that benefit-cost analyses accompany any new proposed or final regulations, federal regulatory agencies have often resisted, and Congress has battled to have these presidential orders weakened.

In addition to the political unease over benefit-cost analysis, there is also more than a little public concern about its use in environmental decision-making. This concern is harder to document, but it shows up often in public meetings, opinion polls, and everyday discussions.

Political Unease

Political reservations about using benefit-cost analysis to help make risk management decisions are based on several concerns.

Distributional Issues

Benefit-cost analysis is in one sense distributionally neutral. That is, a dollar's worth of benefits (or costs) count the same regardless of the economic position, geographic location, or other characteristics of the individuals to whom they accrue. This can spell trouble in political circles.

Consider, for instance, the case of acid rain. Emissions of sulfur and nitrogen oxides from coal-fired utility and industrial boilers, as well as from mobile sources, are believed to be responsible for damages to aquatic ecosystems, forests, agricultural products, materials, and even human health. A variety of control measures are available and reasonably well understood. If risk managers decide to use the "polluter pays" principle, there would be very uneven geographic distributional effects. Because states in the Ohio River valley are emitters of large amounts of sulfur dioxide, they would bear a heavy share of the total costs of controlling emissions. Application of the polluter pays principle could cause electricity bills in those states to increase by as much as 15 to 20 percent. Such geographic concentration of costs has been one of the stumbling blocks to amending the Clean Air Act of 1970 to deal with acid rain.

Imprecise Information

Politicians view benefit-cost analyses of risk management options with suspicion for another reason. Estimates of benefits and costs must rest on a foundation of knowledge of the physical, biological, and engineering systems involved as well as the economic factors determining monetary values. For example, it must be possible to answer such questions as: How much will indoor radon concentrations be reduced by air filtration equipment? How many fewer cases of lung cancer will there be if radon levels are reduced? How much will emissions be reduced by vapor recovery devices on gasoline pumps? What effect will this have on atmospheric ozone levels? What will be the impact on agricultural productivity of reduced ozone concentrations?

None of these questions is easy to answer. We sometimes have no more than well-educated guesses about the answers to these technological, physical, and biological questions. Some critics therefore believe that benefit-cost analysis can be rigged; that it will more often than not be used to justify a risk management decision that is taken not for analytical but rather for political or other reasons.

Myth of Abundance

One of the uses of benefit-cost analysis is to help ration scarce resources among competing ends.

This use is a reflection of the fact that there are more things worth doing than there are resources with which to do them. While this sounds innocuous enough, politicians generally prefer to avoid making explicit such declarations. No politician is likely to gain much support for telling a group that, although they are bearing some environmental risk from problem X, the risk is relatively small and the money necessary to reduce it could be better spent elsewhere. *Even when it knows better,* the public likes to be told that its government is working to eliminate all environmentally transmitted risks. Sensing this, politicians shy away from analytical approaches based on the premise that resources are finite and priorities have to be set.

Public Unease

Politicians aside, the public has additional concerns about applying benefit-cost techniques to risk management problems.

Uncompensated Risk

Benefit-cost analysis is silent on the question of whether the losers from any risk management policy should be compensated. In practice, therefore, even policies that result in aggregate benefits in excess of costs could still leave some people worse off. It would be natural for the losers to oppose the policies. And this opposition could be quite vocal if the losses were concentrated among a relatively small group of people.

Nowhere is the issue of uncompensated risk more clearly visible than in the problem of proposed siting of LULUs (locally undesirable land uses) such as hazardous waste incinerators and low-level nuclear waste disposal facilities. Those around any proposed site will reject the argument that it is in the best interest of society for them to accept the increased risks that these facilities pose. And very often their opposition will prove successful. In an effort to deal with this impasse, analysts have begun to propose mechanisms for compensating the losers. (See the box, "Compensating the Losers.")

Compensating the Losers

No one likes having a hazardous waste disposal facility in his or her neighborhood, even though that location may be the best choice from society's perspective. What about compensating these "losers" for having the site nearby? The benefit-cost criterion for risk management policies only identifies those cases where the gains are large enough so that there is a potential to fully compensate the losers. It is silent on the question of whether compensation should be paid or not.

Robert Cameron Mitchell and Richard T. Carson, in their article "Property Rights, Protection, and the Siting of Hazardous Waste Facilities" in the American Economic Review *(May 1986), have proposed that communities as political entities be granted what is, in effect, a property right—that is, a right to refuse to accept proposed LULUs (locally undesirable land uses). Along with this right would be an obligation for the community to hold a referendum on any proposal to locate a LULU within its boundaries. Any corporation or larger government entity wishing to place a LULU within the community's boundaries would have to offer compensation to the community if it expected to gain approval through the referendum.*

The "Right" To Be Risk-Free

Many citizens feel that they have a basic and inalienable "right" to be free from contaminants in the water they drink, the air they breathe, and the food they eat. They resent these rights being weighed against economic dislocations, balance-of-trade concerns, and other seemingly impersonal factors.

There is a ready response to such objections. First, even those rights guaranteed in the Bill of Rights are not absolute. For instance, one's freedom of speech is restricted when it comes to standing up in a crowded theater and shouting "Fire!" While no formal benefit-cost analysis supported these relatively mild restrictions on our basic rights, they are premised implicitly on the notion that completely unfettered speech or assembly may sometimes do more harm than good. In other words, the benefits of some restraints may be worth the costs.

To those who would argue that we have a right to be free from all environmental risks, the counter-argument would run as follows. First, in a fundamental physical sense, we can never be free of such risks. Primitive woodburning puts harmful particulate matter in the air, and the human digestive system ensures that some wastes will always be with us. Thus, a no-risk world is simply impossible. Even if it were not, some risks would surely be judged to be so small in comparison to the costs of alleviating them that it would be best to accept them.

Expert versus Lay Opinion

Public opinion polls show a steady erosion of public faith in experts. For a variety of reasons—some having to do with erroneous predictions in the past (for example, that nuclear power would become too cheap to meter), some having to do with generally increasing skepticism—the public seems less willing to be reassured that a particular risk, while real, is nonetheless quite small. This means that benefit-cost analyses, which depend critically on expert opinion or findings, will also have detractors among the populace. This becomes all the more likely when the experts themselves represent business concerns or, if university-based, derive part of their funding from corporations or trade associations. In cases where such suspicions are rampant, it becomes difficult to quell fears that experts feel are unwarranted. This divergence between expert and lay opinion cuts the other way, too. The public is often very slow to warm to concerns that experts may place near the top of their list of environmental risks.

Qualitative Dimensions of Risk

Risk analysts are sometimes puzzled when people react strongly to what may seem to be relatively small risks, yet appear to accept, or even seek out, risks such as skydiving and motorcycle racing. Such behavior is understandable. The risks associated with such sports are voluntarily borne, while one has little choice about the air one breathes while outside. Research during the last twenty years or so

Update

After the Republicans took control of the Congress at the start of 1995, there were renewed efforts to institute a requirement for the performance of risk-benefit and benefit-cost analyses of proposed regulations by all federal agencies (see Paul R. Portney, "Cartoon Caricatures of Regulatory Reform," in *Resources* 121, Fall 1995). None of these proposals made it into law. Although no sweeping regulatory reform has been passed, a number of less ambitious bills have been enacted that somewhat expand the role of analysis in regulation. One bill has made it into law—the Small Business Regulatory Enforcement and Fairness Act of 1996—that gives Congress the ability to overturn any major regulation issued by a federal agency. Through 1998, however, Congress had not used this authority.

has demonstrated over and over again that such characteristics as voluntariness, familiarity, and dread influence the way individuals perceive and react to risks.

Facing Facts

Economics is the science of scarcity, and society is surely limited in the resources it can allocate to the control of environmental risks. Thus, it is important to think analytically about which risks we want to address first and how much control we wish to pursue. Like it or not, tradeoffs will be made when these risks are addressed. This follows directly from the observation above that society's resources are limited. Because this is so, we simply cannot eradicate any and all risks.

At some point decisionmakers will have to say to themselves that additional risk reductions will be so expensive that they are probably not worth additional effort. The virtue of economics is that it makes these decisions explicit. In other words, it forces decisionmakers to say openly, for example, that society cannot afford to spend $1 billion to save an additional life through more stringent regulation of substance X. While such acknowledgements are often painful, they do enable the public to see the tradeoffs that their elected officials are making and object if they disagree with them. Pretending that such tradeoffs do not have to be made only means that they will be made implicitly and out of the public eye.

One conclusion, then, is that the public and its political leaders would be well served if the public better understood economic methods and their application to problems of environmental risk management. The fact that this is a familiar refrain does not detract from its importance.

Sauce for the goose, however, is sauce for the gander. Just as it would behoove the public and its political leaders to better understand the economic approach to risk management, so too must economists understand why their message is so often ignored. While benefit-cost and cost-effectiveness analyses have their strengths, they also have weaknesses, some of which are nearly fatal in the political realm. Until economists do more than pay lip service to the importance of distributional concerns in real policymaking, for instance, they will remain peripherally involved in policy formulation at best.

Economists must also understand that the public cares about more than simply the statistical magnitude of risks. It is also concerned about the mechanisms through which these risks are trans-mitted, the degree to which the risks are voluntary, the benefits that accompany the risks, and other dimensions that are often disregarded in standard economic analyses. Until these concerns are acknowledged and incorporated in our economic models, economists may dismiss as irrational responses that make very real sense.

Suggested Reading

American Chemical Society. 1998. *Understanding Risk Analysis: A Short Guide for Health, Safety, and Environmental Policy Making*. Washington, D.C.: ACS.

Arrow, Kenneth J. and others. 1996. Is There a Role for Benefit-Cost Analysis in Environmental, Health, and Safety Regulation? *Science* 272 (April 12):221–22.

Davies, J. Clarence. 1996. *Comparing Environmental Risks: Tools for Setting Governmental Priorities*. Washington, D.C.: Resources for the Future.

Morgenstern, Richard D., and Marc K. Landy. 1997. Economic Analysis: Benefits, Costs, Implications. In Richard D. Morgenstern (ed.), *Economic Analyses at EPA: Assessing Regulatory Impact*. Washington, D.C.: Resources for the Future.

Weidenbaum, Murray. 1998. *A New Approach to Regulatory Reform*. Policy Study 147. August. St. Louis: Center for the Study of American Business, Washington University. Also available at the CSAB Web site, http://csab.wustl.edu/.

Does Environmental Policy Conflict with Economic Growth?

4

David Gardiner and Paul R. Portney

Debates about environmental regulation most often revolve around its economic consequences, particularly its effects on economic growth. Recently, this debate has become sharper. In addition to the "traditional" view that environmental regulation impedes economic growth (most often espoused by those in the business community), an opposing school of thought has developed. According to its proponents, not only can environmental regulation provide health and ecosystem protection, but it can stimulate the economy and enhance U.S. competitiveness at the same time.

Because this debate has extraordinarily important policy consequences, Resources for the Future sought a way to air—and clarify—the issues bound up in it. Accordingly, RFF's Paul R. Portney and David Gardiner of the U.S. Environmental Protection Agency's Office of Policy, Planning, and Evaluation discussed them on December 1, 1993, at one of RFF's regular Wednesday seminars. The two sections that follow adhere faithfully to their opening remarks.

Although Gardiner is a proponent of the new view and Portney is more sympathetic to the traditional one, both agree that the relationship between the economy and the environment is a complicated one. Moreover, both agree that the debate so far quite often has been exaggerated and misleading. This presentation is intended to help shed light on a most important subject.

Originally published in *Resources,* No. 115, Spring 1994.

David Gardiner

Conventional economic wisdom tends to focus on trade-offs as the basis for exploring the relationship between the environment and the economy. It suggests that environmental policy conflicts with economic progress. The U.S. Environmental Protection Agency (EPA) is trying to dispel this false dichotomy by leading discussion away from the somewhat reactive focus on trade-offs and toward a more proactive focus on ways to achieve environmental protection and economic progress at the same time.

Conventional Model of the Economy-Environment Relationship

The conventional approach to exploring the relationship between the environment and the economy is to pit one against the other—as if the real trade-off were between environmental protection and economic progress. By economic progress I mean quantitative and qualitative progress in the context of clean and equitable improvements to socioeconomic systems. Quantitative improvements enable us to meet the essential needs of the present generation without compromising the ability of future generations to meet their own needs. Qualitative improvements reflect our capacity to convert physical resource use into improved services for satisfying human wants.

In general, the conventional approach ignores changes in technology and changes in consumer preferences, and it assumes that everyone out there in the marketplace is fully informed. It also treats expenditures on environmental protection as expenses, rather than investments, and affords no intrinsic or economic value whatsoever to natural resources, such as clean air and clean water.

In reality, none of these assumptions holds true. This is why less-than-optimal outcomes result for both the economy and the environment when decision makers adopt an either/or model of the economy-environment interaction.

One such outcome resulted when U.S. manufacturers in the automobile coatings segment of the paints and coatings industry failed to anticipate

public demand for stronger environmental regulations or opportunities for cost-effective, safe, and clean technological advances. As a result, the manufacture of all water-borne basecoats used in the United States relies substantially on technology developed by European suppliers.

Another example of a less-than-optimal outcome comes from the agriculture sector, where either/or assumptions and market imperfections have left the potential for realizing economic and environmental benefits substantially unmet. A recent cooperative study undertaken by EPA and the University of Missouri indicates that, when compared to conventional systems of farming, cropping systems that incorporate reduced tillage, greater cropping diversity, and more efficient management of commercial pesticides and fertilizers can improve resource conservation, reduce environmental risks, reduce costs of production, and increase short-run profits.

To obviate the false assumptions that lead to less-than-optimal decision making, we must change the very nature of the debate over the relationship between the economy and the environment. This can be achieved, at least in part, by shifting discussions about that relationship away from the either/or model.

Environmental and economic interdependence is strongly linked to the development and diffusion of technology. As noted above, false assumptions about technology, tastes, and environmental investments form the basis of the view that increased pollution reduction can only be achieved at the expense of economic progress or vice versa, that greater economic activity inevitably hurts the environment. In reality, the myriad relationships between the economy and the environment are continually changing.

New Perspective

The key question, then, is not "Does environmental policy conflict with economic progress?" but rather, "How can we get environmental protection and economic progress at the same time?" Clean technologies and management practices have a particularly important role to play in answering this question, as do price and institutional reforms that encourage reductions in all polluting emissions per unit of industrial output. And because the demand for environmental goods and services, or for a clean environment, increases at a slightly greater rate than income in most cases, we know that the demand for a clean environment is going to increase domestically and internationally.

We want to help give direction to that demand on an international level, so that when the market forms we can meet that demand with U.S. technology. Moreover, we want to provide incentives to industry to target its new capital investments in manufacturing practices and processes that are sustainable over the long term. In this way, we can realize environmental and economic benefits from the ongoing process of technology turnover in all industries.

The development and diffusion of environmentally sound technologies can change the way in which goods and services are produced and also generate benefits that can increase human welfare. The most promising areas for realizing the gains of environmental technology today relate to energy use and the development of alternative fuels, to biotechnology and the development of agricultural practices that use fewer inputs and harmful pesticides, and to industrial production processes that reduce or prevent pollution.

It's worth noting that industry's focus on environmental concerns results not only from the need to comply with environmental regulations; firms are also recognizing new business opportunities and realizing economic gains. Indeed, U.S. industry is racing to capture the world market for new and emerging technologies, which the Organisation for Economic Co-operation and Development estimates to be worth $200–300 billion and forecasts to see sustained growth over the next decade. In addition, "environmentally friendly" has become a powerful marketing tool across all sectors, industries, and services, a tool that recognizes consumer preferences for products that have less harmful impacts on the environment.

Examples abound that let us "brag" about the economic and environmental benefits that result when the interdependence of economic and environmental goals are recognized, understood, and strategically advanced. Inform Inc., a New York–based, nonprofit, environmental research organization, reports that in many cases initiatives to reduce pollution at its source have decreased waste streams by 90 percent or more and resulted in significant savings. The savings, tallied for 62 projects, came to $21 million annually.

In one case cited by Inform, a medium-sized resin and adhesives facility in California made operational changes that slashed by 93 percent its major phenol-laden waste stream, which for years had been discharged first to the local sewer and then to an onsite pond. This reduction has saved the company more than $150,000 per year in waste disposal and potential legal costs. In another case, a reagent chemicals plant in New Jersey computerized its materials tracking system, identified twenty-one source reduction initiatives, and cut more than 600,000 pounds of waste to achieve annual savings exceeding half a million dollars.

State governments also have documented some good examples. Minnesota estimates that six manufacturers using recyclable materials have created around 1,700 jobs, $39 million in new wages, and an increase of $100 million in Gross State Product. Maine reports that recycling added nearly $300 million in wages, profits, savings, and secondary impacts, as well as more than 2,000 jobs to its economy in 1992. There are many more examples, and we want to continue to add to them.

New Approaches

One of EPA's driving principles under Administrator Browner is an uncompromising commitment to environmental goals, while allowing flexibility as to how those goals are met. This combination of uncompromising commitment and flexibility is designed to yield innovation and jobs, as well as better environmental results.

The agency recently announced a major initiative to work closely with industry, states, and environmental groups to explore—on an industry-by-industry basis—coordinated rulemaking, permit streamlining, multi-media compliance and enforcement opportunities, and pollution prevention and environmental technology opportunities that offer "cleaner, cheaper" environmental results. Through initiatives such as these, we can expose the false premises that undermine constructive dialogue on the environment and the economy. Moreover, by demonstrating the interdependence of environmental and economic goals, we can create a new model of thinking that encourages decision makers to leverage the positive relationship between environmental protection and economic progress.

Paul R. Portney

I welcome this opportunity to react to David Gardiner's views on environmental regulation and its connection to economic growth. Because of the importance of this connection, and the key role that Gardiner's Office of Policy, Planning, and Evaluation plays in EPA's analyses of such issues, his willingness to exchange views is encouraging.

On several key points, I find myself in substantial agreement with him. For instance, the debate over environmental regulation *has* often made it seem that we must choose—in an either/or fashion—between economic growth and environmental quality. In fact, the two can coexist.

For example, between 1970 and 1990, per capita real disposable personal income in the United States (the best measure of what the average person has available to spend) increased by 42 percent. Meanwhile, concentrations of airborne lead, perhaps the most harmful of all the common air pollutants, fell by 90 percent between 1983 and 1990 alone. In addition, the period 1970–1990 saw significant reductions in ambient concentrations of sulfur dioxide, particulate matter, and carbon monoxide in almost every major metropolitan area of the United States, as well as significant—though much less uniform—improvements in water quality. Strictly speaking, then, we do not face an

"either/or" choice when thinking about economic growth and environmental quality, and it is wrong to suggest otherwise.

I also agree with Gardiner that new environmental regulations do not inevitably lead to plant closures and unemployment. In fact, as he points out, a substantial number of people are now employed in what might loosely be referred to as the "environmental industry." (Total U.S. employment in this industry is about one million people.) This positive side of the "jobs" issue is routinely ignored by critics of regulatory programs.

Finally, I agree wholeheartedly with the emphasis Gardiner places on the importance of concocting cheaper ways of meeting the goals of U.S. environmental policy. Twenty years of careful research have demonstrated that we can meet our present environmental goals for a fraction (perhaps as little as 50 percent) of the $130 billion we now spend each year to comply with federal environmental regulation. Even if the annual savings were as little as 10 percent, or $13 billion, this would be roughly equivalent to all federal income assistance to poor families and nearly three times the amount of federal assistance to schools for disadvantaged children. We have to take advantage of opportunities like this.

Despite these points of agreement, however, I take issue with some of what Gardiner has to say. And I disagree fundamentally with a message I believe is implicit in his remarks: we can avoid painful choices when setting environmental goals and instead "have it all." That's simply not true, and we had better recognize this admittedly unpleasant reality if we are to fashion wise economic and environmental policies.

Keeping Score with Jobs

Gardiner refers several times to favorable job impacts from environmental measures. But we need to keep three things in mind when thinking about jobs and regulation. First, despite much rhetoric from both sides, environmental regulation will never have much of an impact on the aggregate level of employment in the United States. Rather, total employment is determined by much broader forces—such as domestic and international fiscal and monetary policy, attitudes toward saving and investment, and the quality of our labor force. True, regulation can "create" or "destroy" jobs in the short run, but only temporarily; in the long run, the opportunities for productive employment depend on the factors identified above.

Second, the environmental industry is now and probably always will be relatively small in the grand scheme of things. (The one million jobs in the environmental industry represent about eight-tenths of one percent of total civilian employment in the United States.) As economist Richard Schmalensee has pointed out, the year-to-year fluctuation in total U.S. employment is sometimes only slightly smaller than the whole of the environmental industry. This is emphatically *not* to disparage that industry—indeed, the United States enjoys a favorable balance of trade in environmental goods and services, one I hope grows larger still. But if given the choice between world dominance in the environmental industry or, say, a comparably strong position in automobile manufacturing, chemical production, or agriculture, we would be foolish not to choose any of the latter.

Third, even if environmental regulation could affect the overall level of employment in the long term, counting jobs created or destroyed is simply a poor way to evaluate environmental policies. Consider a regulation that resulted in the closure of a large factory employing hundreds of workers. While surely lamentable, this might be a very good policy from an overall social standpoint if the factory simply could not operate without discharging substances very harmful to human health and the environment. Conversely, one could envision a regulatory program that, in the short term, "created" jobs for hundreds or even thousands of workers. Yet if this program did little or nothing to improve environmental quality, it would be foolish to implement it despite its employment effects—the environmental pork barrel is no more benign than that from which other kinds of make-work projects are often plucked.

Separating Wheat from Chaff

How then do we distinguish wise from unwise policy proposals? The answer is at once very simple and very complicated. In my view, desirable regulations are those that promise to produce positive effects (improved human health, ecosystem protection, aesthetic amenities) that, when considered qualitatively yet carefully by our elected and appointed officials, more than offset the negative consequences that will result (higher prices to consumers, possible plant closures reduced productivity). In other words, wise regulations are those that pass a kind of commonsense benefit-cost test.

Three quick points are in order. First, and obviously, this type of evaluation is more easily described than done. Determining when the pros swamp the cons is often terribly difficult for any one of us to figure out; throw in the fact that we all have a different system of weights and measures, and you have the makings of environmental policy quagmires and donnybrooks.

Second, note my emphasis on a *qualitative* weighing of benefits and costs. While this may make me persona non grata among my fellow economists, I do not believe that a full-blown benefit-cost analysis—one in which all favorable and unfavorable effects must be expressed in dollar terms—should ever be the basis for a regulatory decision. In my view, uncertainties about valuation, the choice of a discount rate (and sometimes even whether to discount future effects at all), the appropriate handling of distributional concerns, and perhaps other problems as well, will always militate against policymaking by reliance on quantitative benefit-cost alone.

Third, these first two observations should not—repeat *not*—be taken to suggest that quantitative benefit-cost analysis has no useful role to play in environmental policymaking. Not only can this type of analysis help put on an equal footing many effects that seem incommensurable at first blush, but it can also reveal starkly the *implicit* values we hold that we often are understandably reluctant to express in dollars and cents. Better to make such trade-offs openly and explicitly, where all can see them, than to fuzz them over by pretending that they do not exist.

Moreover, contrary to some assertions, benefit-cost analysis is perfectly capable of supporting stringent environmental regulation. Among other policies, benefit-cost analyses have supported the Clean Air Act Amendments of 1970, the removal of lead from gasoline, and the phase-out of chlorofluorocarbons (CFCs) because of their role in stratospheric ozone depletion. To be sure, benefit-cost analyses have also cast serious doubt on the wisdom of certain other environmental proposals. Sauce for the goose is sauce for the gander, after all. One thing I think we can be sure of is this: environmental statutes that prohibit even the qualitative weighing of benefits and costs in standard-setting ensure uninformed policymaking.

Cost-Free Regulation?

Note my insistence that there *will* be costs to any regulation. Gardiner provides examples that suggest environmental regulation often jogs firms into discovering money-making opportunities about which they were previously ignorant. In these cases, he implies, citizens get the benefits of a cleaner environment while the regulated firm makes out well, too. In such cases, do we not escape the unpleasantness of trade-offs?

I think not. First, while there surely will be cases where complying with a regulation causes a firm to recognize a money-making opportunity it had been overlooking, I think it unlikely that such instances abound or that the associated profits will be very large. While corporations are hardly the paragons of efficiency that economics textbooks sometimes suggest, a kind of Darwinian market discipline does exist that forces firms to search out and take advantage of profitable opportunities.

More importantly, suppose a firm does realize profits rather than incur out-of-pocket costs when complying with an environmental regulation. In this case, surely, the regulation is costless, right? Wrong. While much more subtle, there is a cost here, too—an opportunity cost that takes the form of the returns the firm would have earned had it invested its expenditures on environmental compliance in other areas—say, on expanding its plant, retraining its work force, or intensifying its research and development efforts. In the same vein, incidentally, there is an opportunity cost associated with a firm's investment in any of the latter activities, even if that investment pays off handsomely. This cost is measured by what the firm could have earned had it put the funds to another use.

While opportunity costs are much less obvious than out-of-pocket expenditures for air or water pollution control equipment, cleaner fuels, or waste cleanup, they are no less real. Moreover, since it will never be possible to spend the same dollars on two things at once, a cost will *always* be associated with each environmental regulatory program. In some cases, it may take the form of out-of-pocket expenditures; but even when regulatory compliance helps a firm make money, we must be sophisticated enough to ask how well the firm would have done if it had put that same money to a different use.

In this regard, then, we can never have our cake and eat it too. Spending money in pursuit of environmental goals has been and can be a *very* wise use of society's scarce resources. But there will always be a cost to environmental regulatory programs, and environmental "paradigms" that promise otherwise are misleading and destined to disappoint.

5

Health-Based Environmental Standards
Balancing Costs with Benefits

Paul R. Portney and Winston Harrington

The 104th Congress is hotly debating an increased role for benefit-cost analysis in regulations pertaining to the environment, safety, and health. Portney and Harrington describe two arguments commonly propounded by those opposed to balancing health benefits against economic costs; then they explain why these arguments are not as convincing as they might seem at first glance.

Balancing the pros and cons of a proposed action seems like a commonsense approach to decisionmaking. But often that is not the approach embodied in environmental legislation. In establishing health-based environmental standards under the Clean Air Act, the Safe Drinking Water Act, and several other major environmental laws, for instance, Congress all but explicitly prohibits the U.S. Environmental Protection Agency (EPA) from balancing the benefits of tighter standards against the attendant costs. Given the 104th Congress's strong interest in using benefit-cost analysis for federal regulation, why have previous legislatures excluded such balancing from the most important standard-setting decisions made by EPA?

Below, we identify two basic arguments that have been put forward for disregarding costs in environmental decisionmaking and raise counterarguments to both. While these arguments and counterarguments require a more thorough analysis than we can devote to them here, our hope is that we will stimulate a more open and enlightened debate about them than we have seen to date.

The Right-to-a-Safe-Environment Argument

The right-to-a-safe-environment argument is perhaps the most common response to those (like us) who would seek to balance benefits and costs in standard setting. This argument is, of course, based on the presumption that safe levels of environmental contaminants *can* be found, a presumption that is appar-

Originally published in *Resources,* No. 120, Summer 1995.

ent in our environmental laws. For instance, the Clean Air Act requires EPA to provide "an adequate margin of safety... requisite to protect public health" in setting National Ambient Air Quality Standards. From our perspective, the right-to-a-safe-environment argument has two flaws—the first scientific, the second philosophical.

From a scientific standpoint, the problem is that *no* safe level is likely to exist for most, if not all, pollutants. Rather, lower ambient concentrations of a particular pollutant almost always will imply lower risks of an adverse health effect (see Figure 1). In the case of air pollution, even very low levels of pollutants pose some risk of adverse reactions in children and the elderly with chronic respiratory disease.

If air quality standards are required by law to provide an adequate margin of safety, and if even weak concentrations of pollutants pose some risk to some individuals, it appears that only zero concentrations could be permitted under the law, for only zero concentrations would provide a "margin of safety" against adverse health effects. But totally eliminating ubiquitous air and water pollutants is impossible in a modern industrial society like ours (and would be impossible even in a primitive society, at least as long as fires were allowed!).

The philosophical problem with the right-to-a-safe-environment argument is whether it makes sense to treat risk-free levels of air and water quality—even if they could be identified—as inalienable rights, such as freedom of speech. Those who oppose a balancing approach to environmental standard setting often argue that we did (or do) no such balancing in establishing and protecting the

Figure 1. Two possible relationships between exposure to pollution and adverse health effects.

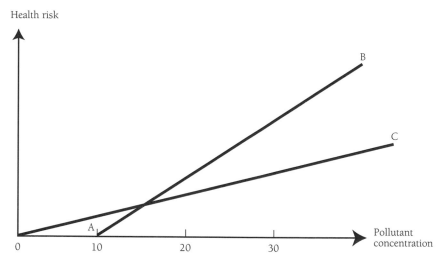

Through clinical, epidemiological, or animal toxicological studies, researchers try to identify a dose-response relationship between exposure to a harmful substance and adverse health effects—for example, incidences of cancer and asthma attacks. The line AB illustrates a case in which pollutant concentrations below ten units cause no such known effects. In this case, setting a pollution standard at some level below ten units would provide a "margin of safety" against adverse health effects. If the true dose-response curve is more like the line 0C, however, then any pollutant concentration above zero units would give rise to some risk. In this case, no safe level is likely to exist for the pollutant in question, making standard setting much more complicated.

basic freedoms that are guaranteed in the Constitution.

But elevating environmental quality to the status of a constitutional right, as some have proposed, would remove neither the necessity for nor the desirability of balancing. Even the basic freedoms that are guaranteed in the Bill of Rights have been subjected to a very crude kind of balancing test. For example, we cannot stand up and scream "Fire!" in a crowded theater; libel laws constrain our ability to write whatever we want to write about a person; and other basic rights are constrained in varying degrees. Such restrictions on the basic rights of Americans reflect a clear balancing mentality—that is, a carefully considered view that some extensions of our fundamental rights could create greater problems (read "costs") than the additional freedoms ("benefits") that the extensions would provide. If the authors and guardians of our Constitution made and continue to make qualitative trade-offs concerning our basic rights, then we see no reason why the freedom to enjoy a clean environment would not be similarly qualified, even after the freedom's elevation to "right"-hood.

Let's suppose that we are to regard environmental quality as a constitutional right. In that case, should we create a constitutional "right" to affordable housing? This amenity is arguably of greater importance to the average citizen than a risk-free environment. What distinguishes the rights guaranteed in the Constitution from those that are not guaranteed? And into which group does the right to a clean environment belong?

We respond to the question about distinguishing rights by noting that the freedoms of speech, religion, and so on are freedoms that people can enjoy extensively without reducing the rights of others. They impose no costs except in those extreme cases where the law already makes restrictions. In contrast, a right to shelter would impose costs on others. In this light, the answer to the second question is clear: in its costliness, a right to a clean environment is more like the right to shelter than the right of free speech. If costly rights were guaranteed in the Constitution, the need for consti-

tutional balancing would be the rule rather than the exception.

To put this argument another way, the need to balance environmental quality against other social objectives will not disappear just because we designate environmental quality a "right," but doing so may make balancing more difficult to achieve. For example, the right to a clean environment would conflict with constitutionally guaranteed rights to use and enjoy private property, as recent congressional debates about "takings" of property attest.

But suppose that environmental quality became a right and that we could identify safe levels of environmental contaminants. The question we would then have to ask is whether society could afford the expenditures that would be required to assure safe air (or water) quality for all citizens. To be sure, we should aspire to this goal; but just as we acknowledge that we have too few resources to accomplish other worthy goals, so too we might collectively decide that we cannot afford to reduce all air pollutants to safe levels everywhere. In view of the costs that might be involved, we might do better to expend at least some of our resources on other important social problems.

We illustrate this assertion using some numbers. By our accounting, we would guess that the nation will be spending at least $25 billion annually to control ground-level ozone by the year 2000. If our rough estimates are correct, we will soon spend about as much each year to comply with the ozone standard as we currently spend on all federal food stamp programs for the poor. Now spending the same amount on ozone control as on food stamps may be perfectly appropriate; after all, people in all walks of life are affected by poor air quality. But we believe that the allocation of resources is a subject about which there should be open and informed debate. In our opinion, we ought not to spend more on ozone control than we do on food stamps (or vaccinations, for that matter) simply because we can find a "safe" level at which to set an ozone standard.

This argument applies to other environmental standards. Even if it were possible to identify a safe level for, say, a drinking water contaminant, it doesn't follow that all communities should be required to meet that level under the Safe Drinking Water Act. Some communities might quite rationally decide to aim at a somewhat less ambitious standard and use the cost savings from doing so to finance another public program. In fact, the current flap in Washington over so-called unfunded mandates—federal regulatory requirements that fall on lower levels of government rather than on corporations—hinges on this point. State and local governments resent being told that they must spend their scarce resources on priorities established in Washington when they face other problems that they sometimes feel are far more pressing.

The Costs-Are-Considered-Anyway Argument

It could be argued that federal regulators inevitably consider costs in real-world environmental decisionmaking, despite the apparent statutory prohibition against doing so. Nonetheless, some people would assert that we should maintain the principle of excluding costs. This argument has two variants.

According to the first variant, we ignore costs in selecting ambient environmental standards, such as standards for the quality of our air and water, but take them into account in writing discharge standards for individual sources of pollution, such as electricity-generation facilities that often put sulfur dioxide into the air or farms that use pesticides that run off into lakes, rivers, and streams. These discharge standards place limits on the amounts of various pollutants that pollution sources can emit into the environment; the limits are intended to bring air and water quality, for example, into attainment with ambient standards. Typically, the discharge standards direct pollution sources to install the "best available technology," when these technological requirements are "affordable" or "economically achievable." In this sense, costs do come into play, ensuring that unaffordable discharge standards will not be imposed.

But what if the affected pollution sources cannot "afford" the technological requirements that would be necessary to meet ambient environmental goals? Short of extending the deadlines for complying with these requirements, EPA has little choice but to close down the affected sources. In short, costs can be taken into account, so long as the ultimate goals of environmental policy will be met; but those costs mean nothing if health-based standards are not met.

Insisting on effective discharge standards is appropriate if truly important health values, ecological values, or both would be compromised. But suppose that all the firms in a particular industry could afford to install the most sophisticated—and, therefore, the most expensive—pollution-control equipment made. Not everyone would agree that they should be required to do so simply because they can afford it—particularly if the health benefits of installing the equipment were deemed to be of marginal significance (that is, would reduce risk very little). While several of our current environmental statutes imply that any affordable environmental goal should be required, we suspect that many people would disagree. And they might ask whether these same statutes are creating a disincentive to succeed by requiring profitable, well-managed firms to meet stringent technological discharge standards, while treating leniently firms or industries that are on the brink of bankruptcy.

The second variant of the costs-are-considered-anyway argument is both frustrating and harder to rebut. According to this variant, we do not have to change environmental laws in order to balance health considerations against economic and other considerations, because such balancing occurs sub rosa each time that EPA sets health-based standards. So why, the argument goes, make balancing a requirement by law?

EPA *does* appear to take economic effects into account in setting some supposedly health-based standards. For instance, in 1978, when EPA promulgated the National Ambient Air Quality Standard for ozone, it stated that finding a literally "safe" ozone level was impossible and that setting a very tight ozone standard would significantly and negatively affect economic and social activities. For this reason, EPA rejected a zero-level standard.

Update

Since the early 1990s, both Congress and EPA have moved cautiously toward greater acceptance and use of cost considerations in legislation and rulemaking. The Safe Drinking Water Amendments of 1996 drastically reduce the number of contaminant standards that must be prepared from twenty-five every three years to an average of one per year; the amendments also require a risk prioritization to guide the regulatory agenda. For new standards, furthermore, EPA must conduct a benefit-cost analysis. Congress also has enacted the Food Quality Protection Act, which repeals the Delaney Clause and allows the EPA to set tolerance levels that balance the risks of pesticide residues against the benefits of pesticide use. On its own initiative, EPA has begun to give consideration to costs and benefits even when such balancing is not strictly allowed by the parent legislation. The recently proposed revisions to the ozone and fine-particulate ambient air quality standards recognize the emerging scientific consensus that there is no safe threshold for these pollutants and, therefore, some kind of balancing is necessary, even if it is forbidden by the Clean Air Act.

According to the documentation supporting the 1978 revision of the ozone standard, public health was the most compelling factor in the revision, but economic impact also was weighed.

If EPA acknowledges that economic impacts play at least some role in its setting of ambient standards under the Clean Air Act, and if this role is recognized and condoned, then it seems to us that Congress should amend the act, and other environmental laws as well, to explicitly allow balancing of health and economic considerations in standard setting. If that is current practice, and there exists general agreement that such practice is appropriate,

then balancing should be explicitly encouraged in the law. Not to do so engenders cynicism about the seriousness of our national intentions as well as contempt for our laws. Moreover, if no "safe" levels of many environmental contaminants can be found (as we suggest above), we cannot understand how Congress can avoid making our environmental laws explicitly require that health effects be balanced against economic and other possible adverse consequences.

Balancing Benefit-Cost Information with Other Information

We do not intend to suggest that establishing ambient environmental standards should be set on the basis of a formal quantitative benefit-cost analysis alone. Several considerations hinder an attempt to do so.

First, despite great progress in understanding how individuals value better health, reduced risks of premature mortality, aesthetic amenities, and other environmental benefits, economists are still a long way from pinning down precisely the marginal benefits associated with proposed changes in ambient environmental standards. In particular, great uncertainty surrounds estimates of how many lives such changes will save, how many illnesses they will prevent, and how much ecosystem protection they will provide.

Second, the costs associated with tighter standards are much harder to estimate than the public—and even some economists—realize. One reason is that regulations can impose costs even when no one must make out-of-pocket compliance expenditures. This would be the case if a regulation led, for example, to the withdrawal from the market of a useful product. Another reason is that regulated parties often cannot foresee technological advances that will reduce their compliance costs.

Third, even if we knew the marginal benefits and costs associated with alternative environmental quality standards, we still would not know whether equating the two would result in the "right" standard. Among other things, we might wish to know just who the winners and losers would be under new standards. For instance, suppose that only millionaires benefited from a tighter air quality standard, while the poor paid all the costs. Even if the added benefits from the tighter standard greatly exceeded the costs, we might resist adopting the new standard unless we could find a way to redistribute some of the net gain. In short, distributional considerations and other nonquantifiable factors having nothing to do with economic efficiency also matter a lot in standard setting.

Objecting to formal benefit-cost analysis as the sole basis for public decisionmaking is easy enough. Determining how such analysis *should* be used is far more difficult. We believe, however, that an analogy drawn from decisionmaking in the private sector can be useful in making this determination.

Before making an important investment decision, a good corporate manager will gather reports on the financial soundness of the venture and the expected future profits. Rather than slavishly basing a final decision on these reports alone, the good manager will temper the analytical information with his or her own judgment and experience. The manager may decide, for example, to overrule an apparently unfavorable financial projection out of a conviction that the long-run health of the company requires entry into new markets that will not pay off for some time. Or he or she may decide that the profit potential does not outweigh the risks of the project. In short, the manager understands that analytical information will rarely be complete or accurate enough to base decisions entirely on it. Giving due weight to and acting on information from all sources is the essence of good decisionmaking, and one of the private sector's strengths is its ability to recognize and reward good decisionmaking.

In the public sector, decisionmaking differs in ways that may make the use of formal benefit-cost methods both more difficult and, arguably, even more important. First, benefit-cost analysis in the public sector will probably be neither as complete nor as precise as its private-sector counterpart.

Second, the public manager may have to weigh additional objectives, such as the distribution of benefits, that do not easily fit into a formal benefit-cost analysis. Finally, success and failure in the public sector are much harder to identify, making any need to take corrective action that much more difficult to discern.

Since feedback from public-sector decisions is often weaker or more ambiguous than that from private-sector decisions, the methods and data used in making decisions become more important. While much of that information will be incomplete or imprecise, it will not be useless as long as its limitations are understood. If public decisionmakers are good at what they do, they will be able to weigh both the content and the quality of information about benefits and costs in the context of available information. Those who believe that decisions would be improved if benefit-cost information were denied to decisionmakers must harbor a pessimistic view of decisionmakers' abilities, a view that sits oddly with a generally expansionist view of the role of regulation.

Taking Economic Issues Seriously

Refusing to admit the need to consider costs may result from our collective desire to believe that difficult trade-offs need not be made. Well, we can't have it all. After more than twenty years of concerted efforts to meet our nation's environmental quality goals, we are still short of the mark in many areas. Moreover, since we now have acted upon the least expensive opportunities to reduce pollution, the remaining options are generally quite costly. Thus, providing all the protection we would like to provide is even less likely than it was two decades ago.

Nothing is wrong with wanting to provide maximum environmental protection to all citizens, just as we would like to provide all the other comforts of a happy and prosperous life. But something is wrong with denying that resources are scarce relative to our prodigious wants and that we must, accordingly, accept unpleasant trade-offs. Since in public rulemakings we openly acknowledge that we cannot find safe levels of environmental contaminants and since we admit the importance of economic considerations, shouldn't we revisit those portions of our environmental statutes that prohibit even the consideration of costs in standard setting? While economic considerations should never take primacy over public health or ecological concerns in policymaking, we believe that the answer to this question is an unambiguous yes.

6

Discounting the Future
Economics and Ethics

Timothy J. Brennan

Do we care more about people alive today than the people who will populate the earth in a hundred years? Should we? Brennan says that battles among economists and environmental advocates over discounting the future benefits and costs of environmental projects arise from confusing these two questions. He argues that separating them can clarify both the cost of serving future generations and our duty to do so.

How much do we care about people whose lives won't begin until long after our own have ended? How much *should* we care about them? These questions come up when we contemplate environmental projects that benefit people who are separated by many years or even by generations from those who pay the costs. Whether the interests of future generations will be at all significant in determining how much we should limit carbon emissions, preserve the ozone layer, or protect endangered species depends on whether a dollar's worth of future benefits is worth less than a dollar's worth of present costs—what economists mean by *discounting*.

Much controversy surrounds the practice of discounting. Divisive caricatures of the discounting wars pit economists, who allegedly view the environment as just another capital asset, against ethicists, who look out for the interests of people born in the future, and environmentalists, who advocate the inherent, noneconomic values in sustaining nature. In reality, discounting battles rage even among economists. Two leading experts on the economics of public projects, William Nordhaus of Yale University and Joseph Stiglitz of the president's Council of Economic Advisers, disagree over the appropriate way to discount the future costs and benefits of climate change.

When an issue has defied resolution for so long, perhaps the difficulty is a misunderstanding of the fundamental questions. Indeed, the difficulty may be that all the seemingly contrary positions on discounting have some validity. One cannot hope to resolve discounting debates among economists or to

Originally published in *Resources*, No. 120, Summer 1995.

allay the intensifying criticisms of discounting from those outside economics, but reflecting on the central arguments and illuminating the relationships between their economic and ethical sides may add a little light to the heat.

What Is Discounting?

One way to understand how discounting works is to compare it with the compounding of interest on savings (see Figure 1). Most people are familiar with the way compound interest increases the value of one's savings over time, in an accelerating way. For example, $100 invested today at 6 percent interest will be worth $106 in a year. Because the 6 percent interest will be earned on not just the initial $100 but the added $6 as well, the gains in the second year will be $6.36. Over time, these compounding gains become substantial. At 6 percent interest, the $100 investment will be worth about $200 in twelve years, $400 in twenty-four years, and $800 in thirty-six years. It will be worth around $3,300 in sixty years and almost $34,000 in a hundred years. A penny saved is more than a penny earned; after a century, the penny becomes $3.40. In 1626, Dutch explorers bought Manhattan for a mere $24; if that sum had been invested at just over 6 percent per year, it would have yielded more than $40 billion in 1990—about the total income generated in Manhattan that year.

Discounting operates in the opposite way. While compounding measures how much present-day investments will be worth in the future, discounting measures how much future benefits are worth today. To figure out this discounted present value, we must first choose a discount rate to transform benefits a year from now into benefits today. If we choose the same discounting rate as the interest rate in the above example of compounding, $106 a year from now would be equal in value to $100 today. Discounting the benefits of a project that generates $200 in twelve years by a discount rate of 6 percent per year would tell us that those benefits are worth $100 today.

To economists, this is the same as saying that $100 invested at an interest rate of 6 percent will

generate $200 in twelve years. For this reason, they often use the terms discount rate and interest rate interchangeably, although *discount rate* properly refers to how much we value future benefits today, while *interest rate* properly refers to how much present investments will produce over time.

The paramount consideration in assessing future environmental benefits is the size of the discount rate: The larger the discount rate, the less future benefits will count when compared with current costs. If the discount rate were 10 percent, $200 in twelve years would be worth only about $64 today; if the rate were 3 percent, the current value would be $140. At a zero discount rate, $1 of benefits in the future would be worth $1 in cost today. Differences in discount rates become crucial for benefits spanning very long periods.

The Obvious Cases for and against Discounting

The close relationship between interest rates and discount rates is the basis for the obvious case in favor of discounting. Suppose that an environmental program costing $100 today would bring $150 in benefits twelve years from now. If other public or business projects yield 6 percent per year, however, those future benefits of $150 would be "worth" only about $75 today after discounting. By investing the $100 today in one of these alternative projects, we could produce $200 in benefits in twelve years, leaving $50 more for the future.

Whether we view the environmental investment in terms of the present value of benefits ($75 as compared with $100) or in terms of an alternative investment that produces benefits of greater value ($200 as compared with $150), it fails the test of the market. Using a bit of economic jargon, we can call this market test the *opportunity-cost rationale* for discounting. Here, opportunity cost refers to the most value we can get by investing $100 in something other than the environment. According to the opportunity-cost rationale, we should discount future benefits from a current project to see if these benefits are worth *at least* as much to people

Figure 1. How much we value future dollars today: The effect of time and the discount rate.

Discounting operates in the reverse direction of compounding. While compounding measures how much present-day investments will be worth in the future, discounting measures how much future benefits are worth today. The illustration below shows how the discounted present value of future benefits can shrink to very small amounts as time goes on. Specifically, it shows how much $100 earned now and in 20, 40, 60, 80, and 100 years is worth today when a 3 percent discount rate is applied.

Along with the passage of time, increases in the discount rate also can dramatically shrink the discounted present value of future benefits. The illustration below shows how much $100 in benefits 100 years from now would be worth today at discount rates ranging from 0 to 6 percent.

When we see how small variations in the timing or discounting of future benefits can make large differences in deciding how much the benefits are worth today, it's easy to understand why discounting can lead to such heated policy debates.

in the future as the benefits they would have if we invested current dollars in medical research, education, more productive technology, and so on.

In effect, the opportunity-cost rationale tells us that our discount rate should be the market interest rate. Consequently, looking at the four factors that produce the interest rates that we see in financial markets will help explain what lies behind discount rates. The first factor is the level of economic activity. If investors want a lot of money for a lot of projects, they will have to pay a higher interest rate for loans; during slow economic times, investors will require fewer loans, leading to a lower interest rate. The second factor is inflation. Future dollars will be discounted if one cannot buy as much with them in the future as one can today. The third factor is risk; a guaranteed bird in the present hand may be worth a chancy two in the future bush. The fourth factor, and the most controversial one in environmental assessments, is what economists call *pure time preference*. This preference refers to the apparent fact that people require more than $1 in promised future benefits in order to be willing to give up $1 in goods today.

Critics of the opportunity-cost rationale often find that discounting leads to a present-day valuation of future environmental benefits that they believe is too low. Threats to life and nature from environmental degradation are notoriously hard to measure and, in the views of many, impossible to compare with the "mere" economic benefits that accrue from investing in a business project. Moreover, the benefits from a business investment might accrue to the wealthy or be frittered away today, while the benefits from an environmental project are likely to be distributed more widely across society and into the future.

Environmental benefits may or may not be overestimated in policy evaluations, and they may or may not be distributed more equitably than the returns from other investments. Those well-known criticisms, however, apply to cost-benefit tests in *any* context. The specific case against discounting fundamentally concerns pure time preference. A principle in most prominent ethical philosophies is that no individual's interests should count more than another's in deciding how social benefits should be distributed. If all men are created equal, as Thomas Jefferson wrote, there can be no justification for regarding the well-being of present generations as more important than that of future generations simply because of the difference in time. Given that principle, are we really justified in refusing to sacrifice $24 in 1995 if that $24 would bring "only" $4 billion—and not $40 billion—to people living in the year 2359? Substituting lives, or the capacity of wealth to save lives, for dollars makes

this question even more vivid and pressing. How could a future life, no matter how distant, be worth less than a present one? Using the language of philosophers and lawyers, we might call the insistence that future lives be valued equally to present ones the *equal standing* argument against discounting future benefits.

Might Cases for and against Discounting Both Be Valid?

Suppose we ask whether present generations should sacrifice short-run economic growth to undertake a particular program to improve the environment and leave more resources for future generations. Proponents of opportunity cost, who would discount future benefits, might say no, but proponents of equal standing, who would not discount future benefits, might say yes.

When a question has two compelling yet contradictory answers, it may really combine two questions in one. A close look at the question "should we undertake this environmental policy now to benefit future generations?" reveals that it asks a question about obligation (what duty do we have to sacrifice today to benefit future generations?) *and* a question about description (if we should sacrifice, do we help future generations more by implementing the proposed environmental policy or by doing something else?).

The economist's opportunity-cost rationale speaks to the question about description. If the goal is to improve the welfare of future generations, we should choose a policy that achieves the largest improvement for a given present cost. Consequently, we should compare the returns to the proposed environmental policy with those to other investments in order to see which are largest. Consider, for example, other investments with the same present-day costs as the environmental policy. If the discounted future benefits from these alternative policies are larger than the those from the environmental policy, we should consider implementing the alternative policies instead. We may be able to do more for future generations by subsidizing basic scientific and medical research or promoting education than by protecting the environment.

An obvious response would be to ask, "Why not invest in environmental protection *and* medical research?" This response brings us to the question about obligation—whether and how much to sacrifice. Unlike the question that asks us to describe and compare the benefits of one program to another, the obligation question asks us to contemplate our duties to future generations. As such, it fundamentally concerns ethical values rather than economic facts. Accordingly, equal standing is a more appropriate perspective from which to answer this question than is opportunity cost.

Proponents of the equal-standing principle have no problem with discounting for inflation or risk. But they find the pure-time-preference component of discounting to be morally controversial, even though the pure-time-preference discount rate is half the 6 percent discount rate drawn from today's markets. While a 19:1 ratio (present value to future value yielded by a 3 percent discount rate) is less philosophically forbidding than the 340:1 ratio (yielded by a 6 percent discount rate), it still is hard to reconcile with the equal-standing principle.

Violating "Hume's Law"

Separating environmental policy questions into questions about description and about obligation uncovers the root of much of the discounting controversy within economic circles and across disciplinary boundaries. This controversy is a consequence of trying to use facts about how people *do* discount to tell us how policymakers *should* discount. This attempt violates a maxim derived from eighteenth-century British philosopher David Hume, who asserted that facts alone cannot tell us what we should do. Any recommendation for what you, I, or society ought to do embodies some ethical principles as well as factual judgments. For example, to recommend policies if and only if their economic benefits exceed their costs would imply

the ethical principle that increasing net economic benefits is the only worthy goal for society.

The fact that we *do* have time preferences may not tell us much about how we ought to regard future generations. Imagine a world where generations do not overlap. In this world, people are like long-lived tulips; every eighty years, a new batch comes to life after the previous batch disappears. Suppose the people in one of those generations happen not to care about any subsequent generations. They would then choose to exhaust resources and degrade the environment without regard for how these actions might lower the quality of life of the people who succeed them. The *fact* of this disregard, however, does not invalidate an ethical principle that people born far in the future deserve a good quality of life as much as people already living.

Using market discount rates to examine ethical questions has made the economics of discounting more complicated than it perhaps needs to be. For example, economists have long argued about whether to calculate pure-time-preference discount rates based on the returns that investors receive before they pay taxes or after they pay taxes and, if after, whether to include corporate income taxes or personal income taxes in the calculation. If pure time preference has only limited ethical relevance in determining how much we should discount, these issues become relatively unimportant.

Divergence between equal-standing and opportunity-cost discount rates would be less important if policies that always did the best from one perspective did the best from the other as well. Unfortunately, this does not always hold. A policy that generates benefits in the short run may have a higher discounted value in an opportunity-cost sense than a policy that produces benefits much later. If we use a lower discount rate—that is, one reflecting more equal standing—the policy with long-term benefits may come out on top. We might need to do more for future generations; moreover, we might be doing the wrong things now. At opportunity-cost discount rates, development of an urban park may be more beneficial than an equally costly plan to reduce greenhouse gas emissions by taxing gasoline. At low or zero discount rates, the gasoline tax may be the more beneficial policy.

Philosopher Mark Sagoff of the University of Maryland suggests that market discount rates may not be a good indicator of the ethical value that people, upon reflection, would place on protecting future generations. Accordingly, we might resolve the discounting issue by having the government set policy based on people's stated ethical views regarding how to weigh current lives and dollars against future lives and dollars. Through a telephone survey of 3,000 U.S. households, Maureen Cropper, Sema Aydede, and Paul R. Portney of RFF determined that the rate at which people apparently discount lives saved is comparable to after-tax returns in financial markets. For example, people discount lives a century from now at about 4 percent per year. Equal-standing advocates can draw scant comfort from such data, which might tell us how a democracy would react if it followed the public's pure time preferences but, according to Hume, don't tell us what the right time preferences are.

Ethically Justified Discounting

Reconciling discounting with ethics may seem impossible, but there is some hope. To say that present and future generations have equal standing in an ethical sense does not necessarily imply that they have the same claim on present resources, because the general level of wealth or well-being may be changing over time. If we follow the ideas of a recent Nobel Prize winner in economics, John Harsanyi of the University of California–Berkeley, we should sacrifice today for the benefit of future generations only if the average well-being of people in the future goes up by more than we lose on average today. If present trends continue, advances in technology and knowledge will make people better off in the future than we are today. In that case, more than a dollar of gains to them would be needed to make up for a dollar lost to us. Any future returns should then be discounted by this difference to ensure that future generations' gains in well-being exceed our losses. According to the view proposed by Harvard University philosopher John Rawls, we might not be justified in making *any* sacrifice for future generations if they would be better off than we are now. If we expect future generations to be worse off than we are, however, Rawls' framework suggests that we should make present-day sacrifices.

More promising justifications for discounting come from critiques of the equal-standing idea itself. Philosophers such as Susan Wolf of Johns Hopkins University and Martha Nussbaum of Brown University have pointed out that to say that everyone has equal standing is to say that no one has special standing—including our families, friends, and fellow citizens. Insistence on equal standing denies the value that special interpersonal relationships hold for us and without which we could not be fully human. This argument may provide some support for asserting that generations closer to us should mean more to us than generations far in the future. (Thomas Schelling of the University of Maryland points out the irony of worrying so much about the welfare of future generations while doing so little to improve the welfare of many of the most destitute among us today.)

As long as resource scarcity makes trade-offs between the present generation and future generations inevitable, no consideration of environmental policies to benefit future generations should ignore economic opportunity cost. Ultimately, decisions to implement or not to implement such proposed policies will be the result of political processes, with all their virtues and imperfections. Justifications for the policies, which are tied in large measure to the degree of discounting, unavoidably involve ethical reflection and judgment. An appreciation of the necessary roles of both economics and ethics should clarify the nature of discounting and promote better understanding of our obligations toward future generations and how to meet them.

Suggested Reading

Arrow, K., and others. 1996. "Intertemporal Equity, Discounting, and Economic Efficiency." In J.P. Bruce, H. Lee, and E.F. Hates (eds.), *Climate Change 1995: Economic and Social Dimensions of Climate Change*. New York: Cambridge University Press.

Lind, R.C., and others. 1982. *Discounting for Time and Risk in Energy Policy*. Washington, DC: Resources for the Future (especially chapters by J. Stiglitz and A.K. Sen).

Parfit, D. 1984. *Reasons and Persons*. New York: Oxford University Press (especially Part Four, Future Generations).

Portney, P., and J. Weyant. In press. *Discounting and Intergenerational Equity*. Washington, DC: Resources for the Future.

Rawls, J. 1971. *A Theory of Justice*. Cambridge: Harvard University Press (especially Chapter V, Distributive Shares).

Schelling, T.C. 1995. Intergenerational Discounting. *Energy Policy* 23(4/5): 395–401.

7

When Is a Life Too Costly To Save?
The Evidence from Environmental Regulations

George L. Van Houtven and Maureen L. Cropper

Some environmental statutes require the U.S. Environmental Protection Agency (EPA) to balance benefits and costs when issuing regulations, while other statutes prohibit such balancing. But do these requirements or prohibitions make a big difference in the regulations that are written? According to a recent study conducted at Resources for the Future, the answer is "no." The study reveals that both benefits and costs appear to have influenced the regulations issued by EPA, regardless of the statutory mandate under which the agency was operating. The study also suggests that the value that EPA implicitly attaches to the prevention of one case of cancer is very high—from $15 million to $45 million. This is much more than individuals appear to be willing to spend to reduce their own risks of death.

Under the various environmental statutes the U.S. Environmental Protection Agency (EPA) administers, the agency is responsible for issuing regulations to protect the public from exposure to pollution. These regulations can include outright bans of certain products—for instance, pesticides and products containing asbestos. They more commonly include limitations on the amount of pollution a factory or vehicle can emit.

Most economists would argue that these regulations should be made, at least in part, on the basis of benefit-cost analyses. That is, they believe that an environmental standard should be set just at that point where the marginal cost of setting a slightly more stringent standard would begin to outweigh the marginal benefit of increased stringency. Congress, however, sometimes limits EPA's ability to engage in such balancing when the agency issues regulations.

For example, under the provisions of the Clean Air Act that pertain to the establishment of maximum permissible air pollution concentrations, EPA cannot take costs into account. When establishing effluent standards under the Clean Water Act, the agency is allowed to consider costs but not benefits. Only two environmental statutes—the Federal Insecticide, Fungicide, and Rodenticide Act (FIFRA) and the Toxic Substances Control Act (TSCA)—actually require that EPA balance the benefits and costs of regulation in setting environmental standards.

Recently, we conducted an after-the-fact analysis of regulatory decisions that EPA has made over the last two decades. Our purpose was to see whether EPA *appears* to have balanced costs

Originally published in *Resources*, No. 114, Winter 1994.

and benefits in issuing its regulations, regardless of whether the law under which the agency is operating directs or prohibits this balancing. In our study, "balancing" is determined by the following question: If we look back at a class of regulations EPA has put in place—for example, emissions standards for toxic air pollutants—do variations in the costs and benefits of all the regulatory options the agency considered at the time help explain the standards selected? If the answer is "yes," we argue that balancing has taken place.

We conclude that EPA *has* acted as if both costs and benefits influence its selection of regulatory standards; specifically, other factors being equal, a more costly standard is less likely to have been selected than a less costly one, and a standard that saves a greater number of lives is more likely to have been selected than one that saves a smaller number of lives.

Intuitively, however, balancing requires more than just paying some attention to costs and benefits. It also requires that the cost EPA is willing to impose on society to save an additional life be regarded as "reasonable." One way to determine what EPA considers reasonable is to see if there is some threshold value for the cost-per-life-saved above which the agency has been reluctant to issue regulations. (For lack of a better term, we call this threshold value the "value of a life" implied by the regulations.) For each class of regulations that we examined, we calculated the value of a life that was implicit in the regulations.

We were especially interested in two issues. The first and most important of these is how the value of a life implicit in environmental regulations compares with society's apparent willingness to pay to save lives: Is this value acceptable to American society? The second issue concerns the way in which the implicit value of a life seems to vary across EPA programs and across population groups: for instance, do environmental regulations pertaining to pesticides place a higher value on a life saved than regulations pertaining to hazardous air pollutants? Also, does EPA implicitly attach more weight to saving the life of a worker exposed to pesticides

or asbestos on the job than to the life of a consumer exposed to these pollutants?

To answer these questions, we gathered data on EPA-estimated costs and benefits associated with three categories of pollutants that EPA regulates:

1. asbestos, sources of which are regulated under TSCA;
2. all cancer-causing pesticides used on food crops that underwent EPA's Special Review process between 1975 and 1989; and
3. all carcinogenic air pollutants for which EPA set National Emissions Standards for Hazardous Air Pollutants between 1975 and 1990.

When we gathered data for each source of these pollutants (each crop in the case of pesticides), we arrived at a total of 39 asbestos regulations, 245 pesticide regulations, and 40 regulations pertaining to four hazardous air pollutants—benzene, inorganic arsenic, radionuclides, and vinyl chloride.

We limited our study to the regulation of carcinogens because quantitative risk data—that is, estimates of the number of lives at risk as a result of exposure to a particular pollutant or product—are available more often for carcinogens than for other substances. The availability of such data for carcinogens implies that the number of lives saved by each of the regulations we examined can be quantified. We also purposely selected some regulations issued under the two statutes (TSCA and FIFRA) that require EPA to balance costs and benefits, as well as those regulations that set emissions standards for hazardous air pollutants under the Clean Air Act, which prohibits such balancing. We included these regulations in our study in order to determine whether the directives given EPA in the enabling legislation made any difference in the way in which the agency appeared to weigh benefits and costs.

It is important to be clear about one thing. We were not privy to EPA's decisionmaking process for any of the regulations discussed here. What we have done is to look back at the information on benefits and costs that the agency had when it formulated the regulations, to examine the pattern of regulatory decisions, and then—using statistical analysis—to ascertain whether these decisions were consistent with the hypothesis that benefits and costs influence the regulatory outcome, regardless of the statutory mandate.

We turn now to a discussion of the specific regulations.

Asbestos Regulations under TSCA

In 1985 EPA announced its intent, under the authority of the Toxic Substances Control Act, to ban the use of asbestos in 39 products. Because TSCA requires EPA to balance benefits and costs, the agency's Notice of Intent to Regulate was followed by a detailed assessment of the health risks associated with exposure to asbestos fibers, as well as the costs that would result from the proposed bans.

Well-documented epidemiological evidence indicates that some forms of asbestos are human carcinogens. This evidence is particularly strong for lung cancer, gastrointestinal cancer, and mesothelioma, a cancer of the lung or abdominal lining. Estimating the number of cancer cases associated with a particular asbestos-containing product (for example, brakes lined with asbestos) requires estimates of the potency of asbestos—that is, of the likelihood that an individual will develop cancer as a function of exposure—as well as of the number of people who are exposed to various levels of asbestos.

In the analysis accompanying its final rule, EPA presented estimates of consumers' and various groups of workers' exposure to each product to be regulated, as well as an estimate of the number of cancer cases then-currently associated with each of these sources of asbestos. It also calculated the number of cancer cases that would be avoided if each product were banned. EPA was able to estimate this number, as well as the cost of the ban, for 31 of the 39 products considered for regulation.

A plotting of the regulatory costs and the number of cancer cases avoided for each of the 31 products for which complete data are available (see Figure 1) is consistent with the hypothesis that EPA considered benefits and costs in issuing its asbestos

decision. Products for which the cost of the ban was low and the number of lives saved was high (tending toward the lower right-hand corner of figure) were almost always banned, while products for which the cost of the ban was high and the number of lives saved was low (tending toward the upper left-hand corner of figure) were for the most part not banned.

Since avoiding cancer cases is the only benefit of the asbestos ban mentioned by EPA (ecological risks, for example, not being mentioned), it is tempting to infer from the plot that there is a threshold value for a cancer case avoided below which all products were banned. For instance, the rule "ban only those products for which the cost-per-life-saved is less than $10 million" (a rule illustrated by the lower of the two diagonal lines in the figure) would explain many of the bans, but it would yield incorrect predictions for some products. Similarly, the rule "do not ban any product for which the cost-per-life-saved is greater than $100 million" (a rule illustrated by the higher of the two diagonal lines in the figure) would be correct almost all the time, but it would yield incorrect predictions for some asbestos-containing products.

To compute the threshold value of a cancer case avoided that is implied by the asbestos regulations, we estimated statistically the line that maximized the number of regulations correctly predicted by the above-noted rule. We found that the implied threshold value of a cancer case avoided is $49 million (measured in 1989 dollars). (This value would have fallen between the two diagonal lines in the figure.) This value seems high—especially in contrast to estimates of the value of life that are based on individuals' willingness to pay for risk reductions.

Consider, for example, the added compensation that workers require to accept jobs that pose increasingly greater health risks, compensation that provides useful information about individuals' risk-reward trade-offs. Based on dozens of studies, the value of life that seems to be implicit in workers' occupational choices is about $5 million, an amount much lower than the value of life implicit

Figure 1. Cost-effectiveness of asbestos ban.

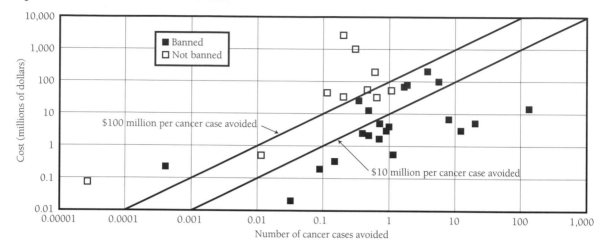

Each of the 31 squares represents a decision by EPA to ban or not to ban the use of asbestos in one particular product. The lower diagonal line illustrates the rule "ban only those products for which the cost-per-life-saved is less than $10 million." The higher diagonal line illustrates the rule "ban only those products for which the cost-per-life-saved is less than $100 million."

in EPA's regulation of asbestos. While labor market compensation is for risks that are voluntarily borne, it is hard to imagine that the additional premium associated with involuntary risks is $44 million. Not coincidentally, perhaps, EPA's failure to give sufficient weight to the costs of regulation in issuing its asbestos bans was cited by the Fifth Circuit Court of Appeals in the *Corrosion Proof Fittings* case, which overturned the ban.

Pesticide Regulations under FIFRA

Under the Federal Insecticide, Fungicide, and Rodenticide Act, EPA is responsible for ensuring that all pesticides used in the United States have no unreasonable adverse effects on the environment. If EPA suspects that a pesticide poses risks to human health or to ecosystems, the pesticide—or, more accurately, the active ingredients used in the pesticide—is subject to what is known as a Special Review.

A Special Review entails a formal risk-benefit analysis of the pesticide, after which EPA can either ban the pesticide for use on specific crops, restrict the manner in which the pesticide is applied, or allow its continued use without modification. Between 1975, when EPA initiated its first Special Review, and December 1989, Special Reviews of 37 active ingredients were completed. We restricted our analysis to those active ingredients that are suspected human carcinogens.

In considering whether or not to ban a pesticide, EPA examines risks of cancer to consumers of food containing pesticide residues and to persons exposed to the pesticide in the workplace—these are the people who mix the pesticides (mixers) and load them into the dispensing equipment (loaders), as well as those who apply the pesticides (applicators). The agency also examines noncancer health risks, such as risks of miscarriages or of fetal damage. In addition, it considers the adverse effects of the exposure of fish, birds, and mammals to pesticides.

Against these risks, EPA balances the benefits of pesticide use—that is, the gains to both farmers and consumers as a result of the increase in agricultural output brought about by pest control. Depending on the relative weight given to these and other factors, EPA might decide that a particular ingredient can no longer be used on a particular crop.

It is tempting to plot the cost of bans against the number of cancer cases avoided for pesticide regulations, as we did for asbestos regulations; however, the resulting diagram would be misleading. Because the avoidance of cancer cases is only one of the benefits of banning a particular use of a pesticide, our inferred threshold value of a cancer case avoided would overstate the value that EPA implicitly attaches to reducing cancer risks. Instead, we estimated a statistical model designed to predict EPA's decisions to cancel (or not cancel) the use of each of the active ingredients in pesticides on each of the food crops for which the ingredients were registered.

Our model, which correctly predicted 87 percent of the 245 decisions EPA made between 1975 and 1989, suggests that EPA considered both the risks and benefits of pesticide use in issuing its pesticide regulations. The benefits of pesticide use were statistically significant and of the expected sign: the higher the benefits of pesticide use, the less likely it was that a pesticide was banned for use on a particular crop. The risks associated with the pesticide were also important in explaining which uses of a pesticide were banned and which were not. Other factors being equal, the higher the risks of cancer to applicators (the group with the highest average exposure to pesticides), the greater the probability that a pesticide was banned. The implicit value of a cancer case avoided among applicators was about $52 million (1989 dollars)—a value remarkably close to the value we found to be implicit in asbestos regulations.

Our analysis was quite surprising in one respect: neither risks to mixers or loaders of pesticides nor dietary risks to consumers seemed to influence EPA's decisions to ban uses of active ingredients in pesticides. One possible explanation for this is that risks to both mixers and loaders and to consumers are lower than risks to applicators, and therefore seen to be a less pressing problem. The

median lifetime cancer risk associated [with] sures to the pesticide ingredients we exa[mined] 1 in 1,000 for applicators but only 1 in 1[...] for consumers of food with pesticide resid[ues].

National Emissions Standards for Hazardous Air Pollutants

In contrast to regulations issued under TSCA and FIFRA, the National Emissions Standards for Hazardous Air Pollutants were, according to the Clean Air Act of 1970, to be set to protect human health without consideration of costs. As we shall see, however, during the mid-1980s EPA did attempt to consider costs in setting emissions standards for sources of hazardous air pollution. In 1987, the Natural Resources Defense Council successfully sued the agency for making costs a factor in the determination of those standards. As discussed below, the ruling in that case had a pronounced effect on EPA's subsequent setting of standards for air pollution.

Section 112 of the Clean Air Act requires EPA to regulate the so-called toxic air pollutants, substances such as benzene, arsenic, asbestos, and mercury. These pollutants are not as ubiquitous as particulates, sulfur oxides, carbon monoxide, and other pollutants for which EPA is to set ambient air quality standards, but they are nonetheless harmful to human health. According to the Clean Air Act, EPA was required to establish a list of toxic air pollutants and then to set emissions limits for various sources of each pollutant. Between 1970 and 1990, only seven such substances were regulated. Five of these air pollutants are carcinogens, but quantitative risk data are available for only four—vinyl chloride, benzene, inorganic arsenic, and radionuclides. We examined the regulation of these substances.

In seeking to regulate the various sources of these four pollutants, EPA considered at least one regulatory option that would reduce emissions of each pollutant, as well as the option of no regulation. For each option, it computed cost, the number of associated cancer cases, and the post-regulation risk to the "maximally exposed individual," the individual who receives the greatest dose of a pollutant from a particular source. For most sources of hazardous air pollution, this individual is not exposed to the pollutant in the workplace, but rather lives near the source of the pollutant (for example, the person whose house is nearest to a copper or lead smelter).

To examine the possible trade-off between benefits and costs in the regulation of hazardous air pollution, we estimated a statistical model to explain which regulatory option was chosen for each of the 40 sources of vinyl chloride, benzene, inorganic arsenic, and radionuclides. Our results suggest that EPA's regulatory choices were consistent with the hypothesis that the agency was balancing the cancer risk reductions due to more stringent regulation against the costs this regulation would entail. In technical parlance, when we used all 40 sources of the four air toxics in our study to estimate our model, the coefficients of both the reduced cancer incidence and the regulatory cost were significant. The implicit value of a cancer case avoided—that is, the value that best enabled us to predict EPA's regulatory decisions—was very high— $153 million, to be exact.

These results look somewhat different, however, if we distinguish regulations issued before 1987 from those issued after 1987, when the U.S. Court of Appeals for the District of Columbia ruled that EPA had improperly considered costs in setting emissions standards for toxic air pollutants. In the so-called Vinyl Chloride decision, EPA was directed to consider the costs as well as the technological feasibility of regulatory options only after an "acceptable risk" level had been achieved.

When we modified our analysis to take this decision into account, our results came out quite differently: they implied that a cancer case avoided was valued at approximately $15 million before the 1987 court ruling and at the same amount after that ruling, so long as maximum individual risk was less than 1 in 10,000. After 1987, however, EPA always elected to regulate the source of a hazardous air pollutant if it posed a greater than 1 in 10,000 cancer risk to the maximally exposed individual; in other

words, the threshold value of a life was infinite. If this risk was less than 1 in 10,000, however, then the threshold value was the same for the post-1987 regulations as for the pre-1987 regulations: $15 million (1989 dollars).

Surprises and Questions

One of the striking findings of our analysis is that, in issuing the asbestos, pesticide, and toxic air pollutant regulations we examined, EPA has been willing to impose substantial costs on consumers and firms in order to save a life. Under each of the two statutes that allow the balancing of benefits and costs, the agency's implicit valuation of a cancer case avoided was in excess of $45 million. Whether members of society would agree with this valuation, which is about ten times greater than individuals implicitly value the risk of death due to occupational hazards, is an important question.

Nevertheless, compensation for risks faced in the workplace is generally for voluntary exposure to immediate risk of death. Exposure to asbestos and pesticides may not be voluntary (even for workers) if people are unaware of the risks they face; this fact may account for the very high implicit value assigned to risk reductions in EPA regulations pertaining to these substances.

It is interesting to note that the value per cancer case avoided that is implicit in regulations pertaining to hazardous air pollutants was about one-third the value implicit in pesticide or asbestos regulations. In a sense, this is quite surprising. Our findings suggest that EPA has, in the past, put in place more stringent regulations under statutes that require it to balance benefits and costs than it does under a statute that directs it to ignore costs and consider health risks only. This does not "prove" that EPA balanced costs and benefits under the Clean Air Act, only that it made decisions that were consistent with the hypothesis that the agency behaved this way.

This in turn raises the question of whether statutes that prohibit consideration of costs in standard setting really make a difference in the regulations that are issued. Our analysis of the setting of the National Emission Standards for Hazardous Air Pollutants suggests that, short of recourse to the courts, prohibitions against consideration of costs may be difficult to enforce. Likewise, Congress may require that the costs of a regulation be balanced against its benefits; but as long as EPA has discretion in the weights it assigns to costs and benefits, the regulations it issues under statutes that allow balancing of benefits and costs may still be very costly.

Suggested Reading

Cropper, Maureen L., William N. Evans, Stephen J. Berardi, Maria Ducla-Soares, and Paul R. Portney. 1992. The Determinants of Pesticide Regulation: A Statistical Analysis of EPA Decision Making. *Journal of Political Economy* 100: 175–97.

Cropper, Maureen L., William N. Evans, Stephen J. Berardi, Maria Ducla-Soares, and Paul R. Portney. 1992. Pesticide Regulation and the Rule-Making Process. *Northeast Journal of Agricultural and Resource Economics* 772–82.

Morgenstern, Richard D., ed. 1997. *Economic Analyses at EPA: Assessing Regulatory Impact.* Washington, DC: Resources for the Future.

Van Houtven, George, and Maureen Cropper. 1996. When is a Life Too Costly to Save? Evidence from Environmental Regulations. *Journal of Environmental Economics and Management* 30(3): 348–68.

8

Assessing Damages from the *Valdez* Oil Spill

A. Myrick Freeman III and Raymond J. Kopp

People who have never been to Prince William Sound and never plan to visit nevertheless may feel a keen sense of loss over the damages it has sustained from ten million gallons of spilled oil. Many economists contend that this loss must he reflected in court damage awards. Moreover, they say, there are methods available for attempting to put a dollar value on it.

When the Exxon *Valdez* hit Bligh Reef in Prince William Sound, Alaska, on March 24, 1989, the massive spill of crude oil triggered several important efforts. The first, of course, has been the attempt to recover as much oil as possible, clean up oiled beaches and shoreline, rescue threatened sea mammals and birds, and restore as much as possible of the affected area to its pre-spill condition. At this writing, the effort is employing more than 4,000 workers and costs may run to the hundreds of millions of dollars. It appears that the Exxon Corporation will be financially liable for cleanup and restoration costs.

The second initiative set into motion is a set of scientific studies of the dispersion, breakdown, and persistence of the spilled oil in the environment and its impacts on the terrestrial and marine ecologies of the region. Shortly after the accident, Exxon agreed to contribute $15 million toward scientific studies of the spill's impact.

And third, economists have taken on the thorny task of quantifying and assigning monetary values to the damages to the natural resources of the impacted region. Under provisions of the Clean Water Act of 1972 and Alaska state law, both the federal and state governments can sue "potentially responsible parties" to recover the economic damages sustained by the publicly owned natural resources of the affected region.

From an economic perspective, natural resource damages resulting from an accident such as the *Valdez* spill stem from the

Originally published in *Resources,* No. 96, Summer 1989. It was adapted from an op-ed piece by Freeman that appeared in the *Wall Street Journal* on May 24, 1989.

reduction in the flow of services from the environment as a consequence of its contamination by spilled oil. These services include such things as the biological productivity of the resource that supports commercial and sports fishing, hunting, and the subsistence efforts of local residents, as well as the visual beauty and amenities that attract tourists. Estimation of the damages involves identifying which valued services are provided by the publicly owned portions of the affected resources, measuring how much these services have been reduced or impaired as a consequence of the spill, and determining how much these lost services were worth to the people who formerly benefited from them.

What Are Damages?

The economic concept of damage is based on the idea of compensation—finding out how much money it would take to make everyone who has been affected by the reduction in service flows as well off as they were before the incident occurred. Although much of the attention of resource economists has been devoted to the analysis of the values of services to those who make direct use of the environment (so-called *use values*), it has been recognized at least since John Krutilla's pioneering article, "Conservation Reconsidered" (*American Economic Review*, 1967), that environmental services could be valued by people who do not make direct use of them (so-called *non-use values*).

In the case of the Prince William Sound area, several types of uses are likely to be impaired by the oil spill. One of the more obvious is the use of the marine ecosystem as a source of fish for the commercial fisheries of Alaska. Prince William Sound houses some of the most productive fisheries in the world and supplies both domestic and international markets. If scientific studies are able to establish a relationship between the spilled oil and the harvest rates of commercial species of fish, quantifying the dollar loss to these fisheries will be relatively easy. This is because market prices—which reflect the value society places on the fishery—can be observed. These prices can provide a basis for mea-

suring this component of the damage caused by the spill. Consumers as well as fishermen and processors are likely to suffer these losses. Consumers' losses may come in the form of higher prices for fish products. Fishermen and processors may suffer because of reduced revenues and higher costs.

By contrast, accurately estimating the damages to some of the other types of uses will be much more difficult. Many of the services provided by the region's resources cannot be purchased in markets. For example, the scenic beauty of the Sound enjoyed by local residents and visitors will be diminished by the presence of the oil on the beaches and the absence of wildlife killed or driven away by the oil pollution. Resident and nonresident sports fishermen may find the region less attractive because of the reduced chance of catching fish or fear of catching contaminated fish. No one is asked to pay a price for the pleasure of enjoying the scenic beauty or the right to fish on the open waters. How then can we place a value on these losses?

Economists have developed techniques for estimating the values of such unpriced resource services. They include drawing inferences from behavior such as willingness to incur travel costs to experience natural environments first hand and asking people directly about the values they place on these environments through what has become known as contingent valuation surveys. Although the absolute accuracy of such estimates cannot be guaranteed, economists do possess a widely accepted framework for measuring these values. The framework applies not only to those services traded in markets but also to those types of individuals' values that are unpriced and therefore not revealed through market transactions.

Non-Use Values

A third category of damage ensues not from the diminution of the quality or quantity of services provided by the Sound, but rather from society's knowledge that a unique natural environment has been injured. Economists refer to these damages as lost *non-use* or *intrinsic values*. The parties suffering lost non-use values may be the same parties who have experienced lost use values but may also include the generally larger group of individuals that had no direct involvement in the area of the spill but nevertheless feel a loss. Generally, economists would argue that non-use values have been reduced if individuals enjoying none of the Sound's use values would have been willing to pay some dollar amount to ensure that a spill of this magnitude would not have taken place or that the probability and/or magnitude of such an occurrence in the future could be reduced.

It would be misleading to suggest that all economists agree about the measurement of non-use values or even agree whether non-use values can be measured at all. However, they do tend to agree that there are features of natural environments like Prince William Sound (such as unusual concentrations of wildlife, including rare and endangered species; unique scenic beauty; and pristine wilderness) that are valued by people who have never been to the Sound or never plan to visit. The origin of these values may lie in a desire to preserve the natural environments for the enjoyment of future generations or from a simple expression of stewardship. Economists generally place diminished non-use values on an equal footing with diminished use values and argue that failure to account for lost non-use values will understate the damage suffered by society. The case of the *Valdez* spill is no exception.

A substantial body of accumulated research dating back to the middle 1960s has led to the development of methods and empirical techniques for quantifying non-use values. Termed *contingent valuation*, the methodology combines knowledge from several disciplines in a consistent conceptual framework to measure non-use values. Contingent valuation is based on surveys used to gather data to help determine society's willingness to pay for public goods. This information about people's willingness to pay can be used as an indication of the existence of non-use values and as a tool for non-use value quantification. The contingent valuation method has been applied in the field on several occasions. It can be used to place dollar values on

> ### Update
> Shortly after this article was written, the state of Alaska assembled a team of experts to conduct a contingent valuation study of the damages caused by the oil spill. Raymond Kopp was a principal investigator on that team. The results of that study placed the dollar value of the injuries at $2.8 billion.

such things as enhanced visibility due to decreased power plant emissions and the diminished non-use values associated with contaminated drinking water.

Measuring non-use values is an admittedly difficult task. By necessity the measurement techniques involve not only economic theory but knowledge of psychology, statistical science, and the techniques of survey research. Moreover, the introduction of these measurement techniques into courts of law and natural resource damage cases can hinge on the courts' interpretation of the rules of evidence as applied to survey results.

In addition, the surveys themselves and the questions they contain may be poorly designed, executed, and analyzed and thus produce results that misleadingly purport to quantify non-use values. Yet a growing body of research shows that well-designed, executed, and analyzed contingent valuation studies lead to valuation estimates that accord nicely with the results from other techniques. Thus there is confidence that contingent valuation can be reliably applied to the estimation of lost non-use values.

Who Should Pay?

Since the *Valdez* spill, it has been widely held that Exxon has behaved irresponsibly and the public should not have to pay for the mistakes of the corporation or its employees. Meanwhile, the prices of crude oil and gasoline have increased in the wake of the spill, prompting outrage and the suspicion of gouging. This suspicion is difficult to substantiate since other factors, such as the loss of some North Sea oil production and lack of refinery capacity in the face of new regulations governing gasoline volatility, appear to have contributed to the price rise.

To the extent that negligence or criminal behavior might have been a factor in the spill, the economic concept of deterrence and simple notions of justice suggest that those with individual or corporate responsibility should bear the costs associated with their actions. But the idea that consumers of petroleum products should bear at least some portion of the costs of environmental risk and damage associated with bringing fuel to the pumps has validity from an economic perspective. Those who use petroleum products do bear some responsibility for the environmental risks and damages that are an inevitable part of the oil exploration, production, and delivery system that serves their demands. An efficient allocation of resources requires that they accept the financial burden that goes along with that responsibility.

Economic Role in Assessment

If the economic damages from the *Valdez* oil spill can be accurately assessed and monetary damages can be collected from the responsible parties, we can expect several economically beneficial consequences in addition to the potential for compensating those who experienced the most serious losses. First, since the damages will become part of the cost of doing business, prices of petroleum products will rise, thus more accurately reflecting the environmental costs associated with their extraction and transportation. Higher prices, in turn, should mean less consumption of petroleum. At the same time, the potential liability for damages due to future spills is also likely to reduce the expected returns to the oil companies from exploration and development in more risky environments. And third, the prospect of liability for large natural resource damage claims should provide the oil industry with incentives to take further actions to increase the safety of their operations and to reduce the risks of future accidents.

On the other hand, if the economic damages are not accurately assessed, or if court-awarded damages are not based on sound economic analysis, we can expect undesirable consequences. An award that is excessive and exceeds the appropriate social compensation will lead to an economically inefficient underutilization of the nation's oil resources, while understating the damages would lead to excessive oil extraction and consumption. Whatever the economic damages are, it is important to do our best to get the numbers right.

The Faustian Bargain

Allen V. Kneese

This chapter was prepared originally in response to a request by the Atomic Energy Commission for comments on one of its documents; the AEC noted that environmental statements for a power reactor should contain a benefit-cost analysis which, among other things, "considers and balances the adverse environmental effects and the environmental, economic, technical and other benefits of the facility." Allen V. Kneese submitted the following remarks to the AEC.

I am submitting this statement as a long-time student and practitioner of benefit-cost analysis, not as a specialist in nuclear energy. It is my belief that benefit-cost analysis cannot answer the most important policy questions associated with the desirability of developing a large-scale, fission-based economy. To expect it to do so is to ask it to bear a burden it cannot sustain. This is so because these questions are of a deep *ethical* character. Benefit-cost analyses certainly cannot solve such questions and may well obscure them.

These questions have to do with whether society should strike the Faustian bargain with atomic scientists and engineers, described by Alvin M. Weinberg in *Science*. If so unforgiving a technology as large-scale nuclear fission energy production is adopted, it will impose a burden of continuous monitoring and sophisticated management of a dangerous material, essentially forever. The penalty of not bearing this burden may be unparalleled disaster. This irreversible burden would be imposed even if nuclear fission were to be used only for a few decades, a mere instant in the pertinent time scales.

Clearly, there are some major advantages in using nuclear fission technology, else it would not have so many well-intentioned and intelligent advocates. Residual heat is produced to a greater extent by current nuclear generating plants than by fossil fuel-fired ones. But, otherwise, the environmental impact of routine operation of the nuclear fuel cycle, including burning the fuel in the reactor, can very likely be brought to a lower level than will be possible with fossil fuel-fired plants. This superior-

Originally published in *Resources,* No. 44, September 1973.

ity may not, however, extend to some forms of other alternatives, such as solar and geothermal energy, which have received comparatively little research and development effort. Insofar as the usual market costs are concerned, there are few published estimates of the costs of various alternatives, and those which are available are afflicted with much uncertainty. In general, however, the costs of nuclear and fossil fuel energy (when residuals generation in the latter is controlled to a high degree) do not seem to be so greatly different. Early evidence suggests that other as yet undeveloped alternatives (such as hot rock geothermal energy) might be economically attractive.

Unfortunately, the advantages of fission are much more readily quantified in the format of a benefit-cost analysis than are the associated hazards. Therefore, there exists the danger that the benefits may seem more real. Furthermore, the conceptual basis of benefit-cost analysis requires that the redistributional effects of the action be, for one or another reason, inconsequential. Here we are speaking of hazards that may affect humanity many generations hence and equity questions that can neither be neglected as inconsequential nor evaluated on any known theoretical or empirical basis. This means that technical people, be they physicists or economists, cannot legitimately make the decision to generate such hazards. Our society confronts a moral problem of a great profundity; in my opinion, it is one of the most consequential that has ever faced mankind. In a democratic society the only legitimate means for making such a choice is through the mechanisms of representative government.

For this reason, during the short interval ahead while dependence on fission energy could still be kept within some bounds, I believe the Congress should make an open and explicit decision about this Faustian bargain. This would best be done after full national discussion at a level of seriousness and detail that the nature of the issue demands. An appropriate starting point could be hearings before a committee of Congress with a broad national policy responsibility. Technically oriented or special-

ized committees would not be suitable to this task. The Joint Economic Committee might be appropriate. Another possibility would be for the Congress to appoint a select committee to consider this and other large ethical questions associated with developing technology. The newly established Office of Technology Assessment could be very useful to such a committee.

Much has been written about hazards associated with the production of fission energy. Until recently, most statements emanating from the scientific community were very reassuring on this matter. But several events in the past year or two have reopened the issue of hazards and revealed it as a real one. I think the pertinent hazards can usefully be divided into two categories—those associated with the actual operation of the fuel cycle for power production and those associated with the long-term storage of radioactive waste. I will discuss both briefly.

The recent failure of a small physical test of emergency core cooling equipment for the present generation of light-water reactors was an alarming event. This is in part because the failure casts doubt upon whether the system would function in the unlikely, but not impossible, event it would be called upon in an actual energy reactor. But it also illustrates the great difficulty of forecasting behavior of components in this complex technology where pertinent experimentation is always difficult and may sometimes be impossible. Other recent unscheduled events were the partial collapse of fuel rods in some reactors.

There have long been deep but suppressed doubts within the scientific community about the adequacy of reactor safety research vis-à-vis the strong emphasis on developing the technology and getting plants on the line. In recent months the Union of Concerned Scientists has called public attention to the hazards of nuclear fission and asked for a moratorium on the construction of new plants and stringent operating controls on existing ones.

The division of opinion in the scientific community about a matter of such moment is deeply disturbing to an outsider.

No doubt there are some additional surprises ahead when other parts of the fuel cycle become more active, particularly in transportation of spent fuel elements and in fuel reprocessing facilities. As yet, there has been essentially no commercial experience in recycling the plutonium produced in nuclear reactors. Furthermore, it is my understanding that the inventory of plutonium in the breeder reactor fuel cycle will be several times greater than the inventory in the light-water reactor fuel cycle with plutonium recycle. Plutonium is one of the deadliest substances known to man. The inhalation of a millionth of a gram—the size of a grain of pollen—appears to be sufficient to cause lung cancer.

Although it is well known in the nuclear community, perhaps the general public is unaware of the magnitude of the disaster which would occur in the event of a severe accident at a nuclear facility. I am told that if an accident occurred at one of today's nuclear plants, resulting in the release of only five percent of only the more volatile fission products, the number of casualties could total between 1,000 and 10,000. The estimated range apparently could shift up or down by a factor of ten or so, depending on assumptions of population density and meteorological conditions.

With breeder reactors, the accidental release of plutonium may be of greater consequence than the release of the more volatile fission products. Plutonium is one of the most potent respiratory carcinogens in existence. In addition to a great variety of other radioactive substances, breeders will contain one, or more, tons of plutonium. While the fraction that could be released following a credible accident is extremely uncertain, it is clear that the release of only a small percentage of this inventory would be equivalent to the release of *all* the volatile fission products in one of today's nuclear plants. Once lost to the environment, the plutonium not ingested by people in the first few hours following an accident would be around to take its toll for generations to come—for tens of thousands of years.

When one factors in the possibility of sabotage and warfare, where power plants are prime targets not just in the United States but also in less developed countries now striving to establish a nuclear industry, then there is almost no limit to the size of the catastrophe one can envisage.

It is argued that the probabilities of such disastrous events are so low that these events fall into the negligible risk category. Perhaps so, but do we really know this? Recent unexpected events raise doubts. How, for example, does one calculate the actions of a fanatical terrorist?

The use of plutonium as an article of commerce and the presence of large quantities of plutonium in the nuclear fuel cycles also worries a number of informed persons in another connection. Plutonium is readily used in the production of nuclear weapons, and governments, possibly even private parties, not now having access to such weapons might value it highly for this purpose. Although an illicit market has not yet been established, its value has been estimated to be comparable to that of heroin (around $5,000 per pound). A certain number of people may be tempted to take great risks to obtain it. AEC Commissioner Larsen, among others, has called attention to this possibility. Thus, a large-scale fission energy economy could inadvertently contribute to the proliferation of nuclear weapons. These might fall into the hands of countries with little to lose or of madmen, of whom we have seen several in high places within recent memory.

In his excellent article referred to above, Weinberg emphasized that part of the Faustian bargain is that to use fission technology safely, society must exercise great vigilance and the highest levels of quality control, continuously and *indefinitely*. As the fission energy economy grows, many plants will be built and operated in countries with comparatively low levels of technological competence and a greater propensity to take risks. A much larger amount of transportation of hazardous materials will probably occur, and safety will become the province of the sea captain as well as the scientist. Moreover, even in countries with higher levels of technological competence, continued success can lead to reduced vigilance. We should recall that we managed to incinerate three astronauts in a very straightforward accident in an extremely high technology operation where the utmost precautions were allegedly being taken.

Deeper moral questions also surround the storage of high-level radioactive wastes. Estimates of how long these waste materials must be isolated from the biosphere apparently contain major elements of uncertainty, but current ones seem to agree on "at least two hundred thousand years."

Favorable consideration has been given to the storage of these wastes in salt formations, and a site for experimental storage was selected at Lyons, Kansas. This particular site proved to be defective. Oil companies had drilled the area full of holes, and there had also been solution mining in the area which left behind an unknown residue of water. But comments of the Kansas Geological Survey raised far deeper and more general questions about the behavior of the pertinent formations under stress and the operations of geological forces on them. The ability of solid earth geophysics to predict for the time scales required proves very limited. Only now are geologists beginning to unravel the plate tectonic theory. Furthermore, there is the political factor. An increasingly informed and environmentally aware public is likely to resist the location of a permanent storage facility anywhere.

Because the site selected proved defective, and possibly in anticipation of political problems, primary emphasis is now being placed upon the design of surface storage facilities intended to last a hundred years or so, while the search for a permanent site continues. These surface storage sites would require continuous monitoring and management of a most sophisticated kind. A complete cooling system breakdown would soon prove disastrous and even greater tragedies can be imagined.

Just to get an idea of the scale of disaster that could take place, consider the following scenario.

Political factors force the federal government to rely on a single above-ground storage site for all high-level radioactive waste accumulated through the year 2000. Some of the more obvious possibilities would be existing storage sites like Hanford or Savannah, which would seem to be likely military targets. A tactical nuclear weapon hits the site and vaporizes a large fraction of the contents of this storage area. The weapon could come from one of the principal nuclear powers, a lesser developed country with one or more nuclear power plants, or it might be crudely fabricated by a terrorist organization from black-market plutonium. I am told that the radiation fallout from such an event could exceed that from all past nuclear testing by a factor of 500 or so, with radiation doses exceeding the annual dose from natural background radiation by an order of magnitude. This would bring about a drastically unfavorable and long-lasting change in the environment of the majority of mankind. The exact magnitude of the disaster is uncertain. That massive numbers of deaths might result seems clear. Furthermore, by the year 2000, high-level wastes would have just begun to accumulate. Estimates for 2020 put them at about three times the 2000 figure.

Sometimes, analogies are used to suggest that the burden placed upon future generations by the "immortal" wastes is really nothing so very unusual. The Pyramids are cited as an instance where a very long-term commitment was made to the future and the dikes of Holland as one where continuous monitoring and maintenance are required indefinitely. These examples do not seem at all apt. They do not have the same quality of irreversibility as the problem at hand, and no major portions of humanity are dependent on them for their very existence. With sufficient effort the Pyramids could have been dismantled and the Pharaohs cremated if a changed doctrine so demanded. It is also worth recalling that most of the tombs were looted already in ancient times. In the 1950s the Dutch dikes were in fact breached by the North Sea. Tragic property losses, but no destruction of human life, ensued. Perhaps a more apt example of the scale of the Faustian bargain would be the irrigation system of ancient Persia. When Tamerlane destroyed it in the 14th century, a civilization ended.

None of these historical examples tell us much about the time scales pertinent here. One speaks of two hundred thousand years. Only a little more than one-hundredth of that time span has passed since the Parthenon was built. We know of no government whose life was more than an instant by comparison with the half-life of plutonium.

It seems clear that there are many factors here which a benefit-cost analysis can never capture in quantitative, commensurable terms. It also seems unrealistic to claim that the nuclear fuel cycle will not sometime, somewhere experience major unscheduled events. These could range in magnitude from local events, like the fire at the Rocky Mountain Arsenal, to an extreme disaster affecting most of mankind. Whether these hazards are worth incurring in view of the benefits achieved is what Alvin Weinberg has referred to as a transscientific question. As professional specialists we can try to provide pertinent information, but we cannot legitimately make the decision, and it should not be left in our hands.

One question I have not yet addressed is whether it is in fact not already too late. Have we already accumulated such a store of high-level waste that further additions would only increase the risks marginally? While the present waste (primarily from the military program plus the plutonium and highly enriched uranium contained in bombs and military stockpiles) is by no means insignificant, the answer to the question appears to be no. I am informed that the projected high-level waste to be accumulated from the civilian nuclear power program will contain more radioactivity than the military waste by 1980 or shortly thereafter. By 2020 the radioactivity in the military waste would represent only a small percentage of the total. Nevertheless, we are already faced with a substantial long-term waste storage problem. Development

of a full-scale fission energy economy would add overwhelmingly to it. In any case, it is never too late to make a decision, only later.

What are the benefits?

The main benefit from near-term development of fission power is the avoidance of certain environmental impacts that would result from alternative energy sources. In addition, fission energy may have a slight cost edge, although this is somewhat controversial, especially in view of the low plant factors of the reactors actually in use. Far-reaching clean-up of the fuel cycle in the coal energy industry, including land reclamation, would require about a 20 percent cost increase over uncontrolled conditions for the large, new coal-fired plants. If this is done, fission plants would appear to have a clear cost edge, although by no means a spectacular one. The cost characteristics of the breeder that would follow the light-water reactors are very uncertain at this point. They appear, among other things, to still be quite contingent on design decisions having to do with safety. The dream of "power too cheap to meter" was exactly that.

Another near-term benefit is that fission plants will contribute to our supply during the energy "crisis" that lies ahead for the next decade or so. One should take note that this crisis was in part caused by delays in getting fission plants on the line. Also, there seems to be a severe limitation in using nuclear plants to deal with short-term phenomena. Their lead time is half again as long as fossil fuel plants—on the order of a decade.

The long-term advantage of fission is that once the breeder is developed we will have a nearly limitless, although not necessarily cheap, supply of energy. This is very important but it does not necessarily argue for a near-term introduction of a full-scale fission economy. Coal supplies are vast, at least adequate for a few hundred years, and we are beginning to learn more about how to cope with the "known devils" of coal. Oil shales and tar sands also are potentially very large sources of energy, although their exploitation will present problems. Geothermal and solar sources have hardly been considered but look promising. Scientists at the AEC's Los Alamos laboratory are optimistic that large geothermal sources can be developed at low cost from deep hot rocks—which are almost limitless in supply. This of course is very uncertain since the necessary technology has been only visualized. One of the potential benefits of solar energy is that its use does not heat the planet. In the long term this may be very important.

Fusion, of course, is the greatest long-term hope. Recently, leaders of the U.S. fusion research effort announced that a fusion demonstration reactor by the mid-1990s is now considered possible. Although there is a risk that the fusion option may never be achieved, its promise is so great that it merits a truly national research and development commitment.

A strategy that I feel merits sober, if not prayerful, consideration is to phase out the present set of fission reactors, put large amounts of resources into dealing with the environmental problems of fossil fuels, and price energy at its full social cost, which will help to limit demand growth. Possibly it would also turn out to be desirable to use a limited number of fission reactors to burn the present stocks of plutonium and thereby transform them into less hazardous substances. At the same time, the vast scientific resources that have developed around our

fission program could be turned to work on fusion, deep geothermal, solar, and other large energy supply sources while continuing research on various types of breeders. It seems quite possible that this program would result in the displacement of fission as the preferred technology for electricity production within a few decades. Despite the extra costs we might have incurred, we would then have reduced the possibility of large-scale energy-associated nuclear disaster in our time and would be leaving a much smaller legacy of "permanent" hazard. On the other hand, we would probably have to suffer the presence of more short-lived undesirable substances in the environment in the near term.

This strategy might fail to turn up an abundant clean source of energy in the long term. In that event, we would still have fission at hand as a developed technological standby, and the ethical validity of using it would then perhaps appear in quite a different light.

We are concerned with issues of great moment. Benefit-cost analysis can supply useful inputs to the political process for making policy decisions, but it cannot begin to provide a complete answer, especially to questions with such far-reaching implications for society. The issues should be aired fully and completely before a committee of Congress having broad policy responsibilities. An explicit decision should then be made by the entire Congress as to whether the risks are worth the benefits.

Part 3

Environmental Regulation

10 Taxing Pollution
An Idea Whose Time Has Come?

Wallace E. Oates

For many decades, economists have argued the case for taxes on pollution. From an economic perspective, such taxes (or effluent charges, as they are often called) serve to correct a serious source of market failure: the absence of a "price" needed to prevent the careless and excessive use of scarce environmental resources. However, the economist's case has not found a receptive audience in the policy arena. Instead, legislators and regulators have ignored this policy approach in favor of the more traditional command-and-control measures under which environmental agencies specify emissions limitations and control technology polluter by polluter. Many studies have documented the unnecessarily high costs and other wastes associated with these programs, but with a few exceptions these studies have not had much effect on the design of environmental policy.

During the past year, a new interest has emerged in taxes on pollution. In the desperate search for new revenue sources to reduce the deficit in the federal budget, legislators both in the House and Senate have begun to give more serious consideration to such taxes. This interest has led to the introduction by Congressman Judd Gregg of New Hampshire of a bill entitled "The Sulfur and Nitrogen Emissions Tax Act of 1987" (H.R. 2497). Hearings on the bill were held by the House Ways and Means Committee in September 1987. Not surprisingly, the hearings brought forth strong opposition from representatives

Originally published in *Resources,* No. 91, Spring 1988. It was based on Oates' testimony in September 1987 before the House Ways and Means Committee on H.R. 2497, the Sulfur and Nitrogen Emissions Tax Act of 1987.

of the coal industry and various public utilities, which appears, for the moment, to have stalled progress on the bill. More generally, however, the Congress continues to be interested in the idea of taxes on pollution as a source of federal revenues.

In principle, taxes on pollution offer not only an effective policy approach to protection of the environment, but also a very appealing addition to the revenue system. While most taxes have harmful side effects on the economy by distorting economic choices, taxes on pollution have socially beneficial side effects such that they cost society less than the revenues they produce.

The Economic Rationale

The economic case for pollution taxes is really just a straightforward corollary of the logic underlying a market system. The proper functioning of a system of free markets depends on the emergence of a set of prices that accurately reflect the cost to society of the resources used in the production of goods and services. Prices are the basic signals in a market system that direct the flow of resources to their most productive use. For most goods and services, the market forces of supply and demand generate the proper price.

Under certain circumstances, however, such prices may not reflect true social cost—and pollution is a classic case of such a circumstance. The basic point is straightforward: the absence of an appropriate price for certain of our scarce environmental resources (such as clean air and water) has led to their overuse, resulting in what the textbooks call a case of "market failure." A producer, for example, whose factory spews smoke on a neighboring residential area is using up a scarce resource—clean air; put slightly differently, the producer is imposing a real cost on those who must live with the dirty air. But the firm need not bear this cost in the same way that it pays for other resources—labor and raw materials—that it employs. While the price of labor and materials encourages the firm to economize on their use, there is no such incentive for the firm to control its smoke emissions. Whenever a scarce resource is made available free of charge (as is the case where individuals have free access to our limited stocks of clean air and water), it is bound to be used to excess. Producers and consumers alike must be made to bear the costs that their activities impose on others if they are to be expected to adjust these activities in light of the costs.

From this perspective, the basic cause of excessive environmental degradation is the absence of an appropriate price for scarce environmental resources. Once accepted, this proposition has a direct policy implication: the need for government to intervene and, in the absence of the interplay of supply and demand, to impose an artificial price, a tax (or effluent fee), on damaging waste emissions. It is easy to show in terms of microeconomic analysis that if this tax is set equal to the value of the damages from an additional unit of emissions, sources will have the proper incentive for controlling their discharges of pollutants. Economic analysis thus suggests the need for taxation of pollution to correct for a serious "failure" in a competitive market system.

Traditional command-and-control programs, by contrast, are much less efficient policy instruments, because it is virtually impossible for the environmental authority that must prescribe control measures to know what the most effective and least costly control technology will be for each polluter. These programs often result in across-the-board requirements for control measures that fail to take account of the particular circumstances of individual polluters. There is now a large body of empirical work that describes the enormous waste associated with these programs. Existing studies of U.S. programs for the management of air and water quality find that control costs to polluters range from double to (in some instances) more than ten times the least-cost outcome! In short, we are paying far more than is necessary to clean up the environment.

Probably even more important than this distortion in the choice among existing control technologies is the lack of compelling incentives for research and development in abatement technology. Over the longer haul, the success of our environmental programs will depend critically upon our capacity to discover and employ more effective and less costly methods of pollution control. The command-and-control approach not only fails to provide these incentives, but may actually discourage such pursuits: a firm that finds a new and better way to control its waste discharges may be "rewarded" by the environmental agency with a tougher set of abatement directives.

In contrast, pollution taxes provide a powerful set of incentives that encourage *both* the selection of the proper control technology among existing options and the search for new and more effective abatement procedures. Such activities would be directly profitable to polluters, since they would reduce costs. For example, a firm that discovers a less expensive way to control its waste discharges could employ this new technique to reduce its discharges yet further. It would realize savings both on its control activities and on its tax bill as a result of lowering the level of waste emissions. And society benefits from the cleaner environment that results in turn. A system of pollution taxes would effectively harness the powerful market force of competition on behalf of environmental protection.

The Revenue Rationale

If the basic economic rationale for instituting pollution taxes is to correct a malfunctioning in the market system—the case set out above—then the level of tax rates should be determined by our environmental objectives, *not* by revenue considerations. Tax rates should be set such that waste discharges are cut back to levels consistent with our targets for air and water quality. Nevertheless, such taxes would naturally produce revenues for the public sector.

Taxes on pollution would constitute a very appealing component of our overall revenue system. Any system of taxation creates a myriad of incentives for individuals and firms, incentives that lead to a variety of adjustments in economic behavior. What is troubling is that the vast majority of these adjust-

ments are harmful: they introduce distortions into the functioning of the economy, with resulting losses in social welfare. A tax on wage or salary income, for example, not only raises revenues but also creates disincentives to work. If 50 or 60 cents of an extra dollar of income is drained off in additional taxes, people have a compelling incentive to cut back on work effort and to substitute untaxed leisure pursuits for income-producing activities.

Moreover, there is now a large body of empirical work suggesting that the welfare losses to society from the distortions caused by our tax system are quite substantial. In a recent survey of this work, Edgar Browning of Texas A&M finds that although there remains great uncertainty concerning the magnitude of the distortion associated with taxes on labor earnings, it is potentially quite large, with estimates ranging from a low of about 8 cents lost per dollar of tax revenues to an upper bound in excess of $1.00. The true costs to society of the taxes we pay are larger (probably much larger) than the amounts of revenue actually collected. Taxes on pollution thus have a major attraction: not only can they provide needed protection for the environment, but they can substitute for other revenue sources that damage the economy.

Admittedly, the revenue potential of pollution taxes, while not trivial, appears modest. David Terkla of Boston University has made a detailed study of the potential revenues from the taxation of sulfur and particulate-matter emissions from stationary sources in the United States. Terkla calculates that annual revenues in 1982 dollars would range from a lower bound of about $1.8 billion to a high of $8.7 billion. The great bulk of these revenues, he estimates, would come from the tax on sulfur discharges. Revenue estimates for this tax alone range from a minimum of $1.7 billion to an upper bound of $8.5 billion. Another recent estimate, this one prepared by the staffs of the Joint Committee on Taxation and the House Committee on Ways and Means, assumes a tax of 45 cents per pound on sulfur and nitrogen dioxide emissions from boilers and furnaces and puts annual revenues from the tax at about $6.3 billion.

These revenue estimates are not large when compared to the overall revenue needs of the public sector. Pollution taxes obviously could not completely replace major taxes, such as income and sales taxes, in our revenue system. Nevertheless, they have a potential role to play. A broad set of pollution taxes encompassing most of the major air and water pollutants—and perhaps other wastes, including those that find their way into landfills—would obviously produce more tax receipts. These monies could relieve pressures for injurious increases (or perhaps permit decreases) in rates of the major forms of taxation.

Revenues from pollution taxes could thus substitute to some extent for revenues from "harmful" taxes. Terkla, in his aforementioned study, also explored the potential "efficiency gains" from the use of pollution taxes. These gains result from removing distortions to the economy caused by income and other taxes. Terkla finds the gains to be of measurable significance. He calculates that if his estimated revenues from the taxes on sulfur and particulate matter were substituted for tax receipts from labor income, there would be a savings to society ranging from $0.6 billion to $3.1 billion. If these revenues were substituted for corporate income taxes, the efficiency gains would be even larger—ranging from $1 billion to $4.9 billion. Such results suggest that the gains from a more efficient economy might offset as much as one half of the revenues themselves.

Implications for Legislation

As the general support for pollution taxes becomes embodied in actual legislation, it is essential to scrutinize carefully the particular provisions of proposed bills to ensure that the essential properties of this approach are not compromised in the legislative process. Certain issues are central to an effective piece of legislation—and the Gregg bill, H.R. 2497, is instructive in this regard.

First, the tax base must be defined correctly. The tax must be levied on the level of waste emissions—that is, it must be a tax per unit of actual waste discharges (for example, a levy of 25 cents per pound of sulfur emitted into the atmosphere). The tax must provide a direct incentive to polluters to cut back on their waste emissions; thus, a tax on the sales or profits of polluting industries will not do—the tax must be directly on emissions. The Gregg bill, incidentally, has the right tax base: it provides for a tax on emissions of sulfur and nitrogen oxides.

Second, legislation should ensure that the revenues from the tax are directed to reducing the federal deficit and later to decreasing levels of other more harmful taxes. There is a strong temptation to direct revenues from a pollution tax into some kind of environmental trust fund. In the case of the Gregg bill, for example, provision is made for the creation of a Sulfur and Nitrogen Emissions Trust Fund from which monies would be made available to assist polluters in their control efforts. This particular measure was designed in part to soften opposition from those subject to the tax. However, as the hearings on the bill made clear, that strategy was not successful and should be strongly resisted. The revenues from pollution taxes should become part of general revenues.

Third, it is desirable to retain some flexibility over the tax rate—unlike the Gregg bill, which prescribes a fixed set of rates. It might well turn out, for example, that the rate prescribed in the bill is insufficiently high to attain our targeted level of air quality. Or, alternatively, the rate could be so high as to lead to "excessive" cleanup at very high cost to society. It is important in such instances to have the capacity to adjust the tax rate to the proper level. This doesn't mean that rates need to be adjusted every month or so, but rather that we should not lock ourselves into a single rate over a long period of time; opportunities should exist to review the rate.

And fourth, for pollution taxes to be effective, they must *not* be accompanied by regulations on pollution control technology. The taxes themselves should provide the incentives for firms to seek out and use the least-cost methods for controlling their emissions. If the environmental agency has pre-

Update

The most recent literature on environmental taxes has brought to light a new aspect of the way in which such taxes impinge on the economy. In addition to reducing pollution and providing revenues that can be used to reduce existing distorting taxes, these taxes interact with the existing tax system (the so-called *tax-interaction effect*) in ways that exacerbate the distortions in economic behavior. Thus, the treatment of such taxes in this article is, in a sense, incomplete. For more about the tax-interaction effect, see chapter 17 of this book.

scribed treatment procedures for polluters, then taxes obviously cannot perform this function. The basic rationale for pollution taxes would be compromised. This point needs special emphasis, for it is fundamental to the proposal for pollution taxes. Such taxes cannot be overlaid on a command-and-control system of regulated technologies without destroying the efficiency-enhancing properties of the effluent tax system.

A Market Alternative

No discussion of market incentives for pollution control would be complete without mentioning an intriguing alternative to pollution taxes: a system of transferable discharge permits. Instead of taxing sources on their emissions, the environmental authority could issue a limited number of permits that authorize the emission of a specific amount of the relevant pollutant (where the number of permits is kept sufficiently low to achieve the environmental target). Sources would then be free to buy and sell these discharge permits. In principle at least, such a system has the same cost-saving properties as a system of pollution taxes—namely, it can achieve the target level of environmental quality at the least cost to society.

Permits may have some advantages over taxes; for one thing, they give the environmental agency direct control over the level of emissions through control of the number of permits (rather than indirect control through regulating the level of the tax rate on discharges). Moreover, since existing regulatory systems make use of permits, making them marketable would appear to be a less wrenching change in our approach to environmental protection than scrapping permit systems altogether for a system of pollution taxes. The Environmental Protection Agency has gone some distance toward such a system of transferable permits with its Emissions Trading Program for the control of certain air pollutants. Some states now have programs that allow sources to exchange emissions entitlements, and many of these exchanges have resulted in large cost savings.

However, taxes have their appeal as well: they present a direct price incentive to sources (without the need to organize a market in discharge permits) and raise needed tax revenues. The choice between these two market instruments for pollution control is a fascinating issue in policy analysis. As economic analysis has shown, where it is crucial to prevent levels of pollution from exceeding certain damaging threshold levels, the permit approach is especially attractive because of the direct control it provides over levels of emissions through limiting the number of permits. But where the greater danger is one of excessive costs of pollution control, taxes or fees have the advantage since sources can always avoid high costs of pollution control by maintaining emissions and paying the associated taxes. Each of these approaches has its own important place in the broader spectrum of policies for management of the environment.

Suggested Reading

Oates, Wallace E. 1994. "Environment and Taxation: The Case of the United States," in OECD, *Environment and Taxation: The Cases of the Netherlands, Sweden, and the United States.* Paris: OECD, pp. 103–43. (A study of the principles of environmental taxation followed by a description and assessment of environmental taxes used by federal, state, and local governments in the United States.)

Organisation for Economic Co-operation and Development (OECD). 1997. *Environmental Taxes and Green Tax Reform*. Paris: OECD. (A review of environmental tax systems with an analysis of issues of implementation.)

11 Regulatory Reform in Air Pollution Control

Tom Tietenberg

As anyone who has tried it knows, regulatory reform concepts that appear disarmingly simple in theory turn out to be distressingly complex when they are applied. Regulations that, from a distance, seem inherently insupportable are discovered, upon closer inspection, to have significant support among special interest groups. Since the status quo engenders so much inertia, many promising ideas languish by the wayside.

In some ways the emissions-trading program—the Environmental Protection Agency's (EPA's) new approach to air pollution control—seems an unlikely candidate for survival. Significant pockets of opposition to the use of economic incentive approaches such as this one exist within Congress as well as within the EPA itself. Nonetheless, the emissions-trading program has become a new wave in environmental policy.

Why has the emissions-trading program survived its critics? And what price has been paid to achieve this survival? With these questions in mind, I spent my 1983–84 sabbatical year looking at the manner in which the theoretical concept had been translated into practice. My objective was not only to gain some understanding of the process of regulatory reform as it was applied in this case, but also to provide some foundation for the further evolution of emissions trading.

Originally published in *Resources*, No. 79, Winter 1985. It was based on an essay that appeared subsequently in *The Colby Alumnus*, vol. 74, no. 2 (March) 1985.

The Nature of the Reform

The 1970 Amendments to the Clean Air Act established an approach to controlling pollution from stationary sources that has become known as the *command-and-control approach*. Stripped to its bare essentials, it involves the specification of a separate emission standard (that is, a legal emission ceiling) for each major pollution discharge point, such as a stack, a vent, or a production process. The enormity of the regulator's task in defining these standards becomes clear when one realizes that the typical industrial facility contains many such discharge points, with some facilities having more than one hundred.

The emissions-trading program attempts to make this approach more flexible by allowing polluters more options in meeting their assigned control responsibilities. The general thrust of the program is to allow polluters to seek alternative, cheaper means of reducing emissions so long as the substitute has an equivalent, or better, effect on air quality.

Specifically, any source reducing its emissions further at any discharge point than that required by law may apply to have this excess reduction certified as an emission-reduction credit. Once certified, the credit may be applied to other discharge points or may be sold to other sources. The conditions under which these credits can be created, stored, transferred, and used are defined by the *bubble, offset, banking,* and *netting* components of the emissions-trading program, as described below.

The Bubble Policy

Despite the fact that studies indicated that the potential cost savings from implementing emissions trading were huge, the program got off to an inauspicious start. Following some five years of industry pressure, the EPA published its first application of emissions trading, called the "bubble policy," in December 1975. Under this policy, an entire plant would be regarded as if it were within a giant bubble, with only the total amount of emissions leaving the bubble being subject to regulation. Its thrust

was to excuse existing plants undergoing expansion or modification from meeting the tough new-source performance standards so long as any resulting emission increases would be offset by decreases elsewhere in the plant. In current parlance, this approach afforded regulatory "relief" rather than regulatory "reform." The specifics of this approach were ruled unacceptable by the courts as being inconsistent with the intent of Congress when it passed the Clean Air Act, though the principles of emissions trading emerged unscathed.

The Offset Policy

A second attempt to introduce emissions trading, which concentrated on regulatory reform rather than relief, was more successful. By 1976, it had become clear that a number of regions would fail to attain the ambient air-quality standards by the deadlines mandated in the Clean Air Act. Thus the EPA was faced with the unpleasant prospect of prohibiting any new sources from entering these regions. As an alternative to out-and-out prohibition, the EPA established an "offset policy," which enables new sources to enter such regions provided that they meet strict emission standards and acquire sufficient offsetting reductions from other facilities so that total regional emissions will be lower after their entry than before. In essence, this program provides a way to improve air quality by reducing emissions at existing sources, but it does so by forcing new sources to find and finance the offsetting reductions.

With the advent of the offset policy, emissions trading had established a precarious foothold in air pollution policy, but at a high price. Although the EPA has achieved its objective of allowing new sources to enter nonattainment areas, it has become very expensive for the sources to commence operations. By shifting all of the financial burden to new sources, existing sources (and existing jobs) are protected—but at the detriment of modernization and technological progress in industrial production that otherwise would have been possible. Compared to an approach that forces new and existing sources to bear a more equitable share of the financial burden of pollution control, the offset policy gives existing sources in nonattainment areas a significant competitive edge over their potential new rivals by perpetuating the traditional regulatory bias against new sources.

The 1977 Amendments

When writing the 1977 Amendments to the Clean Air Act, Congress provided legislative authorization for the offset program. To this day the offset program remains the only component of the program specifically authorized by statute; the others are purely bureaucratic creations, resting solely on general principles articulated in the act. As such, they remain especially vulnerable to hostile judicial interpretation.

Emissions Banking

The next component of the emissions trading program—banking—was added in 1979 as the EPA issued new regulations designed to bring the interim offset program into conformance with the 1977 Amendments. Emissions banking allows sources of emission-reduction credits to store those credits for subsequent sale or use. Prior to the 1979 regulations, banking had been disallowed on the grounds that it was incompatible with the EPA's statutory responsibility to assure that nonattainment areas achieved the ambient standards as rapidly as possible. Confiscation and retirement of emission-reduction credits not immediately used had been seen as a rapid means of improving air quality. Since the 1977 Amendments and the associated implementing regulations provided specific procedures for attaining the standards by the new statutory deadlines that were compatible with emissions banking, these objections were overcome.

Potentially this was an important boost to the program, since in the absence of banking the incentives for controlling emissions beyond the minimum legal requirements are diminished substantially. Without banking, excess control is valuable to its source only if another source simultaneously needs an offsetting reduction. One can easily imag-

ine what would happen in more traditional markets, such as furniture, if the product were confiscated by the state whenever a buyer could not be found soon after the product was finished. Less furniture soon would be available. The same principle holds for emission-reduction credits. The polluter has absolutely no incentive to undertake additional control voluntarily unless he retains an exclusive and transferable property right over the emission-reduction credit until it can be used or sold.

Successful banking programs do exist, as is illustrated by the bank established in Louisville, Kentucky. By May 1984, this bank had some fifteen deposits of emission-reduction credits for total suspended particulates, sulfur dioxide, volatile organic compounds, nitrogen dioxide, and carbon monoxide. This program clearly has been successful in stimulating additional reductions and in facilitating the search for lower-cost means of controlling pollution.

The Revised Bubble Policy

Whereas the establishment of the offset program had been a response to a specific, passionately felt political need to remove the prohibition on growth in nonattainment areas, during the late 1970s interest in expanding the application of the emissions-trading concept grew. Since the 1975 attempt at a bubble policy had been overruled by the courts, the EPA had to proceed cautiously. In view of its need to build a constituency while protecting its flanks from judicial attack, the agency initially proposed heavily circumscribed programs designed to assuage fears and to move slowly. By taking this approach, the EPA sought to assure that the first trades would demonstrate clear, unambiguous benefits and set a useful precedent. At the same time, the number of possible trades would be intentionally limited, giving the states time to adjust to a new program before they were overwhelmed by a flood of applications.

The reincarnated bubble policy allowed stable, existing sources some flexibility in fulfilling their assigned control responsibilities. Whereas the original bubble policy sought to limit the applicability of the regulations, this policy focused on making compliance easier. Instead of forcing each source to produce the stipulated emission reductions at each and every discharge point (as would be required by strict adherence to the previous command-and-control policy), the bubble policy allows each source to choose its own mix of emission reductions so long as the air-quality effects remain equivalent.

The relatively slow pace of trading following these initiatives convinced the EPA that the substantive reforms would have to be accompanied by procedural reforms if the program were to live up to its potential. Originally, the bubble policy could be used only if the approving state included the intended trade in a formal revision to its state implementation plan (SIP). Because the SIP approval process is the primary means by which the EPA exercises its responsibility for assuring state compliance with the Clean Air Act, SIP revisions are bureaucratically cumbersome. For example, when the Reagan administration took office in 1981, a backlog of 643 proposed changes in SIPs awaited the EPA's approval. Because any SIP revision has to fulfill a large number of procedural requirements, state control authorities are reluctant to file revisions unless they are absolutely necessary. Requiring bubble trades to be approved through SIP revisions proved a surefire way to limit the state control authority's interest in the program.

In 1981 the EPA significantly lessened this burden by approving the generic rules that the states intended to use to govern trades. So long as subsequent trades conformed to these rules, no SIP revision was required for each trade. For the first time the state control authorities were able to see the bubble policy as something other than a procedural nightmare.

Resurrecting Regulatory Relief

The design and fate of the netting program—the final component of the emissions-trading package—provides an interesting example of what happens when the irresistible force associated with a bureaucracy committed to regulatory flexibility

runs into the immovable object represented by rigid statutes. Both the offset and revised bubble policies are regulatory reform measures in that they offer flexible ways to comply with the statutes. However, not all areas of pollution policy allow this flexibility. In particular (as the treatment of the first bubble policy by the courts indicates), the statutory language permits little opportunity for expanding or modifying sources to use emissions trading in complying with the new-source performance standards; such sources have to achieve the stipulated reductions at each discharge point. It remained an open question, however, as to whether emissions trading could be used to limit the applicability of the new-source review process, an additional layer of regulatory control imposed by the 1977 Amendments. By providing a means of gaining relief from this additional regulatory burden without relieving the source of the obligation to meet the new-source performance standards, the EPA attempted to open a new channel for emissions trading while respecting the previous court decision.

The Netting Program

Netting allows modifying or expanding sources to escape the burden of new-source review requirements so long as any net increase (counting the emission-reduction credits) in plantwide emissions is insignificant. By "netting out" of review, the facility may be exempt from the need to acquire preconstruction permits as well as from meeting the associated requirements, such as modeling or monitoring the new source's impact on air quality, procuring offsets, and meeting the most stringent emission standards. It is *not* exempt from any applicable new-source performance standards.

While this program could have exempted a large number of modified sources from review, it was successfully challenged in court by the Natural Resources Defense Council. Ruling that exemption of modified sources from review in areas with air quality worse than the standards was inconsistent with the statutory intent to reach attainment as expeditiously as possible, the appeals court voided the netting rules as they applied to sources in those areas. By constantly referring to netting as the "bubble policy," the court cast a cloud over the application of the bubble policy as well as over the application of netting. The U.S. Supreme Court reversed this ruling, but not before a lengthy period had passed, during which the use of netting and the bubble policy was suspended as states awaited the outcome of judicial review.

An Appraisal

There is little doubt that emissions trading has improved upon the command-and-control policy that preceded it. The EPA estimates that more than 2,500 emission trades have taken place since the program's inception. The frequency and significance of these trades has triggered a new set of private support institutions. For example, in 1984 a new brokerage house was established solely for the purpose of facilitating emission-reduction credit transactions.

Many of these transactions have facilitated the modernization and expansion of existing plants as well as the construction of new ones in areas of the country not meeting ambient air-quality standards. Each of these trades represents an affirmation of the basic premise of emissions trading—allowing sources to trade emission-reduction credits reduces the cost of complying with the law.

One of the substantial benefits of this lower compliance cost has been an increase in the number of firms complying with the terms of their permits. It no longer pays for noncomplying sources to engage in expensive litigation to avoid compliance. In many instances, the bubble policy was the means by which previously noncomplying sources were brought into compliance for the first time. In two cases the actions taken under the bubble policy contributed to the state authorities' ability to demonstrate that the ambient standards in the affected nonattainment areas would be achieved.

The enduring role that the EPA emissions-trading program is playing is directly attributable to the fact that it was preceded by a very cost-ineffective regulatory policy. Not only did this create a demand

Update

Following the developments chronicled in this article, the emissions-trading concept has subsequently been applied in several new national, state, and international contexts. In the United States, federal emissions-trading programs have been used to eliminate the lead in gasoline, as well as to reduce acid rain and ozone-depleting gases. States also have begun to adopt emissions-trading programs, with the best known being the RECLAIM program in the greater Los Angeles area. In addition, emissions trading has been enlisted to control greenhouse gases as part of the Kyoto Protocol to the Climate Change Convention and also has been incorporated in domestic air pollution control policies in countries such as Canada and Chile.

While the later U.S. programs share fundamental characteristics with the earlier programs, they differ considerably in their implementation details. Whereas the earlier programs sought to increase the flexibility of traditional regulatory policy, the latest program, the sulfur allowance program, has used emissions trading as a replacement for traditional regulation.

for approaches that offered to reduce costs, but also it provided a ready-made baseline for the trades, smoothing the transition to emissions-trading. Had the command-and-control policy been more cost-effective, it is doubtful that the emissions trading policy could have gained the foothold it has.

Paradoxically, the ability to overlay this program on an existing but cost-ineffective policy was a key to its political success, but it also has diminished the effectiveness of the program in several specific ways:

• In response to command-and-control regulation, a great deal of capital equipment already had been installed prior to the inception of the emissions-trading program. Since much of this installed durable capital was cost-ineffective and its owners were unable to benefit from the emissions-trading program, this has reduced the savings achievable by the program from what would have been possible if the program had started with a clean slate.

- A particularly unfortunate side effect of overlaying emissions trading on a preexisting command-and-control allocation also arose when some sources complied rather rapidly and others proved more recalcitrant. Because the emissions-trading option appeared late in the game, sources that immediately complied with the command-and-control regulations were precluded from using the emissions-trading program to their greatest advantage, while those who were able to fend off the earlier expensive standards could, with the advent of emissions trading, reach compliance at a substantially lower cost. In this way the introduction of an emissions-trading program rewarded slow compliance, which strikes many potential supporters as patently unfair.

- The bias against new sources that characterized the command-and-control policy has persisted, albeit to a lesser degree, in the emissions-trading program. Not only are new or expanding sources required to buy emission-reduction credits to offset any emission increases that remain after the installation of required controls, but new sources typically must meet the prescribed emission standards by installing the control equipment necessary to meet the mandated reductions at each discharge point. In contrast, existing sources are not required to acquire credits to offset their remaining emissions, and they can use emission reduction credits to meet their statutory responsibilities

rather than producing the mandated emission reduction at every discharge point. This bias effectively delays the replacement of older, heavier-polluting facilities with newer, less-polluting facilities.

- The notion that firms might have a property right in surplus emission reductions was not a part of the command-and-control system and has been hard for some control authorities to swallow. In some jurisdictions confiscation of certified credits is a distinct possibility, destroying much of the incentive to create additional emission reductions.

These flaws must be kept in perspective. Although a definite price has been paid for survival, it is not so large as to overshadow the very positive accomplishments of the program. The emissions-trading program loses its utopian luster upon closer inspection, but it nonetheless has made a lasting contribution to environmental policy. The realms of the possible and of the desirable rarely overlap completely; it is comforting to know that the overlap is not inconsequential.

Suggested Reading

South Coast Air Quality Management District. 1998. The RECLAIM Program. http://www.aqmd.gov/reclaim/reclaim.html

Tietenberg, T.H. 1998. Bibliography on the Use of Tradable Permits for Environmental Protection. http://www.colby.edu/personal/thtieten/

Tietenberg, T.H. 1998. "Ethical Influences on the Evolution of the US Tradable Permit Approach to Air Pollution Control." *Ecological Economics* 24(2,3): 241–257.

U.S. Environmental Protection Agency. 1998. The Sulfur Allowance Program for Controlling Acid Rain.http://www.epa.gov/acidrain/trading.html

12 Trading Emissions To Clean the Air
Exchanges Few but Savings Many

Dallas Burtraw

Trading emission allowances is proving both an environmental and economic success, with annual emissions of sulfur dioxide nearly halved at a cost of a third to a half as much as under earlier regulatory approaches. Yet actual trading volume between utilities is unexpectedly low. Burtraw explains why and what it might mean for the future.

The allowance trading program for sulfur-dioxide (SO_2) emissions is a good example of a legislative initiative that is both an environmental *and* an economic success. As the centerpiece of Title IV of the 1990 amendments to the Clean Air Act, the allowance trading program is reducing annual emissions of SO_2 by nearly 50 percent and is doing so for about one-half to one-third of the cost that would have been incurred using the approach taken throughout the first twenty years of federal air pollution control.

At the same time, however, the volume of trading between utilities of SO_2 emission allowances is well below original expectations, with only about 2 to 3 million allowances traded in 1995, the first year of the program's first phase (one allowance equals one ton of SO_2).

How is it that the program has generated tremendous cost savings with few allowances changing hands? The major reason is, in a word, flexibility.

The New Flexibility

The success of the SO_2 program comes as no surprise to many scholars. They predicted that the largest share of economic benefits from a trading program would come not from the trading of allowances per se, but from what economists call "dynamic efficiency"—innovation, competition, and discovery of new ways of compliance. Title IV freed electric power companies from the constraints of traditional regulations, which effectively

Originally published in *Resources*, No. 122, Winter 1996.

spelled out exactly how a requirement was to be met, and instead gave the utilities the flexibility to figure out for themselves how to achieve compliance.

Given the new flexibility, many firms have found ways to reduce the cost of controlling SO_2 emissions that do not rely either directly or very heavily on the allowance trading program. For instance, some utilities have switched entirely to low-sulfur coal, whose price has fallen substantially over the last five years. Other power plants have begun blending coals with varying sulfur content in order to reduce average SO_2 emissions, something thought impractical just a few years ago. Deregulation of the railroad industry has also led to a steep drop in the cost of shipping low-sulfur coal from west to east.

The fact that few emission allowances have been traded may be seen as ironic as well as paradoxical, given the controversy that the program provoked at the time of its enactment. Many environmentalists opposed the program for authorizing (and implicitly endorsing) the sale of the right to pollute. But as groups such as the Environmental Defense Fund point out, traditional command-and-control regulations had been *giving* away the right to pollute for free. Federal air pollution regulations, for example, have allowed SO_2 emissions from electric utilities to grow along with increased production and as new plants are built.

Under the Title IV amendments, however, the utilities face the first cap ever on SO_2 emissions. In Phase I of the program (1995–2000) the nation's 110 dirtiest coal-fired electric power plants are required to reduce SO_2 emissions averaged across these facilities to about 2.5 pounds per million British thermal units (mmBTUs) of electricity generated. In a quid pro quo for the cap, the facilities receive annual allowances for emissions, which they can transfer to other plants within their own systems, sell to other utilities, or save for later use. In the absence of robust inter-utility trading, the program's cap on emissions operates like a performance standard applied to individual utilities.

Few Trades, Low Prices

Just as the volume of trading is lower than what was expected at the time the Title IV program was enacted, so are the prices of allowances. As Table 1 indicates, before passage of the Clean Air Act amendments in 1990, estimates of marginal emission abatement costs were as high as $1,500 per ton, which is the figure stipulated in the act for direct allowance sales by the U.S. Environmental Protection Agency (EPA). In debates surrounding the 1990 amendments, EPA cited estimates of marginal abatement costs about half as high, which became the bases for estimates of allowance prices. After passage of the amendments, estimates plummeted still further. In early 1995, the price of allowances traded privately was about $170 per ton and fell to the low $100s by year end. The marginal price of 1995 allowances in the EPA auction administered by the Chicago Board of Trade ranged from $122 to $140 per ton.

The fall in prices was a product of low demand relative to supply. But the new flexibility that the utilities are enjoying under Title IV and the array of low-cost compliance options they have to choose from are not the only reasons the market is not more active. In trying to explain the low prices, the role that state public utility commissions (PUCs) play has to be considered

Many of the rules that states have imposed on the utilities potentially inhibit allowance trading. Depending on what the rules say, allowances may look less attractive than other available cost-cutting strategies. What the rules say, for example, about the recovery costs of investments (such as the allowed rate of return, the depreciation rate, and the risk that expenses might not be passed on to ratepayers) often differs across compliance strategies and sometimes is designed to create preferences for one strategy over another.

Furthermore, what the rules say is not always enough to go on. Neither the Federal Energy Regulatory Commission nor the PUCs have provided adequate guidance about cost recovery rules yet. Uncertainty about what shape these regulations

Table 1. Marginal Cost Estimates and Realizations for Compliance Options (dollars per ton)

Industry estimates pre-1989	EPA 1990 estimate	Early allowance trades	Current allowance trades	1993 CBOT allowance auction	1994 CBOT allowance auction	1995 CBOT allowance auction
$1,500	$750	$250	Low $100s	$122	$140	$126

Note: CBOT is Chicago Board of Trade.

will take has contributed to the cautious reception allowance trading is receiving.

A related problem is explicit prohibition by legislatures on trades that might undermine local economic activity. Nearly every state with substantial Phase I compliance obligations has enacted rules or incentives to promote the use of local coal, for instance, by offering pre-approval for cost recovery of investments in scrubbers.

Not a Perfect Program

In trying to understand the program's few trades and low prices, many analysts also have criticized EPA's annual auction of 2.8 percent of allowances, which began in 1993 to jump-start the market. As set out in the statute, the auction is a discriminating-price, sealed-bid one that provides bidders and sellers with strategic incentives to underbid their reservation prices. Critics say it is a poorly designed mechanism that generates prices below those emerging from trades between utilities.

Still another reason for low prices is the extra 3.5 million allowances introduced in Phase I. The purpose of this provision in the 1990 amendments was to subsidize utilities that install scrubbers and thus cushion the blow to states producing high-sulfur coal. The effect has been to encourage scrubbing even if it is not really the least-cost option, as well as to increase the supply of allowances and depress the price.

The most important explanation for low allowance prices, however, has to do with changing market fundamentals in coal markets, rail transportation of coal, and equipment suppliers. In particular, falling prices in the coal and scrubber markets have had a profound effect on how the industry has complied with the SO_2 emissions cap in Phase I of the program. In 1990 many analysts projected average prices for low-sulfur central Appalachian coal to reach $40 per ton by 1995, but last year the price was less than $25. According to the U.S. General Accounting Office, scrubber prices fell by nearly half over this same period.

One explanation for this turn of events is the unanticipated degree to which markets have been drawn into direct competition. The result is a decline in prices below those forecast in every potential option for compliance.

Compliance Options of Choice

Using the freedom that Title IV gave them, electric utilities have met their clean air requirements in innovative ways. The process of fuel switching to, and fuel blending with, low-sulfur coal is the most widely used compliance option. The low cost of this strategy is one reason; that it is relatively non-capital-intensive in a period of general change in the industry is another.

Like fuel switching, fuel blending has lower capital costs associated with it than scrubbing. Experimentation prompted by Title IV has led to an improved understanding of the ability to blend coals with varying levels of sulfur contents. Detrimental effects of incompatible blending on plant equipment designed to operate using a particular type of coal are fewer than originally supposed.

SO₂ Program Saves Billions—and Could Save a Billion More

The table below presents two sets of estimates of the relative annual costs associated with three different scenarios for implementing the sulfur-dioxide emission reduction goals of the Clean Air Act: a command-and-control approach, limited allowance transfers within firms, and active allowance trading across firms. EPA produced the first set of estimates, which were used by Congress to develop the program as a key provision of Title IV. The U.S. General Accounting Office (GAO) produced the second set, which summarizes what has happened since Title IV took effect.

Projected Annual Costs under Alternative Implementations for 2001 (in Billions of Dollars)

	Command-and-control baseline	Constrained trading (internal transfers)	Flexible interutility trading
EPA (1989)	—	$3.3–$4.7	$2.7–$4.0
GAO (1994)	$4.3	$2.5	$1.4

Sources: *U.S. Environmental Protection Agency. 1989. "Economic Analysis of Title IV (Acid Rain Provisions) of the Administration's Proposed Clean Air Act Amendments," prepared by ICF Resources Incorporated, Washington, D.C.*

U.S. Government Accounting Office. 1994. "Air Pollution: Allowance Trading Offers an Opportunity to Reduce Emissions at Less Cost," GAO/RCED-95-30, Washington, D.C.

Three important points emerge from the data in the table. First, GAO estimates that, by the beginning of Phase II, costs resulting from limited allowance transfers within a company—the implementation scenario we have observed to date—will be almost 40 percent less than under a command-and-control baseline. The baseline that GAO used was an emission rate applied to each facility, which yielded lower estimates than specific technology requirements would and thus yields a conservative estimate of these cost savings.

The second point emerges in the comparison between estimates for each category of implementation scenario. EPA estimates are relatively low compared with the other projections made before passage of the Clean Air Act amendments in 1990. In part, this is because ICF Resources, which conducted the analysis for EPA, maintains a sophisticated coal market model, and correctly anticipated that low-sulfur coal would play the prominent role in compliance at least through Phase I of the program.

Nonetheless, even under the most optimistic implementation scenario of active trading, EPA's lower bound for the cost of the program was $2.7 billion per year in 2001. In contrast, GAO found that constrained trading conditions would yield a cost of $2.5 billion, which is lower than the most optimistic projection made before the amendments were passed.

By way of comparison, a command-and-control program to reduce SO₂ emissions was estimated to cost as much as $7 billion annually if every utility had been required to install scrubbers, or $4.3 billion annually if uniform emission rates had been applied to individual power plants.

The third point evident in the table data is that sizable savings still remain available through an improved trading program. GAO estimates that potential savings total another billion dollars a year, which is more than 20 percent of baseline estimates.

Many observers of the development of the 1990 amendments foresaw bottlenecks in rail transport that would preclude western coal from playing a big role in the compliance plans of eastern utilities. Thus, forecasts hinged on prices for low-sulfur Appalachian coal locally available to eastern utilities. But these bottlenecks have failed to materialize.

One reason rail has responded so enthusiastically to the potential new markets for low-sulfur coal created by Title IV is that rail itself was deregulated under the Staggers Act of 1980. Lines moving low-sulfur coal out of the Powder River Basin in northeast Wyoming and southeast Montana are now the busiest in the world. Indeed the experience of the deregulated railroads may foreshadow the experience to come of utilities subject to the Clean Air Act, which follows a pattern of regulatory reform that has also touched telephones, airlines, and natural gas over the last two decades.

Title IV has also inspired a reduction in costs within the scrubber industry to stay competitive. For the first time, an incentive exists to improve the efficiency of scrubbers, since each ton of SO_2 saved is one allowance earned. New scrubbers exhibit increased efficiency and reliability. Improvements in scrubber design and use of materials have reduced maintenance costs and increased utilization rates.

Title IV "Star" Is Still in the Wings

Inter-utility allowance trading—the aspect of the Title IV program that observers anticipated would be the leading star—thus far has been the option least commonly used. Illinois Power is the only utility to rely heavily on allowances for Phase I compliance. Only Carolina Power and Light and Georgia Power seem likely to do so in Phase II.

Even in the absence of extensive trading, however, allowances are proving useful. For example, utilities are saving millions by not having to purchase spare scrubber modules for use during maintenance periods and outages; they are relying on allowances instead. Similarly, utilities are able to delay capital investments in scrubbers by relying on allowances. This is particularly useful at a time when many utilities are reluctant to make new investments until more is known about the direction that the regulation of the electricity industry is likely to take.

In the face of uncertainty surrounding compliance strategies, performance standards alone may be inadequate to stimulate innovation because utilities may be unwilling to experiment if failure has a

Update

The prospects for SO_2 emission allowance trading have continued to brighten over the first three years of the program. Each year, the volume of trades between independent firms has virtually doubled, totaling nearly 8 million allowances in 1997. Meanwhile, the program has achieved 100% compliance among affected facilities. A significant allowance bank of nearly 11 million tons is expected by 2000 and this bank will carry over into Phase II of the program. This bank will enable emissions above allowance allocations in Phase II, but it has provided early reductions in Phase I. It also has provided the opportunity for significant cost savings, as firms have used banking to help time their investments and to reduce their costs. Allowance prices remain low, compared to expectations about the marginal cost of compliance. Prices also have been volatile, a signal that the market is still maturing, and an indication of uncertainty about other regulatory changes that could affect the program, including electricity deregulation, NO_x policies, and climate change policies.

high cost. The allowance offers a convenient value as insurance, even if it is not a primary compliance strategy in itself.

What the Future Holds

One question looms: will de facto performance standards that exist in the absence of active allowance trading remain sufficient to keep down the costs of controlling emissions? This uncertainty becomes especially pertinent with regard to Phase II of the Title IV program, which will take effect in 2000 and apply to all fossil fuel power plants greater than 25 megawatts in size. To add to the challenge in Phase II, utilities also will have to cut the total amount of averaged emissions to 1.2 pounds of SO_2 per mmBTUs.

Right now, the availability of low-priced, low-sulfur coal has allowed most utilities to comply with Phase I of Title IV relatively cost-effectively. Even without institutional obstacles to allowance trading, robust trading would not be expected when this low-cost option is commonly available. Obstacles to a more liquid allowance market are not too important in the short term. But that may change.

Whether low-sulfur coal will continue to provide a commonly available low-cost compliance strategy may be the crux issue. Current estimates show costs increasing over time as the result of an expected depletion of Appalachian low-sulfur coal and of allowances banked during Phase I.

If the supply does in fact dwindle, some utilities will turn to other options, such as the installation of scrubbers. Such a move will likely result in significant differences in marginal costs across companies. It is then that utilities may take a second giant step by moving beyond performance-based standards to broad-scale trading of emission allowances among themselves.

For that reason, flaws in how trading is currently implemented in Title IV and the obstacles to it created by state regulators should be addressed before Phase II is under way. Unless corrected, inadequate and parochial regulations that stymie allowance trading will grow in importance, and Title IV may be much less successful than it has been to date.

Will allowance trading ever become the star that its fans expected it would become when the 1990 amendments to the Clean Air Act were being drafted? Projected annual costs using alternative compliance options indicate that significant cost savings may continue to accrue mainly from the flexibility afforded by Title IV. But that doesn't rule out the possibility of an active trading program waiting in the wings. If utility regulators decide to improve the prospects for a more liquid allowance trading market, the savings will be all the more dramatic.

Suggested Reading

Burtraw, Dallas, Alan J. Krupnick, Erin Mansur, David Austin, and Deirdre Farrell. 1998. The Costs and Benefits of Reducing Acid Rain. *Contemporary Economic Policy* forthcoming.

Burtraw, Dallas. 1996. The SO_2 Emissions Trading Program: Cost Savings without Allowance Trades. *Contemporary Economic Policy* XIV (April): 79–94.

Schmalensee, Richard, Paul L. Joskow, A. Denny Ellerman, Juan Pablo Montero, and Elizabeth M. Bailey. 1998. An Interim Evaluation of Sulfur Dioxide Emissions Trading. *Journal of Economic Perspectives* 12(3, Summer): 53–68.

Stavins, Robert N. 1998. What Can We Learn from the Grand Policy Experiment? Lessons from SO_2 Allowance Trading. *Journal of Economic Perspectives* 12(3, Summer): 68–88.

13 Shifting Gears
New Directions for Cars and Clean Air

Winston Harrington and Margaret A. Walls

As deadlines set by the Clean Air Act Amendments of 1990 loom and pass, state and local officials are scrambling to evaluate policy options and adopt programs that will effectively reduce the motor vehicle emissions that can form ozone. Among the mandated policy options are several command-and-control approaches, some of which call for new developments in emission-control technology. But these approaches may not be as cost-effective as other options that rely on economic incentives. Until these economic incentives have been investigated, decisionmakers should be cautious about moving ahead with approaches that could have high costs, or uncertain results, or both.

With the Clean Air Act Amendments of 1970, the United States initiated a bold new approach to air pollution problems. For the most part, this approach seems to have worked, as air quality standards set as a result of the 1970 amendments have now been achieved in most locations. But ground-level ozone (smog) still remains a problem in many urban areas.

That is why the Clean Air Act Amendments of 1990 required stricter control of the emission of ozone *precursors*—mainly oxides of nitrogen (NO_x) and volatile organic compounds (VOCs) from stationary sources such as factories and from mobile sources such as cars and buses. But the amendments also set unrealistically short deadlines for action.

State and local officials are now hastily deciding on policies to control motor vehicle emissions and attain ambient ozone standards. Such decisions could affect the design of cars for decades to come, reshaping the entire car industry (and perhaps the oil industry) and costing motorists billions of dollars each year. Yet, it is not clear that some options for reducing motor vehicle emissions will appreciably affect ambient ozone.

In this article, we explain why many urban areas have not attained ambient ozone standards and how some of the mobile-source provisions of the latest Clean Air Act amendments share many shortcomings of their predecessors. We then present estimates of the cost-effectiveness of various options for reducing motor vehicle emissions and note the reasons to treat these estimates cautiously. We conclude by considering how policymak-

Originally published in *Resources,* No. 115, Spring 1994.

ers could make best use of the various options and offer our views on how policies can be designed to yield the biggest "bang for the buck."

Why Many Urban Areas Still Have Ozone Problems

One of the major reasons why ozone remains a problem is that ozone formation is complex and not well understood. Ozone is not emitted directly; rather, it is formed from precursor pollutants in a series of complex chemical reactions on hot, sunny days. This makes it difficult to relate reductions in precursors to reductions in ozone. The sources of precursors are extraordinarily numerous and, in the case of VOCs, varied; but the most important source—especially in nonattainment areas where ozone levels exceed the standard—is motor vehicles. In 1989, the Office of Technology Assessment estimated that cars, trucks, and buses contributed 45 percent of the VOCs and 30–66 percent of the NO_x emissions in nonattainment areas. Recent studies suggest that these percentages may be even higher.

Rather than directly regulate the driving behavior of millions of motorists, Congress opted to target car manufacturers. In 1970, it began to set increasingly stringent emissions standards (in terms of grams of pollution per mile) for new cars, so that highly polluting vehicles would be replaced by less-polluting ones. It is estimated that these standards make the average purchase price of today's car $500 to nearly $1,200 higher than it would be in the absence of the standards.

The regulations themselves may have contributed to the persistence of the ozone problem. First, even though VOC emissions from *new* cars today are about 95 percent below those from cars in the late 1960s, *average* emissions rates of the U.S. car fleet have not fallen by nearly this much. This is because emissions control systems tend to break down as cars get older, causing emissions to rise. The problem is compounded by the fact that the average age of the U.S. car fleet has increased from 5.1 years in 1969 to 7.7 years in 1990. Second, the

regulations have focused only on emissions rates, ignoring vehicle miles traveled (VMTs). Since 1970, VMTs have increased by 69 percent, partially offsetting reductions in emissions per mile brought about by new-car emissions standards. Third, since the standards were primarily directed at tailpipe emissions, they did not reduce emissions from fuel evaporation, which may account for 10–50 percent of total VOC emissions. Finally, the importance of NO_x in ozone formation was overlooked until recently.

Additional complicating factors include rising roadway congestion over the last two decades and an increase in the average number of trips per household. Congestion increases both evaporative and tailpipe emissions, since VOC emissions rates are higher at low speeds and in stop-and-go traffic. The 22 percent rise between 1969 and 1990 in the number of car trips taken daily by the average household has also increased emissions. Trips increase emissions because a cold vehicle pollutes at a much higher rate than a warm one and because emissions are greatest during cold starts.

Clean Air Act Amendments of 1990

The Clean Air Act Amendments of 1990 continue the practice of using new-car emissions standards as the primary means for reducing overall car emissions. The amendments significantly tighten emissions rates for new cars beginning with the 1994 model year. They also allow states to adopt California's vehicle emissions standards, which are stricter than federal standards and are scheduled to become even stricter in the future.

Unlike the Clean Air Act Amendments of 1970, the 1990 amendments recognize the importance of evaporative emissions. For example, they force the U.S. Environmental Protection Agency (EPA) to establish regulations to control these emissions and require the use of reformulated gasoline in areas with the worst ozone problems. Other provisions push the frontier of automotive technology by requiring the introduction of alternative-fuel vehicles (for example, cars that run on methanol or compressed natural gas) in certain commercial and government vehicle fleets in nonattainment areas and by setting up a pilot program in California where such vehicles will be introduced to the general public.

The 1990 amendments also acknowledge the significance of the disparity between new-car and average-car emissions rates, as well as the effect of increasing VMTs on total vehicle emissions. They do this by requiring "enhanced" vehicle inspection and maintenance programs in the areas with the worst ozone problems and by requiring that local transportation plans conform to Clean Air Act goals. Above all, the 1990 amendments recognize that new-car emissions standards are not the only way to reduce motor vehicle emissions, and they leave to local and state governments many decisions about adopting alternative policies.

Despite these improvements in regulating motor vehicle emissions, the 1990 amendments still suffer from three shortcomings that will make it difficult to attain ambient ozone standards in a cost-effective manner. First, the amendments retain an excessive reliance on emissions standards and technological solutions. Second, they perpetuate EPA's practice of basing emissions-reduction estimates on its computer models rather than on empirical data, despite the fact that the estimates produced by those models can be grossly inaccurate. Third, they tend to target vehicles with *either* high emissions per mile *or* high mileage, instead of vehicles with both.

Because the 1990 amendments set imminent deadlines for attaining ambient ozone standards, they could lead states into hasty and expensive decisions. For example, in February 1994, the Ozone Transport Commission, which coordinates air quality decisions in the northeastern states, requested EPA's permission to adopt California's vehicle emissions standards.

Cost-Effectiveness of Various Options

The 1990 amendments give local areas flexibility to choose among many ways to reduce vehicle emis-

sions. But which approaches to choose? How can state and local officials avoid costly efforts with uncertain results?

One way of evaluating policy options is to estimate the cost-effectiveness of each policy—that is, the cost in dollars per ton of pollutant reduced—and then compare the estimates. With such information, states presumably could adopt the low-cost options first. Various groups, including Resources for the Future (RFF), have studied the cost-effectiveness of individual policy options in reducing the emission of VOCs, and our analysis of these studies is summarized here (see Table 1). The cost-effectiveness of these approaches varies greatly, from $1,650 per ton of VOCs reduced for emissions-based vehicle registration fees to $29,000–$108,000 for electric vehicles. In general, EPA considers any approach that costs less than $5,000 per ton of emissions reduced to be highly cost-effective. Options that reduce VOCs for less than $10,000 per ton are still considered reasonable.

The emissions-reduction options we consider here can be divided into two types. The first type is *command-and-control* approaches that set emissions standards or that specify emission-control technologies. The second type is *economic-incentive* approaches that change prices (such as car purchase prices and gasoline prices) and thereby lead motorists to make decisions that reduce vehicle emissions. For our analysis, we compared options of both types.

Command-and-control approaches include the mandated use of reformulated gasoline, the creation of enhanced inspection and maintenance (I&M) programs, and the replacement of gasoline vehicles by alternative-fuel vehicles. We considered these three approaches, which are required by the 1990 amendments in some nonattainment areas. We also analyzed the group of low-emission vehicles mandated by California.

Reformulated gasoline may be one of the cheaper options for reducing VOCs, according to EPA's estimate of its cost-effectiveness. The overall emissions reductions from reformulated gasoline are small, however.

Table 1. Estimates of the Cost-Effectiveness of Alternative Approaches to Reducing Motor Vehicle Emissions (in $ per ton of VOCs reduced)

Command-and-Control Approaches	
Reformulated gasoline	
Federal	$1,900–3,900[a]
California	$4,100–5,100[a]
Inspection and maintenance	
EPA enhanced	$4,500–6,000
Remote sensing	$2,600–6,000
Hybrid	$4,000–6,000
Alternative-fuel vehicles	
Methanol	$30,000–60,000
Compressed natural gas	$12,000–22,000
Electric	$29,000–108,000
California vehicles	
Transitional low-emission vehicles	$3,700–21,000[b]
Low-emission vehicles	$2,200–27,000[b]
Ultra-low-emission vehicles	$4,200–41,000[b]
Economic-Incentive Approaches	
Accelerated vehicle-retirement	$4,000–6,000
Gasoline-tax increase	$4,500
Emissions-based vehicle registration fees	$1,650[c]

Note: Unless otherwise indicated, estimates are RFF estimates based on RFF studies or other studies. For more details on all estimates, see discussion paper 94-26, "Shifting Gears: New Directions for Cars and Clean Air" by Winston Harrington, Margaret A. Walls, and Virginia D. McConnell.

[a] These EPA estimates are based on reformulation of gasoline according to EPA's recipe (which increases the price per gallon by 3 cents) and according to California's recipe (which increases the price by 8–11 cents per gallon).

[b] These estimates are derived from studies by the California Air Resources Board and the Automotive Consulting Group.

[c] These estimates are based on a draft study by Energy and Environmental Analysis, Inc.

According to RFF estimates, I&M programs are somewhat more expensive, but they may yield larger emissions reductions. The enhanced I&M program mandated by EPA requires that a new, more accurate, but also more expensive, tailpipe-emissions test be used and that inspections be performed at centralized facilities that only test vehicles, rather than at service stations that both test and repair vehicles. Alternative I&M programs include remote sensing, which uses roadside monitoring and detection devices to measure vehicle emissions, and "hybrid" programs, which employ remote sensing but also subject vehicles to enhanced I&M every two to four years.

Alternative-fuel vehicles, according to RFF estimates, are a very expensive option for reducing VOCs. Policies that require these vehicles are poorly targeted because they ignore emissions from vehicles already on the road. Moreover, such policies might cause car manufacturers to increase the price of all vehicles so that they can cover the cost of producing alternative-fuel vehicles. If so, motorists might choose to keep their old vehicles rather than purchase new, less-polluting ones.

These problems may be compounded if states adopt California's phasedown to "ultra-low-emission vehicles" and "zero-emission vehicles."

One of the problems with command-and-control approaches to regulating motor vehicle emissions is that they are targeted to reduce emissions *rates* instead of total emissions. The variability in vehicle use makes emission-rate regulation less promising than regulations targeted at both emissions rates and mileage. (For example, vehicles driven more than 25,000 miles per year make up only 10 percent of all motor vehicles yet account for 30 percent of the total VMTs.)

Because the command-and-control approaches that we studied rely on emissions standards and technological solutions, they are not, in general, as potentially cost-effective as approaches that are well-targeted *and* rely on economic incentives. Economic-incentive approaches include accelerated vehicle-retirement (AVR) programs, gasoline taxes, and vehicle registration fees based on emissions rates.

According to RFF estimates, AVR programs and a gas-tax increase of 4.3 cents per gallon reduce VOCs at about the same cost per ton as I&M programs, but both approaches may yield only limited total emissions reductions. AVR programs are not well targeted since the cars they take off the road have at most only a few years of life remaining. AVR programs will not substantially reduce car emissions unless the programs are large scale, in which case their cost-effectiveness decreases (see Chapter 14).

Gas-tax increases are poorly targeted as a means to reduce VOC emissions because they discourage the use of all cars, not just the most-polluting ones. Furthermore, a large tax increase could lead consumers to purchase cars with greater fuel efficiency—behavior that could offset emissions reductions in the long run. Gasoline taxes nonetheless would be an incentive for energy conservation and therefore lead to reductions in emissions of carbon dioxide and other "greenhouse" gases.

Vehicle registration fees based on emissions rates appear much more promising. Unlike new-car emissions standards, these fees target emissions from *all* vehicles on the road. They also give motorists the proper incentives to maintain (or scrap) their vehicles, since the fee would be higher for a car that pollutes more. Based on a recent preliminary analysis of such fees, we conclude that they are more cost-effective than the other options examined here. Again, we note that there are inefficiencies in approaches that reduce emissions rates rather than total emissions. In the context of emissions-based registration fees, cars that have different emissions per mile would be charged different fees even when their mileages are such that their *total* emissions levels are the same.

By the EPA benchmarks of $5,000 and $10,000 per ton of VOCs reduced, reformulated gasoline, I&M programs, AVR programs, gas taxes, and, in particular, emissions-based vehicle registration fees all appear to be attractive. In contrast, alternative-fuel vehicles and the low-emission vehicles mandated by California appear to be very unattractive.

Caveats

Estimates of the cost-effectiveness of all these options should be interpreted cautiously for several reasons. First, the true cost-effectiveness of any particular option depends on previously implemented options. For example, an I&M program might be less cost-effective if it were implemented after the use of reformulated gasoline than if it were implemented beforehand.

Second, policies that cost-effectively reduce VOCs—for example, accelerated vehicle-retirement and reformulated gasoline—are not necessarily effective at reducing NO_x. In some areas of the country, NO_x reduction is essential for ozone improvements.

Third, uncertainty pervades the emissions-reduction estimates on which calculations of cost-effectiveness are based, and often analysts resort to "best case" outcomes. For example, the estimates for EPA's enhanced I&M program assume that cars identified as high emitters are successfully repaired, but repairs have often been ineffective. In one study, more than half of all vehicles that underwent repairs to reduce emissions had *greater* emissions afterwards!

Fourth, some of the cost estimates are also highly uncertain, particularly those for alternative-fuel vehicles and the low-emission vehicles mandated by California. Some observers believe that technological advances could greatly reduce costs; in the case of alternative-fuel vehicles, however, we feel that it is highly unlikely that any such advances are imminent.

Toward More Efficient and Effective Policy

To ensure that policies to reduce motor vehicle emissions are cost-effective, we must design them with three characteristics in mind. First, we should target policies as precisely as possible to reduce total emissions, rather than emissions per mile, in those places and at those times when ozone creation is at its peak. Second, we must design policies that give motorists incentives consistent with pollution reduction. Third, to the extent possible, we must measure performance on the basis of actual emissions rather than on estimates from computer models.

None of the options considered here is ideal with respect to all three of these characteristics, but each could be improved with relatively minor changes. For example, I&M programs might be more cost-effective if they went after only the very dirtiest vehicles, which tend to be the easiest to detect as well as the most likely to be effectively repaired. Targeting the very high polluters also might mean that a simpler, less costly emissions test could be used instead of the new emissions test developed and promoted by EPA.

Instead of reducing all VMTs through a gasoline tax, better approaches would include congestion pricing of roadways and downtown parking taxes. Though not targeted at high-polluting vehicles, these options can be targeted to areas with ozone problems, and they have the substantial advantage of reducing traffic congestion.

Emissions-based vehicle registration fees appear to be very cost-effective, and, unlike many of the policies considered here, the potential emissions reductions could be made almost as large as desired simply by raising the fee. This policy could be better targeted and thus even more cost-effective if it were based on a car's estimated total emissions during peak ozone periods. This would require information about the car's average emissions rates and its mileage in particular locations at particular times. Here, remote sensing could be valuable.

Mobile remote-sensing units make it relatively inexpensive to measure car emissions when ground-level ozone is at its peak. A program that uses these units could require motorists to pay high registration fees when their cars' average emissions rates—based on several sensor readings—rise above a certain level. Such a program could target total emissions, not just emissions rates, by setting a high fee for cars that not only have high emissions but also pass by the sensors many times.

Remote-sensing devices, however, cannot monitor evaporative emissions, and they are difficult or impossible to use in bad weather. In addition, there may be questions about whether the measured emissions faithfully represent the average performance of cars.

Emissions-based registration fees are promising, but they may encounter the same attitudes that economic instruments for environmental policies always seem to face, including skepticism about their effects on pollution and concern about equity. Nonetheless, the tide in environmental policy has been shifting toward such instruments. Perhaps the time is right for their application to mobile-source emissions control.

Recommendations

The most economically attractive ways of reducing motor vehicle emissions would be directed at cars already on the road and would require extensive use of economic incentives. Emissions-based registration fees hold much promise. Such a policy could achieve substantial emissions reductions at relatively low cost even if based only on emissions rates as determined by a conventional emissions test. To reach its full potential, however, an emissions-based registration fee should reflect mileage during peak ozone periods as well as emissions rates. This may require the use of remote sensing. Further investigation of remote sensing to deal with the real—or to lay to rest the perceived—problems associated with that technology is warranted.

Until emissions-based registration fees and other economic-incentive approaches are investigated, it would be a serious mistake for states to commit themselves prematurely to command-and-control approaches, which may prove to be costly, ineffective, and difficult to back away from. Thus, while I&M programs are promising because they target emissions from all cars (not just new cars), we should avoid a nationally mandated, uniform I&M program until we gather data from demonstration programs.

Until some of the other, cheaper alternatives have been investigated, it would be premature for the rest of the country to adopt California's new-car emissions standards. Given great uncertainty and dubious benefits about the costs of new types of low-emission vehicles, it seems wise to let California experiment with these cars by itself. If the costs prove low, then the kinds of economic policies we advocate will bring them to market, thus achieving the emissions-reduction goals of the California vehicle program but without legislative fiat.

14 Will Speeding the Retirement of Old Cars Improve Air Quality?

Anna Alberini, David Edelstein, and Virginia D. McConnell

Even under increasingly stringent emissions standards, cars are still contributing to urban air pollution. Part of the problem lies with old cars—those manufactured before 1980. Although these cars make up a relatively small percentage of the nation's car fleet, their removal from the road could eliminate a large percentage of some emissions. This is why states, as well as polluting firms that are looking for ways to get emissions-reduction credits, have expressed interest in running accelerated vehicle-retirement (AVR) programs. However, there is much controversy about just how cost-effective AVR programs are likely to be. RFF research suggests that the answer depends on how the cars they enlist differ from the rest of the old-car fleet, as well as on the programs' size, duration, and location.

Despite a substantial decrease in the past twenty-five years, motor-vehicle emissions continue to be a major contributor to air pollution in many urban areas. It is now clear that air quality objectives will never be met simply by setting increasingly stringent emissions standards for vehicles built in the future; something must be done about the emissions of vehicles built in the past.

Given that emissions from newer vehicles have already been drastically cut, policymakers are focusing on emissions from older cars—that is, 1980 and earlier model-year cars. These older cars often do not have advanced emission-control equipment; when they do, it sometimes no longer functions properly. Thus they tend to emit pollutants at much higher levels, on average, than newer vehicles.

Considering that pre-1980 cars make up only 18 percent of the vehicles in use in the United States and account for only 8 percent of total miles driven, they contribute a surprisingly large share of total motor vehicle emissions. On a typical hot summer day, they emit approximately 40 percent of the hydrocarbon, 40 percent of the carbon monoxide, and 25 percent of the nitrogen oxide emissions of the nation's car fleet.

These statistics suggest that a potentially effective way to reduce hydrocarbon, carbon monoxide, and nitrogen oxide emissions in urban areas is to take older, highly polluting cars off the road. Accelerated vehicle-retirement (AVR) programs offer a new market-based opportunity to do this. These programs buy pre-1980 cars from their owners, usually at a price

Originally published in *Resources*, No. 115, Spring 1994.

ranging from $500 to $800, and then scrap the cars.

Because AVR programs remove polluting cars from the road, they are one way for private firms and states to fulfill their emissions-reduction obligations. Through AVR programs, firms that are looking for a lower-cost alternative to reducing their own pollution are given a mechanism to earn credits for short-term emissions reductions. AVR programs are also one of many options states are exploring to meet goals for air-pollution control.

At first glance, AVR programs may appear to be an attractive way to tackle urban air problems. However, there are questions about how effective buying and scrapping old cars will be in reducing emissions and how much AVR programs will cost relative to other policies to cut pollution.

In 1992 the President's Commission on Environmental Quality (PCEQ) commissioned us to conduct a study of a small, one-time AVR program run in Delaware by U.S. Generating Company (USGen), an independent electric-power producer. In that study—developed jointly by PCEQ, Resources for the Future, and USGen—we tried to shed light on some of the controversial issues surrounding AVR programs.

Issues in Evaluating AVR Programs

Firms and agencies considering whether to operate an AVR program will want to determine the amount of emissions that the program has the potential to reduce. Making this determination is fairly complex, as it means forecasting the outcome of three events that cannot be observed directly: (1) the quantity of pollutants that each car would have emitted had it not been scrapped, (2) the quantity of pollutants emitted by cars bought to replace the scrapped cars, and (3) the number of cars that can be purchased through an AVR program at different prices. An uncertainty that affects the emissions-reduction potential of long-term, large-scale AVR programs is the effect such programs have on the purchase price that owners of old cars are willing to accept.

Many factors determine the amount of emissions that a scrapped car *would* have emitted if it had remained on the road. The car's emissions rate, average annual mileage, and expected remaining life all contribute to what we term the *avoided emissions*. Because these variables are impossible to measure, emissions reductions are usually calculated by imputing to the scrapped cars the emissions rates, annual mileage, and remaining life of "average" cars of the same years as the scrapped cars. However, the resulting estimates of avoided emissions are unlikely to be correct because the cars that enter an AVR program are not representative of cars of a certain age. On the one hand, the scrapped cars are at the low end of the used-car market and are therefore likely to have a relatively shorter remaining life span than the average vehicle of their age. They also may be driven less. On the other hand, they may have greater per-mile emissions.

Emissions from cars and other modes of transportation that replace scrapped cars are also difficult to predict. Even if we know what cars have been purchased to replace scrapped cars and thus can measure their emissions, it is unlikely that we will also know what cars have been purchased by the people who sold the replacement cars. Nor is it likely we will know what cars the people who sold *those* cars bought to replace the cars *they* sold and so on. Because these chains of transactions are impossible to track, the emissions rates of replacement cars are often assumed to be equal to the average emissions rate of a region's car fleet.

Predicting how vehicle owners will respond to different purchase-price offers presents still more difficulties. Presumably, the higher the offer, the greater the number of cars AVR programs will enlist. But without information about how many cars each different offer will attract, a firm or agency cannot predict what level of emissions reductions can be achieved. This uncertainty makes AVR programs a gamble, because the firms or agencies that run them often will be required to reduce emissions by a specific amount.

When AVR programs are designed to be a large or steady source of emissions-reduction credits for a region, another set of problems arises. Long-term, large-scale AVR programs may create so much demand for old cars that used-car prices will rise. The retirement of a large number of old cars in the region served by the program is likely to increase the value of the remaining old cars in that region. If so, large-scale programs will have to offer increasingly higher prices to obtain a given number of cars; this means that they will reduce emissions at a higher per-ton cost than small-scale programs, which do not affect the market for old cars.

In addition, ongoing AVR programs may create the wrong incentives for car owners. For instance, people might be encouraged to keep their cars longer than they would normally, so as to have an old car to sell. Also, people living in one region might offer their cars to a program operating in a different region. If so, emissions would not be reduced in the geographic area where they were intended to be reduced.

Study of the Delaware Vehicle-Retirement Program

Given these uncertainties, the role AVR programs should play in reducing air pollution is still unclear. However, our analysis of data collected from USGen's AVR program in Delaware in the fall of 1992 provides some insight about the role these programs *could* play.

As noted above, it is difficult for firms and agencies to determine whether AVR programs are worth starting without knowing how many cars they will attract. Before starting its program, USGen estimated that it would have to buy and scrap 125 cars if it wanted to offset an increase in air pollution caused by transporting coal to one of its power plants.

The company predicted that it would recruit this number of cars if it targeted a select group of car owners to participate in its program. Initially, USGen tried to enlist cars from among the 1,034 cars that had received waivers from Delaware's vehicle inspection and maintenance (I&M) pro-

gram. These waivers allow cars to be driven after they undergo repairs to reduce tailpipe emissions and fail to pass the I&M program's emissions test a second time. Because cars with waivers are likely to be the most polluting cars on the road, they are the ones AVR programs want to enlist so as to obtain the greatest emissions reductions.

Since USGen enlisted only sixty cars from owners of waivered cars, it made offers to about 3,000 owners of cars randomly chosen from the pre-1980 car fleet. From these, it recruited sixty-five additional cars.

The combination of both waivered and non-waivered cars gave us the opportunity to test the feasibility and quantify the benefits of an AVR program that targets highly polluting pre-1980 cars, as well as a program that accepts any pre-1980 car.

Half of the 125 cars USGen purchased were given emissions tests. Using the results of the emissions tests, we estimated the average emissions rate of the scrapped cars. We found that, on average, the waivered vehicles emitted about 60 percent more hydrocarbons from their tailpipes than the nonwaivered vehicles. Using a model developed by the U.S. Environmental Protection Agency, we then predicted the average emissions rate of replacement cars.

We surveyed car owners participating in the program to obtain information on which to base estimates of the annual mileage and the expected remaining life of the 125 cars scrapped. We also surveyed a sample of car owners who were solicited to participate in the AVR program but who declined to do so. By surveying both those who accepted USGen's $500 purchase-price offer and those who refused it, we were able to determine how the scrapped cars differed from the fleet of pre-1980 cars as a whole.

In the surveys, we asked car owners how often and how many miles they drove their cars, what condition their cars were in, how much longer they planned to keep their cars, whether they expected the cars to need major repairs in the near future, and how much effort it took for them to maintain their cars.

To gauge how the purchase price affected participation in the program, we asked respondents to give us their *reservation price*—that is, the minimum offer they would have accepted for their cars.

To get better estimates of avoided and replacement emissions, we conducted follow-up surveys of both participants and nonparticipants one year later. These surveys examined how participants had replaced their cars—by purchasing new cars or relying on public transportation, for example—and how their driving habits had changed.

Study Results

Using data from the surveys and emissions tests, we were able to estimate the potential for emissions reductions from USGen's AVR program, as well as the program's cost-effectiveness compared with other mechanisms for reducing emissions. We also were able to draw some general conclusions about AVR programs.

Our data analysis focused on the relationship between the value of old cars and the cars' expected remaining life. We found that old cars with low values typically have a short remaining life. This finding, which is based on the first empirical evidence ever collected about the remaining life of cars sold at different purchase-price offers, confirms what economic theory suggests. AVR programs that offer $500 will attract cars that would have remained on the road no more than two years on average. This information is essential for evaluating the cost-effectiveness and emissions-reduction potential of AVR programs.

In estimating the cost-effectiveness of the Delaware AVR program and in deriving what economists would call an emissions-supply function (the function that predicts the number of tons of emissions reduced at varying purchase-price offers), we had to make some assumptions about replacement cars' emissions and usage. We based these assumptions on data gathered from the original and follow-up surveys.

Both surveys indicated that the scrapped cars were driven just as many miles (if not more) than the cars not sold and scrapped and that annual mileage was not meaningfully correlated with the age of individual pre-1980 cars. The follow-up surveys showed that, on average, replacement cars were driven no more than scrapped cars had been. We assumed, therefore, that a scrapped car would be replaced by another car with equal annual mileage. The follow-up surveys also showed that the average replacement car was a 1986 model-year car. Because such a car is very similar to the "average" car in the U.S. car fleet, we assumed that the emissions rate of replacement cars was the same as the average emissions rate of the nation's car fleet.

Using these assumptions and the emissions-test and survey data, we estimated a statistical model that correlates the remaining life of a car with the likelihood that its owner will sell it to an AVR program. Based on estimates generated by the model, we determined in two different ways the Delaware AVR program's cost-effectiveness in reducing hydrocarbon emissions. First, we estimated the cost-effectiveness of the entire program, which included both waivered and nonwaivered cars. We found that the program reduced hydrocarbon emissions by about fifteen tons at a per-ton cost of about $5,000. Second, we restricted our analysis to only the waivered cars and found that the per-ton cost of emissions reductions was a little more than $4,000.

Next we estimated the cost of reducing hydrocarbon emissions for a hypothetical AVR program that pays $500 for any pre-1980 car. We found that the comparatively lower emissions of nonwaivered cars in such a program would increase the per-ton cost of reducing hydrocarbon emissions to about $6,000 (see Table 1).

According to these estimates, a program that targets waivered cars appears to be more cost-effective than a program that accepts any older car. If we had taken the value of carbon monoxide and nitrogen oxide reductions into account, the cost-effectiveness of each program would have increased.

Emissions reductions depend, of course, on the number of cars recruited, and participation in AVR programs appears to be very sensitive to the purchase price offered. USGen's offer of $500

Table 1. Predicted Cost-Effectiveness, Participation Rates, and Expected Remaining Vehicle Life in a Small-Scale Accelerated Vehicle-Retirement Program at Various Purchase-Price Offers

Purchase price offer	Average cost per ton of HC reductions	Predicted participation of the pre-1980 fleet	Expected remaining life of a participating vehicle (years)
$400	$5,370	1.8 %	1.5
$500	$5,946	4.3 %	1.7
$600	$6,219	8.0 %	1.9
$700	$6,572	12.8 %	2.1
$800	$6,904	18.2 %	2.3
$900	$7,194	24.0 %	2.4
$1,000	$7,509	30.0 %	2.5
$1,100	$7,838	36.0 %	2.7
$1,200	$8,167	41.7 %	2.8
$1,300	$8,477	47.2 %	2.9
$1,400	$8,800	52.3 %	3.0
$1,500	$9,123	57.0 %	3.1

Note: The program does not target the most highly polluting cars, but rather accepts any pre-1980 car. The shading reflects uncertainty about estimates of cost-effectiveness, participation rates, and expected remaining life at high purchase prices.

attracted only 4.3 percent of the total population of Delaware's pre-1980 cars, and only 5.9 percent of its subpopulation of waivered cars. If the offer price had been increased to $700, we estimate that the company's program would have attracted approximately 13 percent of the total population and 18 percent of the waivered fleet.

Our cost-effectiveness estimates indicate that the expected remaining life of cars purchased by AVR programs is also very sensitive to purchase-price offers. According to our estimates, a car sold to an AVR program for $500 would otherwise have remained on the road for about 1.7 years, and a car sold for $700 would have been driven another 2.1 years. In contrast, the average remaining life for the typical pre-1980 car is about 4.2 years. The significant difference between this figure and our estimates shows the danger of relying on average fleetwide estimates of variables, such as expected remaining life, when projecting the benefits of AVR programs.

Another important conjecture is that high purchase-price offers may adversely affect the cost-effectiveness of AVR programs that attract a large percentage of a region's old-car fleet. Removing a large number of cars from the fleet could increase average vehicle prices and thus influence the willingness of potential participants to accept a given offer price. Participation rates would therefore be lower at all offer prices. If this is the case, our estimates of the per-ton cost of removing hydrocarbon emissions would be too low, and our estimates of participation rates would be too high.

Once we established estimates of remaining vehicle life and participation rates, we were able to derive an emissions-supply function that predicts the number of tons of emissions reduced at varying purchase-price offers (see Figure 1). The emissions-supply curve is an increasing function of these offers: the higher the offer, the higher the number of cars AVR programs will attract and the more emissions savings the programs will realize. The slope of the curve depends on car owners' responsiveness to the offer price, which in turn depends on the number of cars in the targeted fleet that are valued at less than the offer price. Large or

Figure 1. Emissions-supply function: Nontargeted accelerated vehicle-retirement program (1,000 vehicles).

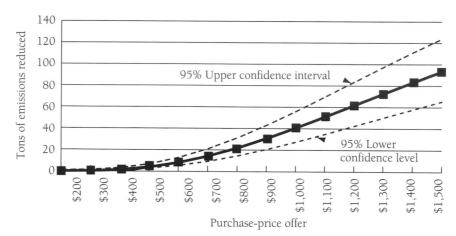

The emissions-supply function predicts the number of tons of emissions reduced at varying purchase-price offers. The solid-line curve is the estimate of this function for a small-scale AVR program that purchased cars representative of the pre-1980 car fleet rather than targeting highly polluting pre-1980 cars. The broken-line curves are the 95 percent confidence intervals, reflecting the uncertainty surrounding the estimates.

ongoing AVR programs would likely make the curve flatter—especially at high offer prices—because high used-car prices would make most owners less willing to participate in an AVR program.

The Future of AVR Programs

Whether AVR programs are cost-effective relative to other means of reducing hydrocarbon emissions will depend on the severity of the air quality problems and the extent of existing pollution controls in the regions they serve. As noted above, the Delaware AVR program, which offered car owners $500 to scrap their cars, reduced hydrocarbon emissions from waivered cars at a per-ton cost of $4,000. By comparison, the U.S. Environmental Protection Agency estimates that the substitution of reformulated gasoline for regular gasoline would reduce hydrocarbon emissions at a per-ton cost of $3,900.

A small-scale AVR program that does not target waivered cars is less cost-effective. It will reduce hydrocarbon emissions at a per-ton cost ranging from about $5,000 (at a $400 purchase-price offer)

to about $7,000 (at a $900 purchase-price offer). However, a small-scale AVR program is more cost-effective than a program to replace cars that run on gasoline with cars that run on natural gas or on methanol, which would reduce hydrocarbon emissions at a per-ton cost of $12,000 and $30,000, respectively.

We conclude that small-scale, short-term AVR programs can be cost-effective for some regions. Programs that target the most highly polluting old cars may be the most promising. However, only a very large-scale program will generate appreciable emissions reductions. Since such a program is also likely to increase used-car prices, its cost-effectiveness is likely to decrease over time. The cost-effectiveness of ongoing AVR programs may also decrease in the long run because these programs are likely to create adverse incentives and unexpected consequences, especially when combined with efforts to target highly polluting vehicles.

Another area of uncertainty concerns the interaction of AVR programs and states' vehicle inspection and maintenance (I&M) programs. If the

Update

Since we wrote this article, there continues to be interest in old-car scrap programs as a way to reduce emissions of hydrocarbons and nitrogen oxides in polluted urban areas. Illinois and Colorado both ran old-car scrap pilot programs in recent years. In the Los Angeles region, the South Coast Air Quality Management District has allowed old-car scrap programs to be part of the private sector emissions-trading program in the region. Industrial or stationary sources of pollution can purchase and scrap old cars in lieu of other ways of reducing equivalent emissions. Since 1993, twenty-two thousand vehicles have been scrapped as part of this program, with the current market-determined price at about $500 to $600 per vehicle. The current rules assume the remaining useful life of these vehicles is three years, which is different from the 1.7 years estimated here for the Delaware program. While there are likely some differences in the fleet distribution and the used-car pricing structure between the two states, it is unlikely to account for all of this difference. This is perhaps the most important component of the RFF study of old car scrappage—that it is important to consider the "selection" problem (old cars recruited for such programs are likely to be those in poor condition with only a short remaining life) in designing and granting emission reductions for these programs. California is considering a major old-car scrap effort in the future, and it will be interesting to see how they deal with this issue.

bination with state I&M programs. For instance, joint AVR-I&M programs might allow owners of cars that do not pass emissions tests to choose to either repair or scrap their cars. Our research suggests that the emissions-reduction potential of AVR programs could be maximized if the programs were set up to complement other means of reducing urban air emissions.

Suggested Reading

Alberini, Anna, David Edelstein, Winston Harrington, and Virginia McConnell. 1994. "Reducing Emissions from Old Cars: The Economics of the Delaware Vehicle Retirement Program." Discussion Paper QE94-27 (April). Washington, DC: Resources for the Future.

Hahn, Robert W. 1995. An Econometric Analysis of Scrappage. *RAND Journal of Economics* 16(2).

Illinois Environmental Protection Agency. 1993. *Pilot Project for Vehicle Scrapping in Illinois.* Springfield, Illinois.

South Coast Air Quality Management District, Rule 1610, *Old Vehicle Scrapping.*

I&M programs become more effective in the future, they may reduce the emissions-reduction potential of AVR programs by identifying and removing from the road many of the most polluting cars. However, there may be many roles for AVR programs in com-

A Voluntary Approach to Environmental Regulation
The 33/50 Program

Seema Arora and Timothy N. Cason

Voluntary pollution reduction gives companies an opportunity to take least-cost actions to reduce pollution and at the same time gain positive public recognition. Given these potential advantages, will voluntary pollution reduction programs attract large numbers of participants and result in large pollution reductions? An analysis of the U.S. Environmental Protection Agency's 33/50 Program suggests that willingness to participate in that program varies greatly among industries and among firms; indeed, only a small percentage of any industry's firms are participating in the program. However, the companies that are participating are responsible for a large percentage of toxic emissions. Thus pollution reductions due to the program could be substantial.

Pollution reduction programs that encourage voluntary participation by companies are gaining currency as a viable approach to environmental improvement. But can voluntary programs be effective in reducing pollution? What kind of company would decide to participate? And what kinds of pollution reductions would be made?

To answer these questions, we conducted a study of the 33/50 Program, a voluntary pollution prevention initiative designed by the U.S. Environmental Protection Agency (EPA) to reduce toxic releases. This program stresses cooperation between regulators and industry and provides positive feedback and awards to participating firms. We evaluated factors that lead to participation in this program by industries and by individual firms. We also compared the 33/50 Program with other voluntary pollution control programs. Before we summarize our findings, however, we present some background on voluntary compliance and the 33/50 Program itself.

The Movement toward Voluntary Compliance

In 1984, a poisonous gas leak from a Union Carbide pesticide plant in Bhopal, India, killed more than 2,500 people and permanently disabled some 50,000 more. Since then, the potential for accidental chemical releases has worried residents of communities near industrial plants, who have wanted to know what chemicals these plants are emitting in order to prepare for such releases. U.S. residents began advocating local community right-

Originally published in *Resources*, No. 116, Summer 1994.

to-know laws, augmenting a movement for worker right-to-know laws that had begun in the late 1970s.

The chemical disaster in Bhopal also catalyzed the movement for a federal community right-to-know law. In 1986, Congress passed the Emergency Planning and Community Right-to-Know Act, which embodies the principle of public disclosure. The act requires all manufacturing facilities to report annually on releases and transfers of more than 320 toxic chemicals. This reporting has resulted in the creation of a national database called the Toxics Release Inventory (TRI).

One of the results of mandated public disclosure has been public pressure for accountability. Such pressure may be exerted by consumer groups, citizen action groups, or the media. Even the mere anticipation of public pressure can lead companies to alter their behavior, as it did in the case of Monsanto.

When the TRI was first publicly reported in 1987, Monsanto discovered that it was one of the largest polluters. This discovery led the company to pledge to reduce its toxic air releases by 90 percent by the end of 1992. Several features of this pledge are striking. First, the pledge was voluntary, as the company was not violating any environmental standards. Second, it came from the highest echelon of the corporation—in fact, from Richard Mahoney, Monsanto's chief executive officer. Third, it set a trend for other polluting firms to follow.

While public disclosure prompted Monsanto to act before consumers, citizen action groups, and the media had time to react to the TRI information, other companies needed more urging. Soon after the first TRI was reported, the *New York Times* published a full-page advertisement, which was sponsored by citizen action groups, highlighting the top ten corporate land polluters, water polluters, and air polluters. Firms that figured prominently in the ad immediately approached EPA and pledged to improve their environmental performance.

By the late 1980s, many companies that had not been at the forefront of environmental stewardship began to adopt a much more proactive environmental stance. Among the results of the compa-

nies' inclination toward voluntary action was the 33/50 Program.

The 33/50 Program

The 33/50 Program gets its name from its two-step reduction goals: a 33 percent reduction of chemical releases and transfers from 1988 levels by 1992 and a 50 percent reduction by 1995. The program encourages firms to develop less-toxic substitutes for highly toxic chemicals, reformulate products, and redesign production processes in order to reduce pollution at its source. It focuses on seventeen of the 320 TRI chemicals that are highly toxic, are produced by industry in large volumes, and present pollution prevention opportunities. The 33/50 Program stresses flexibility, allowing participants to reduce releases of any of these chemicals into any environmental media (air, land, or water). Since about 70 percent of these releases are into the air, however, the 33/50 Program is primarily an air toxics reduction program.

Participation in the program is voluntary and does not change a firm's responsibilities for complying with environmental laws. Indeed, EPA claims that it will not give preferential treatment—such as relaxed regulatory oversight or enforcement of EPA regulations—to program participants. Because participation is voluntary, commitments to achieve pollution reductions are not legally enforceable—in fact, firms are free to renege. Nevertheless, many companies that have decided to participate in the 33/50 Program have submitted detailed timetables and pollution reduction targets.

Incentives for participation in the 33/50 Program include public recognition by EPA, special awards for outstanding achievements in pollution prevention, and, significantly, the opportunity to take least-cost actions to mitigate pollution. Unlike mandatory programs, this voluntary program allows firms the flexibility to make the emissions reductions that are most cost-effective for them. Moreover, EPA provides assistance to the companies making these reductions by conducting regional pollution prevention workshops and by providing access to the agency's Pollution Prevention Information Exchange System.

Voluntary pollution reduction programs such as the 33/50 Program appeal to regulators because the programs require EPA to engage in no costly rulemakings. Furthermore, they save regulators the substantial costs of monitoring and enforcing compliance.

EPA initiated the 33/50 Program in February 1991, when it invited 555 companies with substantial chemical releases to participate. It later extended this invitation to all other firms that release chemicals targeted by the 33/50 Program. As of March 1994, the agency had invited more than 8,000 companies to participate in the program. To date, nearly 1,200 of these firms have done so.

The 33/50 Program has been hailed as a success. It exceeded its 1992 interim goal (a 33 percent reduction in emissions) by more than 100 million pounds—a reduction of more than 40 percent from 1988 emissions levels. According to the projections of participating firms, the 1995 target is also likely to be achieved.

Participation by Industry and EPA Region

Since participation is critical to the success of voluntary pollution reduction programs, we examined the factors that may have led 1,100 of the more than 7,000 firms in our study sample to take part in the 33/50 Program. Our analysis revealed substantial variation in the willingness to participate among different industries and EPA regions. Among industries, this variation may be explained by levels of advertising as well as research and development (R&D) expenditures; the strength and environmental commitment of trade and manufacturer associations; and each industry's market structure. Among EPA regions, the variation may be due to differences in the regions' environmental regulations. We look at each of these factors in turn.

The amount of money an industry spends on advertising and on R&D helps to explain which industries participate in the 33/50 Program. Industries with high advertising expenditures tend to have high levels of contact with consumers. If consumers are environmentally conscious, we would expect that participation in the 33/50 Program would be higher among industries that produce final products, and hence have a lot of consumer contact, than among industries that produce inputs to final products. When we tested this hypothesis using advertising expenditures as a proxy for consumer contact, we found that the greater an industry's advertising expenditures, the greater the likelihood that it participates in the 33/50 Program. Industries with high R&D expenditures are also likely to participate in the program, perhaps because a commitment to developing new products is consistent with the program's goals.

The comparative strength and environmental commitment of trade and manufacturer associations is another factor in industry participation: industries with associations that exert a strong measure of influence on members' actions and that stress environmental stewardship are likely participants. The high participation rate within the chemical industry may be owing in part to the fact that all members of the Chemical Manufacturers Association must join Responsible Care, an initiative with goals similar to those of the 33/50 Program.

The market structure of each industry may also help explain which industries participate in the program. Recent trends in "green" marketing and in consumer awareness of environmental issues, as well as theoretical work on firms' environmental performance, provide a basis for the expectation that firms compete on environmental variables, particularly when they are part of an industry in which competition is great and individual market shares are small. We confirmed this intuition in a study of a small sample of firms for which we were able to combine financial (or economic) information with toxic release data. The study indicates that unconcentrated industries, in which firms have many competitors (and hence small market shares), are more likely to participate in the 33/50 Program than concentrated industries.

Within EPA's ten regions, the variation in willingness to participate may be a result of differences

Figure 1. Participation in EPA's 33/50 Program by industry.

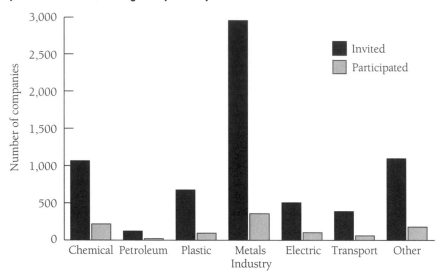

among the regions' environmental regulations. In some regions, EPA may mandate pollution prevention laws or toxics reduction laws that complement 33/50 Program goals. In regions where this is the case, willingness to participate may be relatively high. Moreover, regional variation may reflect the varying stringency of environmental regulations in individual regions. It may also be a measure of the effectiveness of EPA's regional coordinators in recruiting firms to join the 33/50 Program.

Participation by Individual Firms

Our research revealed many determinants of the willingness of individual firms to participate in the 33/50 Program. Overall, we found that only a small percentage of the invited firms in any one industry chose to participate (see Figure 1). However, the firms that did participate were responsible for a large percentage of their industry's toxic emissions (see Figure 2). Specific determinants, such as the volume and number of 33/50 chemicals and other TRI chemicals that a firm emits, a firm's size and financial health, and the intensity with which EPA tries to recruit it, are considered next.

Firms that use high volumes of the seventeen chemicals targeted by the 33/50 Program (as well as of other TRI chemicals) obviously have the potential for making the largest aggregate reduction in releases of these chemicals and are more likely to participate in the 33/50 Program. By voluntarily reducing these releases, these firms may benefit from consumer goodwill.

In certain circumstances, however, the larger a firm's release intensity (as measured by the volume of chemicals emitted per volume of sales), the more unlikely it is to participate in the 33/50 Program. Firms with high release intensities will incur high costs per volume of sales if they switch to alternative chemicals and production processes.

The number of chemicals a firm releases is also a significant determinant of its willingness to participate in the 33/50 Program. Firms that emit a large number of chemicals are more likely to participate, perhaps because these firms possess greater opportunity and flexibility to develop less toxic chemicals.

Holding other factors constant, large firms, as measured by number of employees, are also likely to join the program. These firms may enjoy greater benefits from participation than small firms because they typically serve a larger market demand and because improved environmental performance may generate employee goodwill. Compared with small firms, large firms may also feel more pressure to participate in the 33/50 Program. Large firms have more shareholders, and shareholder pressure for environmental consciousness could spur program participation.

While large size increases the likelihood that a firm will join the 33/50 Program, the fact that a firm has a large number of facilities does not. This finding is contrary to our expectation, since firms could theoretically benefit from public recognition, even if just one of their facilities participated in the program.

Financial health and profitability is another determinant of participation. Increased earnings provide opportunities for firms to invest in pollution prevention. While profitability increases the likelihood of participation, our analysis showed that its effect on the firms in our study sample was not significant.

A significant determinant of a firm's willingness to join the 33/50 Program is the intensity of EPA contact. EPA consulted extensively with the 555 companies it initially invited to join the program. At one point, participation among these companies was as high as 60 percent. By contrast, the

participation rate among the approximately 6,000 companies EPA later invited to join the program has been less than 15 percent. With these companies the agency had comparatively little contact.

Distinguishing between TRI and 33/50 Program Emissions Reductions

Once we knew something about the industries and firms that participated in the 33/50 Program, we wanted to know whether emissions reductions made by program participants were attributable to the 33/50 Program or to the disclosure requirements of the Toxics Release Inventory.

Our research indicates that program participants are not free-riding on the reductions that they made in response to TRI disclosure requirements, which went into effect in 1988. Instead, the 33/50 Program has induced firms to modify their toxic emissions, as is clear from the changing pattern of toxic releases since the program began.

Our analysis suggests that, between 1988 and 1990, releases and transfers of the seventeen chemicals targeted by the 33/50 Program fell by 16 percent, while releases and transfers of other TRI chemicals fell by 24 percent. This pattern changed

Figure 2. Releases of toxic emissions by companies participating in the 33/50 Program as a percentage of emissions for their industries (1990).

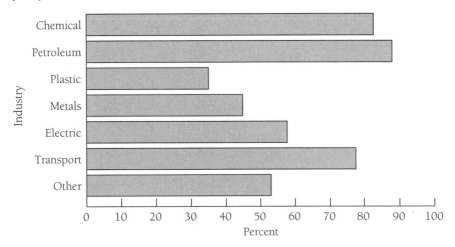

dramatically after the 33/50 Program was initiated. Between 1990 and 1991, releases and transfers of 33/50 Program chemicals fell by 21 percent, while releases and transfers of nonprogram chemicals fell by only 8 percent. The 1992 data reveal that reduction rates for the program chemicals are four times those reported for other TRI chemicals. A breakdown of these data by program participants and nonparticipants reveals that both groups have increased their reductions of chemicals targeted by the 33/50 Program. This suggests spillover effects from the program. The availability of more environmentally friendly products and chemical substitutes has made it easier for even nonparticipants to achieve emissions reductions.

But could reductions in chemicals targeted by the 33/50 Program be "crowding out" potential reductions or even increasing emissions of other chemicals? The answer is probably "no." We found that releases and transfers of nonprogram chemicals by program participants have fallen more than 12 percent. This finding suggests that the 33/50 Program has been successful in setting priorities with respect to the chemicals targeted by firms in their pollution control efforts. In addition to encouraging reductions in emissions of some of the most toxic chemicals, the program may also bring about reductions in emissions of other toxic chemicals.

33/50 Program and Other Voluntary Pollution Control Programs

Our evaluation of the 33/50 Program raised three additional questions: Does a firm's participation in another voluntary pollution reduction program affect its likelihood of participating in the 33/50 Program? Does a firm's participation in the program affect its compliance with environmental regulations? Do firms that participate in the program get preferential treatment in terms of relaxed regulatory oversight and enforcement of EPA regulations?

To answer the first question, we examined the relationship of the 33/50 Program with EPA's Green Lights Program. Participants in the Green Lights Program sign a memorandum of understanding

with EPA in which they agree to install energy-efficient lighting to reduce emissions of greenhouse gases. As with the 33/50 Program, the major incentive for participating in the Green Lights Program is positive public recognition. Of the more than 1,000 participants in this program, ninety are corporations that release chemicals targeted by the 33/50 Program. Our analysis reveals that participation in the Green Lights Program significantly increases the likelihood that a firm will participate in the 33/50 Program. This observation suggests that "environmentally conscious" firms seek to improve their reputation by participating in several voluntary pollution reduction programs at the same time.

Our second question was prompted by fears that firms can use participation in the 33/50 Program to circumvent some environmental regulations under the Clean Air Act. Skeptics of the program argue that this participation may be a way to obtain an extension for complying with certain of the act's requirements. While such an extension may be obtained through participation in the 33/50 Program, it is more appropriately obtained by participation in the Early Reductions Program. Any reductions in hazardous air pollutants documented under the Early Reductions Program may be credited under the 33/50 Program and vice versa. Unlike the 33/50 Program, however, the Early Reductions Program is more stringent and is, in fact, enforceable.

If firms could obtain extensions for compliance with regulations under the Clean Air Act through participation in the 33/50 Program, the success of the program as an alternative policy tool would be diminished. The ability to obtain such extensions would suggest that firms' participation in the program was not really motivated by the desire to gain positive public recognition. However, there is no evidence to support this theory.

Our third question was prompted by the concern that firms participating in the 33/50 Program might get preferential treatment from EPA, despite the agency's claim that it would not relax regulatory oversight or enforcement for program participants. Our examination of enforcement decisions made and penalties proposed in 1993 under the Toxic Substances Control Act (TSCA) provides some evidence that supports EPA's claim. Of the twenty-three companies that were fined under TSCA during that year, eight were participants in the 33/50 Program. These eight companies also received the highest fines. Even within the toxics unit of EPA's enforcement program, participation in the 33/50 Program does not seem to reduce substantially inspections or penalty settlements.

In the enforcement of other environmental laws and programs, EPA intervention on behalf of participants in the 33/50 Program is probably even less likely. Since the 33/50 Program is federal and since most of EPA's enforcement takes place at the state level, widespread intervention in state enforcement programs on behalf of program participants is unlikely. However, participants might believe that they can get preferential treatment, even though EPA's enforcement behavior does not appear to corroborate this belief.

Implications of the 33/50 Program

Our research reveals that the companies with the largest amounts of toxic releases are most likely to take part in the 33/50 Program. This suggests that this voluntary program may achieve substantial pollution reductions because it targets firms with the greatest pollution reduction potential.

Our research also indicates that a voluntary approach to pollution reduction could augment existing command-and-control regulation, under which mandated pollution reductions and prescribed technologies for achieving those reductions give firms little flexibility to control pollution in a cost-effective way. The potential for voluntary programs to augment such regulation is increased when their progress can be tracked through publicly available information that introduces accountability for pollution control and rewards pollution reduction efforts beyond those required by law.

Indeed, public awareness of the pollution reductions achieved through innovative voluntary programs can increase the programs' effectiveness. Regulators can use this awareness to increase participation in such programs, thereby spurring competition in environmental quality. Of course, public disclosure is not a costless exercise for firms, which under the requirements of the Superfund Amendments and Reauthorization Act must report their releases and transfers of chemicals. Estimates of doing so have ranged from EPA's conservative estimate of $4,000 per TRI chemical to the Chemical Manufacturers Association's estimate of $7,000.

The benefits, in terms of consumer goodwill, might outweigh the costs of such disclosure when a firm can document substantial pollution reductions through participation in voluntary pollution control programs. To help ensure these benefits, EPA should provide substantial public recognition and awards to firms achieving such reductions. Greater public awareness of firms' participation in voluntary pollution control programs is key to achieving the program's goals.

Suggested Reading

Arora, S., and T.N. Cason. 1995. An Experiment in Voluntary Environmental Regulation: Participation in EPA's 33/50 Program. *Journal of Environmental Economics and Management* 28: 27–86.

Arora, S., and S. Gangopadhyay. 1995. Toward a Theoretical Model of Voluntary Overcompliance. *Journal of Economic Behavior and Organization* 28: 289–309.

Arora, S., and T.N. Cason. 1996. Why Do Firms Volunteer to Exceed Environmental Regulations? Understanding Participation in EPA's 33/50 Program. *Land Economics* 72 (4): 413–32.

Hamilton, James H. 1995. Pollution as News: Media and Stock Reactions to the Toxics Release Inventory Data. *Journal of Environmental Economics and Management* 28: 98–113.

16 Cleaning Up Superfund

Paul R. Portney and Katherine N. Probst

Controversy over Superfund has come to the fore as the statute comes up for congressional review. Critics of the law express concern about the amount of money being spent for Superfund cleanups, question whether such spending is directed toward cleanups of sites that pose serious health and ecological risks, and bristle at the apparent unfairness of Superfund liability provisions. Defenders of the law point to the increased care with which hazardous materials are now handled and to the large number of privately funded cleanups under way. Although no changes were made the last time Superfund was reauthorized, significant reforms in the law may be enacted by Congress. These reforms hinge on two questions central to all disagreements over Superfund: What is the appropriate extent of cleanup at Superfund sites? How shall the costs of cleanups be apportioned?

Over the last twenty-two years, Congress has enacted seven major laws under which the U.S. Environmental Protection Agency (EPA) has been delegated regulatory responsibility. Six of these laws—the Clean Air Act, the Clean Water Act, the Safe Drinking Water Act, the Toxic Substances Control Act, the Resource Conservation and Recovery Act, and the Federal Insecticide, Fungicide, and Rodenticide Act—could be called "forward looking." That is, under these statutes EPA writes regulations that proscribe the current and future generation, transportation, use, and disposal of a variety of products or pollutants that might endanger human health or the environment.

In contrast with these laws, the seventh major environmental law takes a largely retrospective view. The Comprehensive Environmental Response, Compensation, and Liability Act (CERCLA, better known as Superfund) was enacted to deal with the legacy of sites contaminated by hazardous materials as a result of mining, petroleum refining, manufacturing, waste disposal, and a variety of other economic activities dating back, in a few extreme cases, to the nineteenth century. Superfund has been reauthorized twice since its passage in 1980 and is up for renewal again; as is generally the case, reauthorization provides Congress the opportunity to think about what changes, if any, it wants to make in the law.

There is no shortage of suggestions. Virtually everyone affected by Superfund—from the businesses that feel its economic sting to the citizens who fear nearby contamination—has complaints about the way the waste remediation program estab-

Originally published in *Resources*, No. 114, Winter 1994.

lished under Superfund has worked during the past thirteen years. This makes it very difficult for any-one—in Congress, in the Clinton administration, or in the business or environmental communities—to craft a set of reforms that would be pleasing to all. And despite criticisms, most experts would agree that Superfund has created powerful incentives to reduce the generation and improve the management of hazardous substances. The desire to preserve the most beneficial effects of the law further complicates discussions of reform.

It is our purpose here to provide a brief sketch of the Superfund program as it has evolved to this point, indicate the reforms that various groups have proposed, and offer our views as to the likely outcome of the congressional debate over reauthorization.

Overview of the Program

At the risk of oversimplifying a very complicated statute, we may say the Superfund law has tried to accomplish several things. First, it has provided a mechanism through which contaminated sites that pose serious threats to human health and the environment are identified and ranked. The most troublesome sites are placed on the National Priorities List (NPL)—that group of sites for which federal moneys can be used for cleanup. There are currently 1,286 such sites on the NPL.

Second, the law has established a process for determining which possible remedies (cleanup approaches) are feasible and appropriate for each site on the NPL and for selecting the desired remedy.

Third, it has created several new federal taxes—one falling on petroleum and on chemical feedstocks used in manufacturing, the other a more general corporate income tax—that help stock a trust fund. This fund is used on an emergency basis to finance cleanups of sites posing immediate risks to health and the environment and on a nonemergency basis to finance long-term cleanups at sites where no "responsible parties" can be found and made to clean up the site(s) in question.

Last, but by no means least, the Superfund law has created a mechanism through which EPA can identify these responsible parties, apportion liability among them, and require them to pay for the remedy that has been selected.

Although controversy surrounds each of these provisions, we believe two questions lie at the heart of virtually all serious disagreements over Superfund. First, what is the appropriate extent of cleanup at each of the sites on the NPL? Second, how shall the costs of these cleanups be apportioned? Before we turn to these questions, it is useful to put Superfund in perspective with other federal environmental regulatory programs.

Superfund Costs and Controversies

According to EPA, individuals, government agencies, and businesses had to spend about $130 billion in 1993 to comply with all federal regulations written under the seven environmental laws listed above. Given the attention it has received lately, one would think the Superfund program was responsible for a significant share of these expenditures. In point of fact, however, this is not the case.

As best we can determine, total annual spending pursuant to Superfund requirements in 1993 is likely to be on the order of $6 billion. This figure includes $3 billion in combined expenditures by the U.S. Department of Energy for cleanups at its nuclear weapons plants and by the U.S. Department of Defense for cleanups at its military bases. Of the remaining $3 billion, about $1.5 billion was spent by EPA. This implies that all private firms, state and local governments, and individuals spent less than $2 billion last year on waste remediation under Superfund. Since private parties are not required to report their annual cleanup expenditures, there is great uncertainty about this last figure.

If Superfund currently accounts for a relatively small fraction of total annual expenditures on environmental compliance, why has it become so controversial? Three explanations are likely. First, although current spending for Superfund cleanups may not yet be significant, this amount will change with time. If all 1,286 sites on the NPL are cleaned up at the current average per-site cost of $30 million, for instance, total expenditures will eventually grow to nearly $40 billion. And if EPA continues to add sites to the NPL at the current rate of 50 per year for the next decade, total expenditures will increase by $15 billion. These are sums worthy of serious attention.

Second, there is great concern—particularly among those in the business community—that the moneys being expended on site remediation are not being directed toward very serious risks. These critics allege that EPA systematically overestimates the health and ecological risks arising from site contamination, often by making unreasonable assumptions about the likely human exposures to contaminants. Many of those helping to pay for site cleanups say they are willing to pay to address serious risks, but resent squandering scarce resources on what they regard as often trivial problems. On the other hand, environmentalists contend that serious contamination at some sites is going unaddressed.

Third, controversy surrounds the allocation of cleanup costs. When Superfund was passed in 1980, Congress, having no appetite for increasing federal spending to pay for site cleanups, created a liability system to make it relatively easy for the government to link private parties to sites and make them pay for remediation.

Under Superfund, liability is referred to as retroactive, strict, and joint-and-several. It is retroactive in the sense that it applies to activities that took place before—occasionally long before—Superfund was enacted. Strict liability is that which is unrelated to the care or the negligence responsible parties may have exhibited in the past. Joint-and-several liability implies that any one party at a site can be required by the government to pay for the entire cleanup, regardless of the share of wastes it contributed. (That party can then in turn sue other contributors, but it must incur the legal costs associated with bringing these suits.)

Each of these liability provisions has been exceptionally controversial. For instance, it is galling for one responsible party at a site to be told

that it must shoulder a disproportionately large share of cleanup costs because none of the other contributors can be found, or because they are insolvent or otherwise incapable of paying for cleanup. Similarly, firms that took pains to manage hazardous substances in a responsible way in the past bristle at the fact that their efforts are no defense against Superfund liability.

These controversies lead back to the two questions raised above: What is the appropriate extent of cleanup at Superfund sites? How shall cleanup costs be apportioned? We turn now to these two fundamental issues.

How Clean is Clean?

Section 121 of Superfund spells out the criteria governing site cleanups. Importantly, the law calls for a cleanup that "utilizes permanent solutions and alternative treatment technologies . . . to the maximum extent practicable" at each site. This seemingly innocuous wording is the source of much of the controversy over the Superfund statute.

Consider a site located in or near a residential area. There are houses nearby, but the site—once an industrial dump, say—is fenced off and currently vacant. The soil at the site is contaminated but is not contributing to the contamination of an underground aquifer. Some might find it appropriate in these circumstances to cap the site in order to contain the contamination and to build a more secure fence around it, but then to do little more than continue to monitor it carefully. In view of the risk posed by the site, some might deem this a reasonable response. In fact, they would argue, it would be wrong to go much beyond the measures described here, as doing so would eat up scarce resources that might better be deployed elsewhere.

But would this be a "permanent" remedy? To many in the environmental community and in Congress, the answer is no. Critics of such a risk-based approach—in which the extent of the remedial action depends upon the seriousness of the current health risks a site poses—believe that a permanent remedy is one that goes well beyond containment, extending perhaps to the excavation and incineration of contaminated soils or the pumping and treatment of contaminated groundwater. They would balk at a remedy that would reduce exposure to contamination without removing the contamination itself.

This very brief discussion suggests the basic nature of the debate. On the one side are those who argue that scarce resources necessitate the ranking of cleanup priorities, an activity that implies some sites should receive much less extensive remedies than others. To these individuals, chain-link fences and "Keep Out" signs will constitute appropriate remedies for at least some low-risk sites.

On the other side are those who argue that while a precise balancing of risks against cleanup costs may be a nice conceptual approach to remedy selection, it is a practical impossibility in light of the almost total lack of reliable data on the actual health and environmental risks at sites, as well as the great uncertainties about the costs of various possible remedies. In the view of these people, site-by-site balancing of risks and cleanup costs would drag the cleanup program into the next millennium. In addition, they point to the congressional preference for permanence in remedy selection.

Are there hybrid approaches that might placate both camps? Perhaps. One change that has been suggested would be the establishment by EPA of maximum permissible concentrations of contaminants in soils and groundwater at Superfund sites. Any remedy would be required to meet these standards, but with one important twist: different standards would be established for different sites, depending upon the intended future use of a site. Thus, for instance, a Superfund site that would be redeveloped as an industrial park would have to meet less stringent cleanup standards than a Superfund site on which a housing development or a school would be built. In this way, a crude form of balancing would take place; this hierarchical approach would reflect the fact that humans' exposure to remaining contaminants would be much less likely at the industrial park than at the school playground.

Whether or not such a change is politically feasible remains to be seen. Perhaps surprisingly, tailoring cleanup to intended land use may make less difference than one might suspect from a strict reading of the law. We say this because if one looks closely at the cleanup remedies EPA has selected at Superfund sites all around the United States, it is hard to see any uniform pattern suggesting strict adherence to the concept of permanence. Rather, remedy selection seems to depend at least in part upon which EPA regional office is in charge of a given site, the amount of press attention devoted to the site, and the extent of public involvement there. Interestingly, remedy selection also seems to depend upon the seriousness of the health and environmental risks the site poses. Thus, although the Superfund statute seems to discourage risk-based cleanups, these considerations do seem to be part of the decision-making calculus. (See "Cleanup

Decisions Under Superfund: Do Benefits and Costs Matter?" by Shreekant Gupta, George Van Houtven, and Maureen L. Cropper in the Spring 1993 issue of *Resources,* as well as Chapter 7 of this book.)

Who Pays for Cleanups?

As noted above, Superfund contains expansive liability standards. Not surprisingly, these standards have been the source of tremendous controversy. Equally unsurprising, a number of groups have sought changes in Superfund that would relax these standards in one way or another.

Banks and other lending institutions fear being held liable for cleanup costs under Superfund because they lent money to, and sometimes foreclosed upon, firms that were found to be responsible parties at Superfund sites. Claiming that Congress never intended for "nonpolluters" such as themselves to be caught up in the liability net, these institutions have pressed for elimination of what has come to be known as lender liability.

Also unhappy are municipal officials. A fair number of Superfund sites are landfills that municipalities either operated or contributed wastes to at one time. Like industrial contributors to these sites, they have been named as responsible parties by EPA or have been sued in "contribution actions" by one or more firms that have been stuck by EPA with the cleanup bill. They, too, are claiming that the Superfund law was never meant to impose significant economic costs on them; they generally argue that municipal waste (or garbage, as it used to be called) is much less likely to be among the risky contaminants at municipal landfills. (We should point out, though, that the paint thinners, used motor oils, car batteries, and other liquid and solid wastes that households once casually discarded with their trash can be every bit as toxic as some industrial wastes.)

Private firms, both large and small, are also unhappy with Superfund. They have argued forcefully that they should be neither held liable retroactively nor held liable for contamination caused by others. While there is now no doubt about their legality, retroactive and joint-and-several liability provisions do strike many observers (including some with no financial stake in the matter) as being somewhat unfair. Several recent proposals put forward by coalitions of responsible parties would change the liability provisions of Superfund to address these perceived inequities. For instance, one proposed change would require EPA to pay for any "orphan" shares at Superfund sites; these are the cleanup costs that would be assigned to firms or other responsible parties that either cannot be located or are financially insolvent.

By far the most unhappy bearers of Superfund liability are insurance companies. Although they are never held to be responsible parties by EPA, they, too, have been ensnarled in the recovery of cleanup costs. This is because many of the industrial and other private firms that are responsible parties have sued their insurance companies under the comprehensive general liability policies they have taken out for many years. The responsible parties have contended that these policies do cover the costs they incur to clean up Superfund sites.

For their part, the insurance companies point to the standard language in these policies, language to the effect that coverage pertains to "sudden and accidental" damages. The insurers then argue that the gradual leakage of contaminants at a site is not at all sudden and maintain for this reason that they are not liable. Because insurance is regulated almost exclusively at the state level in the United States, suits brought against insurance companies by responsible parties have been played out in many different state courts; these courts have sided in about equal measure with each of the parties—a confusing situation, to say the least.

Two types of relief for insurers have been suggested. Under one proposal, retroactive liability would be eliminated—that is, no one would be liable for cleaning up wastes disposed of before 1981. This would benefit both responsible parties and their insurers. Under another proposal, Congress would absolve insurers from having to reimburse the cleanup costs of the firms they insured. In either case, the "quid" for this "quo" would be the creation of a separate fund—financed in part or in toto by a tax on insurance companies—to help pay for cleanup costs at those sites where liability is removed.

Nearly two years ago we completed a report (*Assigning Liability for Superfund Cleanups: An Analysis of Policy Options,* Resources for the Future, 1992) that looked carefully at several ways in which the liability standards in Superfund could be changed, including some possible approaches that resemble proposals currently being put forward. At that time we concluded that, while the current liability standards in Superfund are unfair in several respects and result in a lot of litigation, any changes in them will create some new inequities even as they ameliorate others and thus will provide new incentives to sue. We also concluded that EPA has the power under the current Superfund statute to address many of the criticisms being raised, though doing so would not be easy.

Guessing on Changes

It is very difficult to predict what will happen to the Superfund law. The Clinton administration appears to be inching its way steadily toward a set of changes that it will propose to Congress. Although the administration's proposed changes will start the debate in earnest, Congress may well elect to reauthorize the law in its present form for another five years or so, thus putting off—as it did in the prior reauthorization—debate over significant changes. One never goes broke betting on the status quo.

Nevertheless, we hazard the guess that Superfund will be changed in several important ways in this or the next session of Congress. First, it seems likely that Congress will modify Section 121 of Superfund—wherein the standards for cleanups are spelled out—to allow for different degrees of cleanup of Superfund sites depending upon their intended future use. In addition, Congress may restrict "treatment" at NPL sites to highly contaminated hot spots. It also appears likely that Congress will require EPA to pay for orphan shares at NPL sites using an enhanced Superfund trust fund. If so, responsible parties who have complained about joint-and-several liability should be appeased.

Regardless of what happens to Superfund in this or coming years, the United States will be dealing with contaminated sites for decades to come. We will be surprised if the controversy does not outlive the cleanup program.

Suggested Reading

Barnett, Harold C. 1994. *Toxic Debts and the Superfund Dilemma.* University of North Carolina Press: Chapel Hill and London.

Probst, K.N., D. Fullerton, R.E. Litan, and P.R. Portney. *Footing the Bill for Superfund Cleanups: Who Pays and How?* Washington, DC: Brookings Institution and Resources for the Future.

Revesz, R.L., and R.B. Stewart. 1995. *Analyzing Superfund: Economics, Science and Law.* Washington, DC: Resources for the Future.

17 Reducing Carbon Emissions
Interactions with the Tax System Raise the Cost

Ian W.H. Parry

Reducing the amount of carbon dioxide Americans pump into the atmosphere will involve economic costs. These costs are larger than previously thought because emissions reduction policies are likely to aggravate economic distortions created by the tax system. But most of this added cost can be avoided if the policy chosen to reduce emissions raises revenues for the government and these revenues are used to cut other taxes.

Continued accumulation of heat-trapping gases in the atmosphere raises the prospect of future global warming and associated changes in climate. Many countries will attend a conference in December 1997 in Kyoto, Japan, to consider steps to reduce emissions of carbon dioxide (CO_2)—the most important heat-trapping gas. Introducing emissions targets may produce important benefits in terms of avoided future climate change.

Nonetheless, it makes sense to consider which policy approaches might reach these objectives at the lowest economic cost to each country. Recent research suggests that much will be at stake in this respect: the costs of even modest reductions in CO_2 emissions may differ substantially under different types of regulatory policies. To understand why requires a look at how these policies may interact with taxes that already exist in the economy.

The Tax-Interaction Effect

Government spending in the United States is financed primarily by taxes on labor and capital income. Putting aside the potential benefits from these spending programs, the taxes tend to "distort" economic behavior. That is, they reduce employment and investment below levels that would maximize economic efficiency. For example, because personal income taxes reduce take-home pay, the partner of a working spouse may be discouraged from joining the labor force, an older worker may retire earlier, or a worker with one job may be discouraged from

Originally published in *Resources*, No. 128, Summer 1997.

working additional hours in a second job. Employers are likely to hire less labor if social security taxes make employees more expensive. Similarly, capital gains and corporate income taxes reduce the incentives for individuals to save and for firms to invest in new production capacity.

Environmental taxes and regulations tend to discourage economic activity because they raise the costs to firms of producing output. Typically, this leads to a lower overall level of employment and investment in the economy. These "spillover" effects of environmental policies in labor and capital markets add to the distortions created by the tax system. The resulting economic cost has been termed the *tax-interaction effect.*

What would happen if a tax on carbon emissions (as proposed by the European Union) were introduced? The new tax would increase the costs to firms of purchasing coal and oil in particular, which in turn would increase the cost of electricity and gasoline. Most likely, firms would scale back their production activities a little in response to these higher costs, leading to a fall in the level of investment and employment (as happened, for example, in the 1970s when the price of energy increased). But employment and investment are already "too low" because of pre-existing taxes in the economy. This aggravation of distortions created by the tax system would be part of the overall economic cost of a carbon tax.

This is not the end of the story, however, because a carbon tax would raise revenues for the government. These revenues could be used to reduce other taxes in the economy, such as personal and corporate taxes, and thereby reduce the distortion in the level of employment and investment. The economic gain from this so-called revenue-recycling effect could reduce the overall economic costs of a carbon tax significantly. (See the box, "Recycling Revenues: Other Ways to Benefit.")

The Rise and Fall of a Hypothesis

Considerable confusion has arisen recently about the implications of tax distortions in the economy

for the costs of carbon and other environmental taxes. In particular, a number of analysts have mistakenly argued that there would be a "double dividend" from environmental taxes. These analysts have correctly pointed to the potential benefits from the revenue-recycling effect but have failed to recognize the cost from the tax-interaction effect.

Essentially, the double dividend hypothesis asserts that environmental taxes can both reduce pollution emissions and reduce the overall economic costs associated with the tax system. At first glance, this hypothesis seems to be self-evident, if the revenues raised are used to reduce other taxes that discourage work effort and investment. In some European countries, where high taxes, among other factors, have contributed to double-digit unemployment rates, the double dividend hypothesis has been particularly appealing. If environmental tax revenues were used to reduce taxes on labor income, so the hypothesis goes, unemployment and pollution might be reduced simultaneously. More generally, some people have argued that it is better to finance government spending by taxing economic "bads," such as pollution, rather than economic "goods," such as employment and investment.

Economists generally agree that revenue recycling would reduce the net economic cost of environmental taxes. However, recent studies suggest that environmental taxes are likely to *increase* rather than *decrease* the costs associated with the tax system overall. As Lans Bovenberg (Netherlands Bureau for Economic Analysis), Lawrence Goulder (Stanford University and RFF), and others have demonstrated, the adverse effects on employment and investment caused by environmental taxes are generally not fully offset, even if the tax revenues are used to reduce other taxes. That is, the tax-interaction effect dominates the revenue-recycling effect.

Thus, if there were no environmental benefits, it would be better to finance public spending by taxes on, for example, labor income rather than on pollution emissions. Why is this? A tax creates economic costs by inducing households and firms to consume and produce less of the taxed activity and more of other activities. The greater the shift away from the taxed activity, the greater the cost of the tax. Taxes on labor income can only be avoided by people working less and spending more time at home. In contrast, environmental taxes have a much narrower focus and are easier to avoid. A carbon tax can be avoided by an overall reduction in the level of production and employment. However, it can also be avoided by a change in the composition of production away from goods that use a lot of electricity (such as electric ovens and heating appliances) to ones that do not (such as natural gas ovens and heating). Tax economists have long argued that the economic costs of raising revenues are smaller under taxes that have a broad coverage compared with taxes that have a narrower focus.

Of course, this does not mean that environmental taxes should not be implemented. Instead, the environmental and revenue-recycling benefits should be weighed against the costs of reduced production and the loss from the tax-interaction effect. Indeed, recent research generally supports carbon taxes so long as the tax rates are not too high (that is, so long as they do not exceed the incremental value of environmental benefits).

CO$_2$ Permits versus a Carbon Tax

Instead of imposing a carbon tax, the government may reduce CO$_2$ emissions by requiring that firms have a permit for each unit of CO$_2$ emitted. By controlling the total quantity of such permits it gives to firms, the government could limit total CO$_2$ emis-sions to a target level. This permit program would cause a similar reduction in production, employment, and investment, as would a carbon tax. The reduction in employment and investment would add to the distortions created by the tax system, leading to the same cost from the tax-interaction effect.

Whether CO$_2$ permits could also produce the benefit from the revenue-recycling effect would depend on whether the permits were auctioned by the government or given out free to existing firms. If the permits were auctioned off, the government could use the revenues to reduce other taxes in the economy. But if the permits were given out free, as in the case of the existing permits program for sulfur dioxide emissions, no revenue would be collected and there would be no potential for a revenue-recycling effect.

Table 1 summarizes the benefits and costs of carbon abatement policies. The benefits from a carbon tax consist of the potential gains from reducing future climate change (the environmental benefits), and the revenue-recycling effect. The costs consist of the reduced production from industries affected by the tax and the costs of exacerbating tax distortions in the labor and capital markets, or tax-interaction effect. Economists have traditionally focused on (1) and (3) and neglected (2) and (4). This has led to some overstatement of the benefit-to-cost ratio from carbon taxes because the tax-interaction effect generally dominates the revenue-recycling effect.

CO$_2$ permits would produce three of the same effects as the carbon tax: namely, the environmental benefits from reduced future climate change, the

Table 1. The Benefits and Costs of Carbon Abatement Policies

Policy	Environmental benefits (1)	Revenue-recycling benefit (2)	Loss of production (3)	Tax-interaction effect (4)
Carbon tax	+	+	−	−
CO$_2$ permits	+	?	−	−

Note: + denotes benefits; − denotes costs.

cost from reduced production, and the tax-interaction effect. However, the benefit from the revenue-recycling effect could only be obtained if the government auctioned the permits.

Can the Policies Make Society Better Off?

Recent collaborative work by Lawrence Goulder, Roberton Williams, and myself suggests that the tax-interaction effect can raise the overall cost of policies to reduce emissions by a potentially substantial amount. For example, we estimate that the economic costs to the United States from using (non-auctioned) permits to reduce CO_2 emissions by 10 percent below current levels increases by 400 percent when the cost of the tax-interaction effect is taken into account! If instead the permits were auctioned—or a carbon tax were levied—and the revenues were used to finance cuts in other taxes, we estimate that the overall cost of this policy would be reduced by 75 percent.

On top of this, we estimate that the overall economic costs of a free CO_2 permit program would outweigh the environmental benefits—unless these benefits exceeded $25 per ton of carbon reduced. Estimates by William Nordhaus (Yale University) suggest that the benefits from reducing carbon emissions may be *below* $25 per ton, although there is much dispute on this point. If so, *even though the policy would correct a market failure associated with carbon emissions, the benefit would be more than offset by the costs of adding to distortions caused by the tax system.*

In contrast, a policy to reduce emissions that produces the revenue-recycling effect can produce a favorable benefit-to-cost ratio as long as environmental benefits per ton are positive. Thus, a CO_2 emissions reduction policy might produce an overall benefit to society only if it raises revenues for the government.

Other Considerations

It is important to keep in mind, however, that the benefit estimates from reducing CO_2 emissions are

Recycling Revenues: Other Ways to Benefit

Are there other ways that carbon tax revenues might be used to reap economic benefits besides cutting other taxes? Yes, if the revenues were used to reduce the federal budget deficit. In that way, less tax revenue would be required in future years for interest payments and repayment of principal on the national debt. As a result, taxes could be lower, implying less distortion of employment and investment. Of course, in this case the benefits from revenue recycling would occur in the future rather than the present.

The answer is "it depends" if the revenues were used to finance additional public spending. The huge bulk of government expenditure in the United States consists of transfer payments, such as pensions, or expenditures that substitute for private spending, such as medical care and education. Loosely speaking (and ignoring distributional impacts) the benefit to people from a billion dollars of this type of spending is a billion dollars. If instead the revenue were used to reduce other taxes—say the personal income tax—the economic benefits would be greater. Not only would people get a billion dollars but the lower tax rates would favorably alter relative prices in the economy. The rewards for work effort and saving would increase, thereby encouraging more employment and investment. In contrast, increased public spending would not alter relative prices.

However, governments also provide "public goods" that, for various reasons, the private sector may not provide, such as defense, crime prevention, and aid to needy families. People may (or may not) value an additional billion dollars of spending on these goods at more than a billion dollars. If they do, the benefits from this type of revenue recycling may be as large as (or even larger than) the benefits from reducing taxes.

highly speculative at this stage. They do not take into account the (hopefully small) possibility of drastic changes in climate should global warming disturb some unstable mechanism within the climate system. Nor are the potential ecological impacts well understood. We simply do not know enough yet to judge whether global warming will turn out to be a very serious problem or not.

Moreover, there are other factors to consider in the choice of policy instruments to reduce CO_2 emissions. For example, affected industries may oppose a carbon tax that requires them to reduce emissions *and* pay taxes to the government more than a free CO_2 permits program. Other important considerations include the potential impact of a given policy instrument on the private incentives to develop more energy-efficient technologies.

Nonetheless, minimizing the economic costs of any action to reduce CO_2 emissions that might be agreed to in Kyoto is desirable, not only for its own sake, but also for the likelihood of the agreement to stand the test of time. Recent research warns that even modest emissions reductions might be especially costly if the policies used do not raise revenues for the government that are returned to the economy in other tax reductions.

Suggested Reading

Parry, Ian W.H., and Wallace E. Oates. 1998. *Policy Analysis in a Second-Best World*. Discussion Paper No. 98-48. Washington, DC: Resources for the Future.

Part 4

Environmental Federalism

18 Thinking about Environmental Federalism

Wallace E. Oates

Environmental federalism is a complicated and contentious issue. And it is at the center of debates both in this country and in the European Union, where moves are afoot for the harmonization of environmental standards across the member nations. It is helpful in thinking about this issue to go back to some basic "principles." Doing so may not resolve the issue, but at least we can better understand the nature of the argument.

First, the issue is not a simple one of centralization versus decentralization of environmental management. Our governmental systems consist of several levels, and it is clear that there are important roles for nearly all levels of government in environmental protection. The issue is one of aligning specific responsibilities and regulatory instruments with the different levels of government so as best to achieve our environmental objectives.

Second, there exists a body of "principles" (or, perhaps better, "rough guidelines") for making this assignment. In brief (and with some simplification), the central idea emerging from the literature in public economics is that the responsibility for providing a particular public service should be assigned to the smallest jurisdiction whose geographical scope encompasses the relevant benefits and costs associated with the provision of the service.

The rationale for this principle is straightforward. Such decentralization of public decisionmaking allows outputs of public services to be tailored to the particular circumstances—the tastes of residents, the costs of production, and any other

Originally published in *Resources*, No. 130, Winter 1998.

peculiar local features—of each jurisdiction. It is easy to show formally that such a decentralized outcome increases social well-being as compared with a centralized solution requiring more uniform levels of public services across all jurisdictions. In Europe, this is known as the "principle of subsidiarity," and it is enshrined in the Maastricht Treaty for the European Union. In the United States, we think of it more colloquially—as an aversion to the "one size fits all" approach.

Applying this general framework, we can envision a system of environmental management in which the central government sets standards and oversees measures for explicitly national pollution problems and intervenes where pollutants (like acid rain) flow across state and local boundaries; in addition, the central government would support research and the dissemination of knowledge on environmental issues, which benefit people everywhere. At the same time, the states and localities would set their own standards and would manage environmental quality for matters that are contained within their own borders (such things, perhaps, as drinking water, refuse disposal, and air pollutants with solely local effects).

Is this, in fact, the way we do things? Not exactly (as they say in the Hertz ads). Under the Clean Air Act, for example, Congress has directed the Environmental Protection Agency to set uniform national standards for air quality—applicable to every point in the United States. Such standards apply irrespective of whether there is any transporting of pollution across jurisdictional lines. In contrast, under the Clean Water Act, Congress has given the states the responsibility (but subject to EPA approval) for setting their own standards for water quality. Environmental policy in the United States (and Europe as well) is characterized by a certain ambivalence on this matter.

What is the objection to decentralized environmental management? One objection (and this is where things get more complicated) is that state and local governments, in their eagerness to promote economic development through attracting new business investment and creating jobs, will set excessively lax environmental standards to keep down costs of pollution control. What results (so the argument goes) is a "race to the bottom" with states and localities competing with one another to reduce environmental standards. We thus need centralized standard setting and environmental management, as one author has put it, to "save the states from themselves."

But is this true? Note that this is really part of a more general and quite fundamental indictment of all state and local governance that says that economic competition will lead these governments to misbehave—to underprovide public services so as to keep taxes and expensive regulations at excessively low levels. This is curious in one respect. We generally applaud the work of competitive forces in the private sector, where Adam Smith's invisible hand guides self-interested decisions into socially beneficent outcomes. But here we are told that competition is socially harmful in the public sector.

The theory on this is not entirely clear. Certain economic models, for example, find that competition among governments (as in the private sector) encourages precisely the right kinds of decisions. There is no race to the bottom. Active competition for new economic activity in these models provides precisely the correct signals for decisions on public expenditures and taxation. At the same time, it is not difficult to introduce elements (and not unrealistic ones) into these models that generate distortions—in some instances in the form of excessively lax environmental standards. But the theory gives us no sense of the likely magnitude of the potential distortions. Unfortunately, at this juncture we cannot resolve this matter by an appeal to the evidence; existing studies of state and local competition, while of some interest, do not answer our question. At any rate, there exists little systematic evidence that supports the case for a race to the bottom.

My own sense is that there remains a strong case for extensive decentralized environmental management encompassing the setting of standards as well as their enforcement. There has been an impressive growth in both the analytical and administrative capacities of state and local agencies.

Moreover, it simply doesn't make economic sense to insist that all jurisdictions adopt the same set of centrally determined standards for environmental quality. Circumstances differ, and we should take advantage of the opportunities that this provides for a more flexible approach to environmental management. The problems of air and water quality management, for example, are very different between Southern California and Omaha (or Venice and Oslo, in the European setting), and these differences should manifest themselves in the stringency and the form of environmental regulations.

Suggested Reading

Oates, Wallace E. 1998. "Environmental Policy in the European Community: Harmonization or National Standards?" *Empirica* 25(1): 1–13. (An exploration of the principles of environmental federalism with applications to the European Union.)

Pfander, James E. 1996. "Environmental Federalism in Europe and the United States: A Comparative Assessment of Regulation Through the Agency of Member States." In John Braden et al., eds., *Environmental Policy with Political and Economic Integration: The European Union and the United States.* Cheltenham, U.K.: Edward Elgar, pp. 59–131. (A valuable comparative study of the institutions of environmental federalism in Europe and the United States.)

Revesz, Richard L. 1992. "Rehabilitating Interstate Competition: Rethinking the 'Race-to-the-Bottom' Rationale for Federal Environmental Legislation," *New York University Law Review* 67(Dec.): 1210–1254. (An illuminating and provocative treatment of interjurisdictional competition and environmental federalism by a legal scholar.)

U.S. Congressional Budget Office. 1997. *Environmental Federalism and Environmental Protection: Case Studies for Drinking Water and Ground-Level Ozone.* Washington, DC: U.S. Government Printing Office, November. (An insightful discussion of the principles of environmental federalism plus two interesting case studies.)

19 Environmental Federalism

Robert M. Schwab

Nearly all major federal environmental legislation divides the responsibility for controlling pollution between the federal government and the states. The exact division of responsibilities varies substantially from statute to statute.

Consider the Clean Air Act. It requires the federal government to set air quality standards for each of six pollutants and to establish performance standards for new stationary sources of pollution (such as factories and utilities) and mobile sources (such as automobiles). States are required to develop implementation plans, which must be approved by the federal government, to reach these goals; states also share the responsibility for enforcement. In sharp contrast, under the Clean Water Act the individual states are directed to set water quality standards. We thus have uniform national standards for air quality but state-specific standards for water quality.

Which approach is better? Should we move toward greater centralization, thus giving more responsibility for controlling pollution to the federal government? Or should we encourage further decentralization and allow state and local governments a greater voice in environmental policy?

This question is part of a much broader issue. The debate over the proper division of governmental responsibilities is long-standing. It was a key issue in the framing of the Constitution in the 1780s; it lies at the heart of the debate over the Reagan administration's call for a "new federalism" in the 1980s. Many of the questions about environmental federalism should be familiar to those who have thought about other issues in federalism.

Originally published in *Resources*, No. 92, Summer 1988.

Decentralized Policymaking

The case in favor of decentralized decisionmaking is simple, yet powerful and persuasive. For the moment, let's set aside the question of who should set policy and look at the principles we would like to see embedded in policymaking decisions. "Optimal" environmental policy—from an economic standpoint, at least—requires us to continue to reduce pollution in all parts of the country up to the point that the cost of further reducing pollution (what is known as marginal abatement cost) equals the benefits from that reduction (that is, marginal social damage). Focusing only on the economics of the issue, if it costs society $100 to reduce emissions by one additional ton but the benefits from doing so were valued at $200, we would be wise to cut back on pollution further; if, on the other hand, benefits were valued at only $50, we have set environmental standards too high. If we continue with this line of argument, the logic behind the economic principle that the optimal standard is such that marginal abatement cost equals marginal social damage becomes clear.

While it is true that we should apply the same *principle* in every region, we should not set the same environmental *standard* everywhere unless the costs and benefits of pollution control are the same in each region. There are excellent reasons to believe that this is not the case. For example, different industries are clustered in different regions, and it is much more costly for firms in some industries to reduce emissions than others. Thus differences in regional industry mix imply differences in regional abatement costs. Differences in meteorology, topography, and land-use patterns also play important roles in abatement costs. Few would disagree with the proposition that it will be much more costly for Los Angeles to reach any given level of air quality than most other cities; by one estimate, the Los Angeles area would have to reduce gasoline consumption by 82 percent in order to meet the air quality standards set under the Clean Air Act.

Similar considerations arise on the benefit side of the equation. Benefits, after all, are to some extent subjective. It could well be the case that the residents of some states would be willing to pay $200 for a slightly cleaner environment, but that residents of others would be willing to pay only $50. Population size also plays a role. Everyone in an area suffers the damages pollution causes. Therefore, everything else being equal, the greater the population in an area, the greater the benefit from a cleaner environment.

Ongoing research sponsored by RFF and funded by the National Science Foundation points to the validity of this argument. Paul R. Portney, Albert M. McGartland, and Wallace E. Oates have assembled data on damages and abatement costs for total suspended particulates (TSP) for Baltimore and St. Louis. Their preliminary estimate is that the optimal standard for TSP concentrations for Baltimore is nearly 50 percent less stringent than the optimal standard for St. Louis. That is, if we were to equate the costs and benefits of reducing TSP in both cities, the permissible level of pollution would need to be nearly twice as high in Baltimore.

How should we divide the responsibility for setting environmental policy in order to reach this optimal outcome (again recalling that we are only focusing on economic issues)? Federal legislation is rarely sensitive to regional differences in costs and benefits. Typically, federal legislation implies uniform standards; for example, under the Clean Air Act, the federal government sets uniform maximum allowable levels of pollution. Therefore, vesting the federal government with the authority to set environmental policy is unlikely to lead to an optimal outcome.

This line of reasoning should sound familiar to anyone who has been involved in the debate over fiscal federalism. It would make little sense to provide the same menu of public services in every community, or for the federal government to try to tailor policies to meet the local circumstances in each community. Thus we allow state and local governments to decide how much to spend on education, when trash is to be collected, and which beats police officers will patrol. The logic leading to this conclusion also suggests that state and local governments may be in a better position than the federal government to choose the correct level of environmental quality.

Possible Problems

Many people would accept the basic argument in favor of decentralization, but would continue to urge a strong role for the federal government in environmental matters nonetheless. They would contend that while the outcome under federal uniform standards is unlikely to be optimal, it is better than the outcome that would emerge if we were to allow state and local governments to set environmental policy.

Though it might not be immediately obvious, most of the arguments against decentralization turn on a single principle that has a strong basis in economic logic. If there are costs or benefits associated with the environmental policies adopted by one region that are borne by the residents of other regions, then it is unlikely that decentralized decisionmaking will lead to an optimal outcome. We can apply this principle in a range of circumstances.

The clearest cases are those where pollution generated in one state crosses political boundaries into another. For instance, if Illinois firms are forced to reduce pollution, we would expect the citizens of Illinois (and, hopefully, Illinois policymakers) to take into account the benefits from a cleaner environment that Illinois citizens would realize. Thus, Illinois would require its firms to reduce pollution to the point that marginal abatement cost equaled marginal damage suffered by Illinois residents. Under such a decentralized approach, we would not expect Illinois citizens to necessarily take into account the benefits Indiana residents receive. An optimal policy would require Illinoisans to go further and continue to reduce pollution until marginal abatement cost equaled the sum of damages to both Illinois and Indiana residents.

Clearly, in many cases this argument against decentralization is quite persuasive. The problems in coordinating efforts by Delaware, Virginia, Maryland, the District of Columbia, and Pennsyl-

vania to clean up the Chesapeake Bay are well documented (and the Chesapeake is often offered as one of the more successful cooperative undertakings). In some cases, whether or not a pollutant is local is a policy issue; damages from sulfur dioxide emissions can be contained in a small area or dispersed over a much larger area if firms are required to use tall stacks. Some pollutants are international, acid rain being a good example, as the strained relations between Canada and the United States can attest. Some are truly global, carbon dioxide's impact on the climate (often called the greenhouse effect) and the impact of chlorofluorocarbons on the ozone shield being extreme examples of environmental issues that cannot be addressed by states or even nations. In such cases, the benefits from controlling pollution extend well beyond traditional political borders.

Many pollutants, however, cause damages over a much more limited area. Lead, carbon monoxide, and TSP are good examples. In many cases, the control of hazardous waste sites and the establishment of drinking water standards also fall into this category.

Similar issues arise repeatedly when assessing fiscal federalism. Suppose the residents of Community A build parks which are used not only by residents of A but by those of Community B as well (an example of a "fiscal spillover"). Under these conditions, we would expect that Community A had built too few parks. The standard policy solution in this case would be a grant from the federal government to Community A to encourage it to build additional parks.

We can analyze a second argument against decentralization in this framework. People throughout the country derive benefits from the quality of the environment in unique national sites such as national parks and wilderness areas; the residents of New York care about clean air in Yellowstone, as do the residents of Wyoming and Idaho. In many ways this issue is identical to the interjurisdictional pollution problem discussed above, though here the damages from pollution spill over state boundaries even though the pollution itself does not. Again, optimal policy almost certainly requires some federal role.

As a third argument, there are sometimes gains from uniformity and coordination. (Anyone who doubts this should try playing a tape made on a Beta videocassette recorder on a VHS machine or using a piece of software written for an Apple computer on an IBM.) In the context of environmental policy, it would be very expensive to demand that Detroit make cars that meet different standards in Montana, New Jersey, and so on. Clearly there are some benefits from uniform standards; automobiles are probably the most important example. But even in the case of automobiles, the benefits to be gained from a policy on uniformity do not rule out the possibility that the "correct" number of standards may be a happy medium between the current number—two (the Clean Air Act allows California to set its own standards for auto emissions)—and fifty, that is, one that applies to each state.

Interjurisdictional Competition

A fourth argument against allowing lower levels of government to set environmental policy focuses on the effects of interjurisdictional competition. The fear is that in their eagerness to attract new business and jobs, local officials would set environmental standards that are too lax. Thus, so this argument goes, the federal government must set environmental policy in order to "save the states from themselves." Such a stance is consistent with a more general view that nearly all forms of competition among jurisdictions are destructive. The Advisory Commission on Intergovernmental Relations (the focal point for interest in federalism issues within the federal government) for years argued that state and local governments should be encouraged to cooperate with one another.

In the last several years, however, there has emerged a strong sense that competition among governments can be beneficial, just as competition among firms can be beneficial. According to this view, competition forces local government to make decisions that are in the best interest of their con-

stituents. Thus competition stops local officials from overspending; if they were to try to set public budgets that were too high, firms would vote with their feet and move to competing communities.

The following simple example captures the spirit of this argument in the context of environmental policy. Suppose Illinois were about to set a standard for some pollutant. Who would bear the costs associated with this new policy? Since Illinois competes against other states to attract and hold firms, clearly the burden cannot fall on Illinois firms; they have the option to move to another state. Therefore the cost of reducing pollution must fall on Illinois residents in the form of fewer jobs and lower wages.

Under these circumstances, what level of environmental quality should Illinois policymakers set? They might reason as follows. Illinois residents will pay the costs of reducing pollution. Therefore if Illinois policymakers act in the best interest of their constituents, they should set a standard such that the costs of reducing pollution further are just balanced by the benefits. If they set a less stringent standard, Illinois residents would be willing to sacrifice additional jobs and wages in return for a cleaner environment; if they set a more stringent standard, the benefits from less pollution would not be worth the cost. This is exactly what optimal environmental policy requires: marginal abatement cost should equal marginal social damage.

Undoubtedly the real world is much more complicated than the above situation. Firms cannot instantaneously move from Illinois to Hawaii; Illinois firms may have quite a bit of influence with Illinois lawmakers; the costs of pollution control will be concentrated among workers in certain industries while the benefits of a cleaner environment are diffuse; residents of other states will realize some of the benefits if Illinois firms generate less pollution. But this illustration does make an important point. Competition among jurisdictions implies that the costs of government would be borne largely by those who realize the benefits from those policies and will therefore encourage efficient public decisions.

Children and Grandchildren

Finally, as some people have asked, if we vest state and local governments with the authority to set environmental policy, will they properly take into account the interests of future generations? Will communities be inclined to follow policies that yield benefits now (such as more jobs) but would result in substantial damages in the future (such as increased exposure to long-lived pollutants like hazardous wastes)? Problems of this kind would seem to be more acute at the state and local levels than at the federal level. We would presumably all want the federal government to take into account the interests of our children since it is likely that they will live somewhere in this country. But at least some of our children will live in communities elsewhere, and therefore their environmental heritage will depend on the decisions of others. Geographic mobility, so this argument goes, could result in myopic local decisions and suboptimal environmental quality for future generations.

This line of reasoning is not altogether persuasive. There is substantial evidence that many of the factors that determine the quality of life in a community are reflected in community property values—that is, people are willing to pay more to live in a community with good schools, a low crime rate, and a clean environment. Suppose a community were considering allowing a new firm to locate in that community, and suppose further that this firm would cause $100 worth of environmental damages in the future. If environmental quality is reflected in property values, the value of homes in this community would fall by (the present value of) $100 if the firm were allowed to enter. At least in this simple case, current generations could not escape the future costs and benefits of their environmental decisions. Here again, the example ignores some important issues; in particular, it assumes that future residents are aware of and can evaluate the damages from long-lived pollution. But it does suggest that some of the concerns about decentralized policymaking need to be reevaluated.

Sensible Solutions

These considerations are not meant to suggest that we should abolish the U.S. Environmental Protection Agency and ask state and local governments alone to protect the environment. But I would argue that we should seriously consider moving some policymaking responsibility to lower levels of government, while still leaving the federal government with an important role in environmental issues. There are some strong arguments in favor of decentralization. An environmental policy that is correct in one region is unlikely to be correct in all. Federal regulation is rarely sensitive to these differences; instead, it often implies a single uniform policy in all regions.

It is important not to minimize the problems associated with this proposal. In particular, the spillover problem is difficult; the damages from some pollutants do not respect state boundaries. But sensible solutions may be available. Federal matching grants to state and local governments are often used to deal with the problem of fiscal spillovers. Perhaps similar policies (for instance, grants from the federal government to a state in return for an agreement to control pollution more tightly) would be effective in the interjurisdictional pollution problem. Further, we might envision a regulatory structure where states (and possibly lower levels of government) assume the responsibility for controlling some pollutants while the federal government retains the responsibility for others. It is not clear that we can solve all of the problems in environmental federalism, but the issues are sufficiently important that they deserve careful attention.

Part 5

Resource Management

20 Environmental Values and Water Use

Kenneth D. Frederick

During the first seventy years of this century, investments in water resources were driven largely by a desire to control water flows and ensure their availability for domestic, industrial, and agricultural users. In contrast, water related investments and legislation during the last quarter century have been driven largely by a desire to protect and restore the nation's water resources and aquatic environments. The first approach took a heavy toll on the environment and instream water uses; the second is imposing high economic costs, in many instances with little indication that the social benefits exceed those costs. The balancing of the economic and environmental values associated with water use has become a central issue for environmental legislation.

Philosophies guiding the development and use of water resources in the United States have changed greatly during this century. In the early 1900s, rapid construction of dams, reservoirs, and canals proceeded in line with the utilitarian view that dominated water development and use decisions. In that view, leaving water resources unused would be wasteful if those resources were capable of producing crops, power, and other valued products. By the early 1970s, however, growing concern about the effects of rapid growth in water use and development on water quality, fish and wildlife, and natural habitats began to be reflected in environmental legislation. Today, the conflict between water development and environmental protection is evident in controversies surrounding the implementation of much of that legislation. But that conflict is not merely a product of the greater environmental awareness of the last few decades; it has existed throughout the history of water development projects in this century.

Below I examine that history, paying particular attention to the environment versus development contests that have arisen in the context of the Endangered Species Act. I also examine two specific instances in which the difficulty of balancing the economic and environmental interests associated with water use is perhaps most evident today—the development of the Columbia River basin and the relicensing of hydroelectric power plants.

Originally published in *Resources*, No. 117, Fall 1994.

Transforming the Nation's Waters

The nation's first major struggle between environmental interests and water development began in 1901, when the city of San Francisco initiated plans to dam the Tuolumne River in the northern part of Yosemite National Park. San Francisco's plans to develop water supplies and hydroelectric power would flood Hetch Hetchy Valley, extolled as one of the nation's most beautiful and inspirational sites. But efforts to preserve the valley eventually proved futile when Congress approved the flooding of Hetch Hetchy in 1913.

Although preservationists were successful in thwarting subsequent proposals to flood sections of national parks and monuments such as Yellowstone, Glacier, Kings Canyon, and Echo Park, builders encountered little opposition to most dam projects for the next half century. Water projects were promoted and often subsidized to encourage development of the U.S. West; to increase employment, first during the Great Depression and then during the economic transition following World War II; to reduce flooding; and to provide cheap power, transportation, and abundant, reliable water supplies for homes, farms, and factories. Large-scale projects in particular were touted as examples of enlightened use and development of the nation's water resources; they represented a triumph of technology and human enterprise over the uncertainties of nature. And as the demands for the outputs provided by water development mounted, ever-larger projects were proposed to reduce conflicts associated with management of a multiple-purpose dam.

The rate of dam construction accelerated to a frenetic pace following World War II. More than 35,000 new dams were completed between 1945 and 1969, nearly four per day over the twenty-five-year period. The United States' water infrastructure now includes about 75,000 dams; 869 million acre-feet of reservoir storage; 25,000 miles of inland and intracoastal navigation channels supported by more than 200 locks and dams; tens of thousands of groundwater pumps; and millions of miles of canals, pipes, and tunnels for transporting water.

This hydrologic transformation produced many benefits. Streams that once alternately flooded their banks and dried up were controlled to provide dependable sources of supply. Tens of thousands of recreational reservoirs were created, former wetlands and flood-prone areas were developed for urban and agricultural purposes, and deserts were converted into vast urban areas spotted with green lawns, golf courses, and lakes. Virtually everyone had access to relatively inexpensive water at the turn of a tap. By 1980, water was being withdrawn from the nation's surface and groundwater sources at a daily rate of 440 billion gallons (more than 1,900 gallons per person). Nearly one-third of the value of the nation's crop production was being produced on 50 million irrigated acres, and hydropower provided about 11 percent of the nation's electricity and 4 percent of its total energy.

Construction of dams, reservoirs, canals, and so on, supplemented by research to develop new technologies (such as desalinization and weather modification), were widely accepted as the way to provide for growing water demand. And as water became scarcer, development schemes became more grandiose. The North American Water and Power Alliance, conceived in the 1950s and enthusiastically promoted in the 1960s, proposed transporting 110 million acre-feet of water annually (about eight times the average natural flow of the Colorado River) from Alaska and northern Canada to the western United States and northern Mexico. The Bureau of Reclamation's Pacific Southwest Water Plan, presented to the president in 1964, recommended seventeen projects and programs, including a plan to pump Colorado River water over the mountains into central Arizona for Phoenix and Tucson, two big dams on the Trinity River in northern California, a tunnel to divert water from the Trinity to the Sacramento River, a wider aqueduct to deliver more water from northern California to the central and southern parts of the state, and two large hydropower projects at Bridge and Marble Canyons, which are located at opposite ends of Grand Canyon National Park on the Colorado River.

Introducing Environmental Values

Threats to the Grand Canyon and other national parks helped galvanize resistance to these and other large water projects and focused attention on the increasing financial and environmental costs associated with the traditional approach to meeting water demands. At the start of this century, large quantities of water could be developed at relatively low cost. In addition, water was sufficiently plentiful relative to demand that extracting it for one use had little effect on the availability of water for other uses. Finally, there was little concern about the loss of free-flowing streams as a consequence of the construction of dams and the diversion of water flows. By the 1960s, however, the financial costs of developing additional water supplies had increased sharply because the best reservoir sites had already been developed, and the quantity of water controlled (or the safe yield produced) by additional increases in reservoir capacity on a river was subject to sharply diminishing returns. Moreover, as water became scarcer, the implied trade-offs among alternative water uses became more stark.

The strongest objections to proposed new water supply projects were based on their environmental impacts. By 1970, the legacy of environmental degradation associated with past water developments and uses was extensive, and water projects were less likely to be acclaimed as examples of wise resource use. Such projects were increasingly criticized as expensive proposals to quench the insatiable thirst of farmers, cities, and factories, and to provide hydroelectric power at the expense of instream flows and the fish and wildlife habitat and recreational opportunities they support. Thousands of miles of once free-flowing streams had been lost; Grand Coulee Dam alone eliminated a thousand miles of salmon spawning streams in the Columbia River basin. The quality of many of the nation's rivers and lakes had deteriorated to the point that they were unusable for most purposes.

Water projects also contributed to the sharp decline in the nation's wetlands, which store floodwater, control erosion, provide fish and wildlife habitat, improve water quality, and furnish recreational opportunities.

In response to this sad legacy and to the nation's growing desire to protect and restore water quality and aquatic environments, the rules governing the use and development of water resources began shifting against water developers. This shift is evident in federal legislation such as the Wild and Scenic Rivers Act of 1968, the National Environmental Policy Act of 1970 (NEPA), the Federal Water Pollution Control Act Amendments of 1972 (commonly known as the Clean Water Act), and the Endangered Species Act of 1973 (ESA). Development activities that would alter significantly an area's natural amenities now are precluded on thousands of miles of rivers and streams that are preserved under the Wild and Scenic Rivers Act. NEPA requires all federal agencies to give full consideration to environmental effects in planning their programs. As a result, critics of a water project no longer have to prove that the project would have major adverse environmental effects. Instead, project proponents must demonstrate that a water project is environmentally benign, or they must undertake efforts to mitigate the project's adverse effects.

The Clean Water Act affects water development in several important ways. First, together with the Safe Drinking Water Act of 1974 and other legislation regulating the use and cleanup of toxic materials, it has made water quality rather than water supply the driving force behind the nation's water-related investments. The United States has spent more than $500 billion on water pollution control since 1972. Second, the U.S. Environmental Protection Agency (EPA) has used section 404 of the Clean Water Act to veto on environmental grounds more than a dozen water projects. Third, as a result of a May 1994 Supreme Court ruling, the Clean Water Act gives states broad authority to impose minimum stream flow requirements in order to protect water quality.

Development and the Endangered Species Act

Like the Clean Water Act, the Endangered Species Act has become an important factor in many water management decisions. The act's potential influence over water use achieved national attention in 1978, when the Supreme Court issued an injunction halting construction on the Tellico Dam in Tennessee (even though more than $100 million had already been spent on the project, which was 90 percent completed) because the dam threatened the only known habitat of the snail darter. More recently, the U.S. Fish and Wildlife Service evoked the ESA in halting the $590 million Animas–La Plata project in the Colorado River basin one day before construction was scheduled to begin, because the project might harm the endangered Colorado squawfish. The ESA has resulted in delays, modifications, and cost increases in scores of other proposed dam projects.

The ESA has also been used to alter the management of existing projects. The operation of dams on the main stem of the Missouri River has been modified, to the detriment of some water users, in order to protect the nesting grounds of the endangered least tern and the threatened piping plover. In California, the Bureau of Reclamation has sent large quantities of water around the power-generating turbines at Shasta Dam, resulting in millions of dollars in forgone power revenues, in order to provide colder water for the spawning of the threatened winter-run chinook salmon.

In March of this year, the U.S. Fish and Wildlife Service granted special protection to nearly 2,000 miles of the Colorado River and its tributaries in order to protect four species of endangered fish—the Colorado squawfish, the bonytail chub, the humpback chub, and the razorback sucker. This particular action is likely to require the Bureau of Reclamation to alter the operation of its dams and to make new water projects in the Colorado River basin even more difficult to undertake.

The Delta smelt, found only in California's Sacramento–San Joaquin Delta, is under consideration for ESA listing. If the smelt is granted protection, the ability to export water from the Delta to the millions of people who depend on its supply for domestic, agricultural, and industrial uses could be severely limited.

ESA Impact on the Columbia River Basin

The ESA is having its greatest impact on water use within the Columbia River basin, where dams have produced cheap power and enhanced recreational opportunities, irrigated millions of acres, and provided towns located hundreds of miles inland with ports accessible to the ocean. These achievements, however, have come at the expense of the salmon stocks that once inhabited the region's rivers in great numbers.

The Northwest Power Planning Council has spent more than $1.7 billion in taxpayer and ratepayer money since 1980 to rebuild salmon stocks. Measures include making more water available during critical migration periods (and thereby forfeiting power revenues), retrofitting dams with screens that guide fish away from turbines and into channels that lead them past dams and into water below the dams, barging fish around the dams, and constructing fish ladders. Nevertheless, three stocks of salmon that spawn in the Snake River (the principal tributary of the Columbia River) are listed as threatened or endangered, petitions have been filed for listing several other stocks, and as many as eighty-five salmon stocks throughout the Columbia River basin are so weakened that they could be granted protection under the ESA.

While almost everyone agrees that stronger measures are required to protect salmon, no consensus has been reached as to what measures would adequately protect the fish, how much the measures would cost, and who would pay. A review of proposals to facilitate one stage in the life cycle of salmon—smolt migration from the Snake River to the ocean—suggests that the costs and impacts on current water users of restoring the Snake River salmon would be high. The length of time it takes the smolts to migrate downstream is considered critical to the number and health of the juveniles eventually reaching the ocean. Currently, slackwater pools behind the dams can disorient the fish, leave them more exposed to predators, and delay their journey to the ocean. Passage through turbines further reduces their numbers and leaves the survivors weakened.

One proposal for helping the smolts is to increase substantially the rate of water flow and to allow more water to bypass the turbines during the critical migration months. This approach would cost tens of millions of dollars in forgone hydropower revenues and might require irrigators to reduce water withdrawals during low water-flow years. As an emergency measure, in May of this year federal officials opened the spillways on eight dams in the Columbia–Snake river system to push chinook salmon smolts over the tops of the dams. But the experiment was terminated when it appeared that increased nitrogen levels caused by the spills were detrimental to the fish.

Another proposal for helping smolts is to increase streamflow velocity (rather than the volume of flow) by dropping reservoirs to their spillway crest levels during critical downstream migration periods. Preliminary estimates of the costs of modifying the four lower Snake River dams to implement the drawdown strategy range from $600 million to $1.3 billion. In addition, hydropower production, navigation, and reservoir recreation on the Snake River would be adversely affected during the drawdown period.

The most extreme proposal, and perhaps the one that would offer the Snake River salmon the best prospects for recovery, is to remove the dams hindering passage between their spawning grounds and the ocean. All the region's water and energy users would be affected by this strategy.

Dams on the Columbia and Snake rivers are not likely to be removed to protect the salmon, but dam removal for environmental purposes is receiving increased attention. The U.S. Department of Interior, with congressional support, proposes to remove two hydroelectric dams on the Elwha River on Washington's Olympic Peninsula. These dams, which were constructed early in this century, eliminated the river's native fish population. A major obstacle to the proposal will be finding $140–$235 million to remove the dams and restore wild salmon to the river. This estimate does not include the cost of forgone hydropower revenues associated with removal of the dams.

Relicensing of Hydroelectric Plants

Conflicts over environmental and developmental objectives may be more complicated when hydroelectric power is involved because of hydroelectric power's dual environmental effects. On the one hand, hydroelectric power provides a renewable and clean source of power. Unlike fossil fuels, for instance, it does not pollute the air or contribute to the atmospheric concentration of greenhouse gases. On the other hand, the construction and operation of dams to produce hydroelectricity transforms aquatic environments in ways that adversely affect indigenous fish and wildlife and perhaps other water users.

The United States has more than 2,300 hydroelectric power plants with a total capacity of 73,500 megawatts and an annual production of more than 300 billion kilowatt hours. Most of these plants operate under federal licenses issued as many as fifty years ago, when fewer questions were raised about the effects of alternative uses of water resources on fish and wildlife habitat. As the licenses expire (234 lapsed during 1993 alone), the utilities are faced with a complex, costly, and time-consuming relicensing process under the Electric Consumers Protection Act of 1986.

This act requires the Federal Energy Regulatory Commission (FERC) to give power benefits and nonpower benefits (such as the provision of fish and wildlife habitat and recreation) equal consideration in its licensing and relicensing decisions and to award new licenses to the applicant with the plan that is best adapted to the broad public interest. Consequently, an application is likely to require a detailed environmental assessment that includes an evaluation of a power plant's impacts on fish and wildlife habitat, water quality, recreation, land use, local communities, and cultural resources. The relicensing process provides new applicants, as well as environmental interest groups and other parties, an opportunity to voice objections and propose other options.

FERC is considering two new policies that might expand the alternatives to this relicensing process. One policy would identify the circumstances under which a dam should be decommissioned, and the other policy would introduce ways to assess the cumulative effects of several dams in a single river basin. From the perspective of the utilities and power users, the relicensing process itself may result in significant costs, and new licenses (assuming they are eventually granted) may be encumbered with restrictions that diminish the flexibility and productivity of hydroelectric power plants. But the relicensing process might contribute to an improved use of the nation's water resources. Moreover, the environmental and other benefits stemming from any new license restrictions at least might offset the costs associated with any inefficiencies the restrictions introduce in power production.

The Remaining Challenge

The National Water Commission's final report to the president and Congress in 1973 criticized past failures to incorporate adequately ecological processes and environmental values into decisions affecting water development and use. The commission believed that developmental and environmental values frequently can be accommodated with careful planning. But in instances where these values necessarily conflict, the nation needs to develop procedures for striking a balance that serves the public interest fairly and promptly, thereby avoiding the social, economic, and environmental costs attending delays in reaching decisions.

Since publication of the commission's report, the federal government and most state governments have elevated environmental concerns to a prominent role in decisions affecting the development and allocation of water. Investment in new dams and reservoirs has virtually ground to a halt while hundreds of millions of dollars have been spent to improve water quality. Although the ambitious goals of the Clean Water Act have not been fully met, thousands of miles of streams have been protected, and the condition of many of the nation's streams and lakes has improved significantly in spite of increased pressures from an expanding population and economy.

But the nation has not lived up to the National Water Commission's challenge to create procedures that provide for an expeditious balancing of environmental, social, and development values. In some instances, environmental values are introduced preemptively through the Endangered Species Act. In other instances, environmental values are introduced through long and costly judicial or administrative proceedings that may or may not serve the public interest. And in other instances, these values continue to be ignored or shortchanged by institutions rooted in an era when water left in a stream was assumed to have no value. Reauthorization of the Clean Water Act and the ESA provides Congress with another opportunity to address the commission's challenge of introducing environmental values into water use and investment decisions in a balanced and expeditious manner.

Suggested Reading

Frederick, K.D. 1991. "Water Resources: Increasing Demand and Scarce Supplies." In K.D. Frederick and R.A. Sedjo, eds., *America's Renewable Resources: Historical Trends and Current Challenges*. Washington, DC: Resources for the Future.

Frederick, K.D. 1993. "Changing Water Resource Institutions." In Water Science and Technology Board, National Research Council, *Sustaining Our Water Resources*. Washington, DC: National Academy Press.

21 Water as a Source of International Conflict

Kenneth D. Frederick

Efficient use of shared water resources has long been challenged by the reluctance of some neighboring nations to share "their" water, with conflict the most likely result. Rising water costs alone necessitate greater efficiencies: integrated management practices and some market-based means offer avenues for both reducing conflict and curbing costs.

From Canada to Mexico, from Africa to the Middle East, from Asia to Europe, conflicts and the potential for conflicts are growing over the availability of water. While sharing water resources has long been divisive, today's rising environmental, social, and financial costs of managing Earth's most abundant and renewable natural resource exacerbate these perennial tensions. Easing such tensions becomes imperative at a time when demands for water are rising. The greater efficiencies achievable by integrated resource management, developing water markets, and price incentives may prove the best ways to achieve this end.

The Roots of Conflict

Several factors underlie virtually all international conflicts over water and pose problems for managing and allocating it efficiently and equitably. These include the variability and uncertainty of supplies, the interdependencies among users, and the increasing scarcity and rising costs of freshwater. Because water is a "fugitive" resource—naturally flowing from one location and one state (liquid, gas, or solid) to another—individuals and countries have incentives to capture and use the resource before it moves beyond their control but little, if any, incentive to conserve and protect supplies for downstream users.

Also at the root of conflict, however, are other human elements: the vulnerability of water quality and aquatic ecosystems to human activities, the failure to treat water as an economic

Originally published in *Resources,* No. 123, Spring 1996.

resource, the desire for food security and self-sufficiency in arid and semiarid regions of the world, and the importance of water to public health and economic development.

These human factors are making conflicts over water resources within countries increasingly common. When water is shared by two or more countries, the obstacles to achieving efficient, equitable, and conflict-free management are even greater. Such are the situations—described later in this article—between India and its neighbors Pakistan and Bangladesh and among most of the nations of the Middle East

Efficiency vs. Equity

From the standpoint of integrated resource management, these human factors contribute to the inefficient division of an otherwise natural hydrological unit. Efficient management techniques require treating all the water in a given river basin, aquifer, or watershed as a unitary resource: overcoming the tendency among neighbors to exploit water unilaterally would provide a cost-effective way to increase freshwater supplies. The institutional obstacles to achieving this, though, can be considerable.

Even within the United States, multistate water laws, independent water management systems, and institutional inertia impede the introduction of more efficient management systems. Greater obstacles to integrated regional water management are likely when different countries and cultures, and even historical animosities, are involved. Consequently, achieving a sense of equity, perhaps through formalizations of historic patterns of use, among all parties may be a more realistic short-term goal than efficiency in settling international disputes.

Ultimate Market Efficiency

Water will surely become increasingly scarce, however (see the box "Once and Future Water Shortages"), and questions of efficiency ultimately should grow to assume greater significance in

resolving conflicts. Developing markets and market-based prices allows the peaceful transfer of most resources among countries.

Under some very restrictive conditions, markets lead to an efficient distribution and use of a resource: under a wide range of conditions, the market process contributes to a more efficient allocation and management of these resources. Markets can provide individual people as well as countries with increased opportunities and incentives to develop, transfer, and use a resource in ways that would benefit all parties.

Two conditions must be satisfied for the development of efficient markets. There must be well-

Once and Future Water Shortages

Are water supplies, whether considered globally or regionally, sufficient to our needs? Shifts in population (including increased urbanization) and in industrial and economic growth and development can increase both demands on, as well as contamination of, water supplies. In the developing nations, these factors are especially acute and are exacerbated by projected population growth: by 2000, nearly 900 million people (by U.N. figures) will likely be in regions where adequate basic drinking water and sanitation services are already strained, insufficient, or altogether lacking.

The pressures in most developed nations differ as much by degree as by kind: for this and other reasons, there is sound basis that supplies can meet demands. Demands, however, for water for environmental uses— wildlife, fisheries, and recreation—are increasing, and agricultural uses (irrigation demands are particularly heavy in the United States) and contamination continue to be problems.

The unknown factor of possible climate changes may be waiting in the wings for all nations. Water supplies may theoretically increase or decrease across regions under global warming, and present-day uncertainty over where and when these changes might occur makes calculating for climate change a task for the future.

Table 1. Dependence on Imported Surface Water: Countries and Their Percent of Total Flow Originating Outside Their Borders

Egypt	97	Syria	79	Yugoslavia	43
Hungary	95	Congo	77	Bangladesh	42
Mauritania	95	Sudan	77	Thailand	39
Botswana	94	Paraguay	70	Austria	38
Bulgaria	91	Czechoslovakia	69	Pakistan	36
Netherlands	89	Niger	68	Jordan	36
Gambia	86	Iraq	66	Venezuela	35
Cambodia	82	Albania	53	Senegal	34
Romania	82	Uruguay	52	Belgium	33
Luxembourg	80	Germany	51	Israel*	21
		Portugal	48		

*Although only 21 percent of Israel's water comes from outside current borders, a significant fraction of Israel's fresh water supply comes from disputed lands.

Source: Peter Gleick, *Water and Crisis: A Guide to the World's Fresh Water Resources,* Oxford University Press, 1993.

defined and transferable property rights in the resource being transferred, and the buyers and sellers must bear the full benefits and costs of the transfer. Both conditions are now commonly violated for water resources. The fugitive nature of the resource makes it difficult to establish clear property rights, and the interdependencies among users might cause externalities or third-party impacts when the use or location of water is changed.

Regions of Potential Conflicts

Rivers and lakes that border multiple countries, rivers that flow from one country to another, and aquifers that underlie more than one country are international resources: the use of the resource by one country affects the quantity or quality of the resource available to another country. Such situations are numerous: about 200 river basins are shared by two or more countries. Thirteen are shared by five or more countries, and four basins— the Congo, Danube, Nile, and Niger—are shared by nine or more countries. Shared watersheds comprise about 47 percent of the global land area and more than 60 percent of the area on the continents of Africa, Asia, and South America. (Groundwater resources are also frequently shared by two or more

countries.) Table 1 illustrates some of these interdependencies.

The Middle East

The competition for water in the Middle East is so intense that lasting peace in the region is unlikely in the absence of an agreement over shared water use. Indeed, negotiations over water have a separate role in the ongoing peace talks between Israel and its neighbors. Outstanding issues and potential sources of conflict include the allocation and control of the Jordan River, the use of the aquifers underlying the West Bank, and Jordanian objections to the construction and operation of Syrian dams on the Yarmuk, the major tributary of the Jordan River. Water has already been the source of armed conflict in the region between Syria and Israel, once in the 1950s and again in the 1960s.

Several times during the past thirty years, disputes among Turkey, Syria, and Iraq over the development and use of the Euphrates River have nearly ended in armed conflict. Disputes arose in the 1960s when Turkey, where 90 percent of the water originates, and Syria started to plan large-scale withdrawals for irrigation. The conflicts heated up in 1974 when Iraq threatened to bomb the dam at Tabqa, Syria, and massed troops along the border because of the reduced flows Iraq was receiving in the Euphrates. The threats were renewed in the spring of 1975. With the completion of the Ataturk Dam in January 1990, Turkey is in a pivotal position to influence the downstream flow of the river. Potentially, the dam could benefit all countries within the basin by reducing the variability of the river's natural flows. But the dam gives Turkey a potential water weapon that could be used against the downstream countries. The Ataturk Dam and related water projects could reduce flows as much as 40 percent to Syria and 80 percent to Iraq. The threat of reduced water flows has been used in an attempt to force Syria to withdraw support of the Kurdish rebels operating in southern Turkey. Border security and water sharing have been linked in recent negotiations between the two countries.

The Indian Subcontinent

When the Indian subcontinent was partitioned between India and Pakistan in 1947, long-standing conflicts over the Indus River became overnight an international issue between two hostile countries. The partitioning divided the basin physically and split an established irrigation system between the two countries without specifying how the waters were to be divided. India was left with control of the waters supplying Pakistan's irrigation canals, and in 1948 India diverted those waters away from Pakistan. Although the canals were later reopened, the dispute threatened to lead to war.

With help from the World Bank, negotiations over water issues between the two countries began in 1952. Concerns over sovereignty stymied the bank's attempts to develop and manage the basin as a unitary system for the mutual advantage of both countries. The Indus was divided between the two countries, with India receiving the three eastern and Pakistan the three western tributaries. This division deprived Pakistan of the original source of water for its irrigation system. In compensation, India paid for new canals to bring water from the rivers allocated to Pakistan and a consortium of countries financed the construction of storage dams

to ensure Pakistan a reliable supply. At a price, the treaty defused a major source of potential conflict and allowed each country to develop its share of the basin's waters.

Bangladesh, which gained its independence from Pakistan in 1971 with the aid of India's army, is now threatening to cancel its Treaty of Friendship with its former liberator because of conflicts over water. Most of Bangladesh's rivers flow from India, which has shown little concern about the impacts of its water developments on its downstream neighbor. A major diversion from the Ganges River just a few miles from the Bangladesh border has increased salinity levels and reduced water supplies in the Padma River (as the Ganges is known in Bangladesh), threatening the livelihood of millions of Bangladeshis.

The Former Soviet Union

The breakup of the Soviet Union has also converted some formerly domestic water issues into potential international conflicts. Water scarcity and conflicts are particularly acute in the five former central Asian republics of the Soviet Union that share the flows of the Amu Dar'ya and Syr Dar'ya rivers. These two rivers originate in the high mountains to the southeast and flow through deserts to the Aral Sea. As recently as 1960, the Aral Sea was the world's fourth largest lake in area. Since then, water diverted primarily for cotton production has altered the water balance of the Aral; between 1960 and 1989 the sea declined precipitously in level, area, and volume.

Mismanagement of the region's water and land resources has produced an ecological and human disaster with few parallels: the once productive fishing industry has disappeared; rising soil salinity has depressed agricultural yields; and pesticides applied to the cotton have contaminated drinking water supplies in the lower reaches of the river basins, with tragic impacts on human health. Reversing decades of mismanagement and abuse of the region's water supplies may be more difficult now that it requires the cooperation of five struggling newly formed countries.

Potential Benefits of Integrated Management

The lack of clear property rights to international water resources is an obstacle to more efficient resource management and to the resolution of water conflicts. Two extreme and opposing doctrines have been proposed for establishing property rights over international waters. The doctrine of *unlimited territorial sovereignty* states that a country has exclusive rights to the use of waters within its territory. This doctrine, which allows a country to deplete and pollute with no obligation to compensate adversely affected parties, was asserted by the U.S. Attorney General in 1895 in rejecting Mexico's claims to waters originating in the United States. Although the United States subsequently modified its stance on shared waters, this view characterizes India's approach for development of the Ganges River.

The contrasting doctrine of *unlimited territorial integrity* states that one country cannot alter the quantity and quality of water available to another. This doctrine, which greatly constrains how the upstream country can use the resource, is reflected in Egypt's threats against countries proposing water development projects that would reduce the waters of the Nile reaching Egypt.

In the absence of bargaining, both of these doctrines are likely to lead to inefficient outcomes. Under the first doctrine India has no incentive to mitigate the impacts on Bangladesh regardless of the relative magnitude of the damages imposed and the costs of abatement. On the other hand, under the second doctrine upstream countries on the Nile risk the wrath of a more powerful downstream neighbor unless they forgo potentially profitable development projects regardless of how high the costs of mitigation and how small the impacts.

In practice, international water disputes generally have moved away from the extreme positions implied by these two doctrines and toward a doctrine of *equitable and reasonable use*. Although this narrows the likely range of disagreement, it does not provide clear property rights. In the absence of enforceable property rights, the strongest, most clever, and most advantageously positioned countries can claim and use the resource with little concern for the impacts on others. Opportunities for coordinated management may be lost in the acrimony over rights to the resource and obligations to mitigate any adverse impacts imposed on others.

Inefficient management of and conflicts over international water resources reflect problems in the management and allocation of water resources within individual countries. Individual water rights might be limited or poorly defined, constraining the ability to transfer water among competing users and uses in response to changing conditions. Water prices commonly reflect only a small fraction of the social costs associated with use, reducing incentives to conserve and protect the resource. The institutions controlling water use are often rooted in a "pre-scarcity" era when transfers were viewed as unnecessary or unimportant. Similarly, cultural and religious considerations may result in water being viewed as too important or too sacred to have its use determined by the impersonal outcome of markets.

Equity considerations and historical use have been more important than efficiency in the management of both domestic and internationally shared water resources. Relatively few precedents demonstrate the potential advantages of efficient integrated management of an entire hydrologic unit. Yet, as water becomes increasingly scarce, the potential benefits of integrated management—and of institutions that enable scarce resources to be transferred among competing uses in response to changing conditions—will grow. Institutions that perpetuate inefficient water use will become increasingly costly and unstable. Inflexible and inefficient international agreements, which must be self-enforcing, may not be sustainable.

Conclusion

Much of the world's most accessible freshwater supplies are located within basins and aquifers that cross international borders. Inflexible, inefficient, and often inequitable agreements for managing international waters contribute to rising water costs, growing concerns over the adequacy of supplies, and potential conflicts. More flexible allocation mechanisms and efficient management practices are critical for avoiding future conflicts over international supplies and curbing the rise in water costs. Introducing markets and market-based prices, which provide peaceful transfer of other resources among countries, might help promote a more efficient and flexible allocation of water resources located in international basins.

Suggested Reading

Gleick, P.H. 1994. "Water, War and Peace in the Middle East." *Environment* 36/3 (April):6–42.

Beschorner, N. 1992/3. "Water and Instability in the Middle East." Adelphi Paper 273.

Wolf, A.T. 1996. "Middle East Water Conflicts and Directions for Conflict Resolution." Food, Agriculture, and the Environment Discussion Paper 12. Washington, DC: International Food Policy Research Institute.

22 Marketing Water
The Obstacles and the Impetus

Kenneth D. Frederick

As water grows more precious, so do the incentives—and the innovations—to try to apply market principles to its use and management.

Water is becoming increasingly scarce in the United States. Demand is rising along with population, income, and an appreciation for the services and amenities that streams, lakes, and other aquatic ecosystems have to offer. In contrast, the options for increasing supplies are expensive relative to current water prices and often environmentally damaging. Furthermore, contamination and unsustainable rates of groundwater use threaten current supplies in some regions.

Ordinarily, Americans count on prices and markets to balance supply and demand and allocate scarce resources. When demand increases faster than supply, higher prices provide incentives to use less and produce more. And, as conditions change, markets enable resources to move from lower- to higher-value uses. Market forces, however, have been slow to develop as a means of adapting to water scarcity. Both the nature of the resource and the institutions established to control its use help explain why.

Market Obstacles

Efficient markets require that buyers and sellers bear the full costs and benefits of transfers. But interdependencies among the many users of a stream or aquifer make that difficult to do. Selling water rights, for example, is likely to alter the quantity of water in a stream or the location of a diversion or returnflow (water withdrawn from a stream or aquifer that is returned to a location where it can be used again). Third parties—people

Originally published in *Resources*, No. 132, Summer 1998.

benefiting from the water other than the buyer and seller—will be affected by the change. Third-party impacts might include a change in the recreational amenities provided by a free-flowing stream or the erosion of a rural community's tax base when a farmer sells water to a city.

Efficient markets also require well-defined, transferable property rights. But riparian rights, which are still the principal basis of water law in the Eastern United States, are poorly defined because water use is subject to regulatory or judicial interpretations as to what is reasonable or might unduly inconvenience others. Moreover, these rights are not directly marketable because they are attached—and their use is restricted—to the lands adjacent to a stream.

In the West, where streams are less common and flows are smaller and less reliable, "prior appropriation" quickly displaced riparian rights as the primary basis of water law. Appropriative rights are established by withdrawing water from its natural source and putting it to beneficial use. During drought, supplies are allocated according to the principle of "first in time, first in right." This principle provided a powerful incentive for the quick diversion of streamflows and allowed irrigators to acquire the highest priority rights to much of the water. While appropriative rights can be transferable, they are commonly attenuated in ways that limit how and where water can be used.

Water has traditionally been treated as a free resource to be harnessed to serve cities, factories, and farms. Anything less was seen as wasteful. Thus, subsidized water storage and distribution systems and irrigation projects contributed to a nine-fold rise in water withdrawals from 1900 to 1970. They also contributed to the loss of tens of thousands of miles of once free-flowing streams and, eventually, to a shift in national policy. To protect streamflows and recover forgone environmental and recreational values, Congress passed legislation such as the Wild and Scenic Rivers Act of 1968, the National Environmental Policy Act of 1969, and the Endangered Species Act of 1973.

In recent decades, these environmental laws have been used to block construction of many dams and in some cases to challenge previously established rights to divert water from streams and lakes. Domestic, industrial, and agricultural users continue to vie for water that is withdrawn from reservoirs and streams, and now all three groups must also vie with environmentalists and recreationists over how much water can be diverted. Conflicts also arise over the priority that dam managers should give to flood control, water supplies, hydropower production, fish habitat, and recreational opportunities. These conflicts are now generally played out in the courts or administrative proceedings rather than in the marketplace.

Overcoming the Obstacles

If water has been slow to be bought and sold like other commodities, the incentives to do so are strong. Most of the senior water rights in the arid and semiarid West are held by farmers and irrigation districts. They pay nothing for the water itself and generally only a modest amount to have it delivered to their farms. As a result, enormous amounts of water are applied liberally to relatively low-value crops and the marginal value of the water is likely to be well under $50 an acre-foot (af)—the quantity of water that will cover one acre to a depth of one foot. In some cases the value of the water could be increased simply by leaving more in the river to provide hydropower, fish and wildlife habitat, and recreation rather than diverting it for irrigation. In many other instances, the value of water would rise by selling some of it to urban areas that are spending more than ten times as much to augment supplies through recycling or other costly water projects.

Despite the obstacles, the impetus to move from lower to higher value use is driving some water transfers. Temporary transfers are becoming increasingly common to respond to short-term fluctuations in supply and demand. Precisely because they are temporary short-term leases, options to purchase during dry periods, and one-time purchases through water banks blunt a principal third-party concern that a transfer will permanently

Trading Water

The Bureau of Reclamation's Colorado–Big Thompson project brings an average of 230,000 acre-feet of water annually from the Colorado River Basin across the continental divide to northeastern Colorado. Rights to proportional shares of this water are traded actively within the Northern Colorado Water Conservancy District unencumbered by third-party concerns.

Under western water law, downstream users generally own the rights to the returnflows. But in this case the district is able to retain ownership of the returnflows because the water originates in another basin. As a result, rights to the water are traded within the district much like stocks in companies. This arrangement does not eliminate the third-party impacts associated with returnflows, only the need to consider them in transfer decisions. The benefits of being able to transfer water readily among agricultural, municipal, and industrial users exceed any likely third-party costs.

However, limiting sales to within the conservancy district precludes opportunities for even more profitable transactions. For example, an acre-foot of water in perpetuity has sold for $3,500 more in the neighboring Denver suburbs than in the conservancy district.

undermine the economic and social viability of the water-exporting area.

Transfers among farmers within the same irrigation district are common and relatively easy to arrange because the third-party impacts are likely to be small and positive when the water stays within the community. But when farmers want to sell water to cities, irrigation districts resist, fearing the loss of agricultural jobs and income that accompany rural water use.

A water bank provides a clearinghouse to facilitate the pooling of surplus water rights for temporary rental. If well-defined, its rules and procedures can reduce the costs and uncertainties associated with a transaction and increase the opportunities for both buyers and sellers.

California established emergency Drought Water Banks in 1991, 1992, and 1994 to reallocate water among willing buyers and sellers. Water purchased largely from farmers willing to idle land or pump groundwater rather than divert surface water for irrigation was sold to cities and farms or used to protect water quality in the state's delta region and meet instream fish needs. Any adverse third-party impacts on the water-exporting communities were probably insignificant compared with the overall benefits of moving water to higher-value uses. Sales exceeded $68 million in 1991; they averaged less than $11 million in the latter years when drought conditions subsided. Idaho and Texas have established permanent water banks and other states are now considering establishing them as well.

Transferring Permanent Water Rights

Temporary water transfers are particularly useful for adapting to short-run changes attributable to such things as climate variability. They are less effective in dealing with long-term imbalances between supply and demand resulting from changing demographic and economic factors, social preferences, or climate. At some point, the historical allocation of water becomes sufficiently out of line with current conditions to warrant a permanent transfer of rights.

The process of resolving the third-party issues associated with the transfer of a long-term shift in water use is often slow, costly, and contentious. Proposed transfers face the hurdle of proving the negative, that a change will not harm others. This requirement stifles the development of markets in water rights. The Colorado–Big Thompson project (described in the box, "Trading Water"), which has been able to avoid third-party issues, is the exception. The ongoing efforts of the coastal region of Southern California and the city of Las Vegas are more indicative of the obstacles to acquiring additional water.

Both of these geographic areas face the challenge of meeting growing demands for water at a time when their traditional sources are declining and environmental considerations restrict the development of new ones. Los Angeles has already been forced to reduce the amount of water it takes from the Mono Lake region and, to comply with a mandate to improve environmental conditions in Owens Valley, will have to further reduce the city's supplies. In addition, the Southern California Metropolitan Water District (MWD), a large water supplier servicing more than fifteen million consumers including the residents of Los Angeles, is losing access to surplus water (that is, unused entitlements of other states) from the Colorado River. Las Vegas, meanwhile, has been depleting its groundwater stocks, causing subsidence within the city.

Under a 1989 agreement, Southern California's MWD has invested more than $100 million in lining irrigation canals and other water conservation projects in the Imperial Irrigation District. In return, MWD received the right to use the conserved water, approximately 106,000af per year, for at least thirty-five years. Provisions were introduced to assure that neighboring irrigation districts in the United States did not lose their water rights as a consequence. But the impacts on irrigators across the border where groundwater recharge declined were ignored because the Mexicans lack a legal claim to the water.

San Diego receives about 90 percent of its water from the MWD and, as a junior claimant, is the first to be cut back in time of drought. To increase the quantity and reliability of its supplies, the San Diego Water Authority has agreed to fund additional conservation efforts in the Imperial Irrigation District in return for the conserved water. As originally proposed, 20,000af would be transferred in 1999, with the annual quantity increasing to 200,000af after ten years. Disputes with MWD over use of the Colorado River Aqueduct to transport the water, however, have delayed completion of the transaction.

Las Vegas, which is already using most of Nevada's legal entitlement to the Colorado River, is seeking to buy more shares of the river from states with unused entitlements. Legal issues have undermined earlier proposals for interstate and interbasin sales of Colorado River water and enabled Southern

California's MWD to take unused entitlements for free. Rising water values, however, are creating new interest in such sales in Nevada, which lacks rights to surplus flows, and in states wanting to benefit from their unused shares.

In 1996, Arizona established a Water Banking Authority to purchase their own unused Colorado River water for storage in groundwater basins and possible sale to California and Nevada. Interstate sales, however, are tightly restricted; they are limited to 100,000 af/year and only when there is no use for the water in Arizona and there are no shortages on the Colorado River.

Las Vegas is also interested in buying water from Utah, which has not been using its full entitlement. However, a transfer between an upper basin state (Utah) and a lower basin state (Nevada) could require renegotiation of the 1922 Colorado River Compact dividing the river between the two basins.

The Federal Role

State institutions are primarily responsible for allocating waters within their borders. But the federal government—manager of much of the West's surface waters, supplier of water to about 25 percent of their irrigated lands, the source and enforcer of environmental legislation affecting water use, and trustee for Native American water rights—also has a critical role in breaking down the institutional obstacles to permanent water transfers. Some steps in this direction have been taken.

- In 1988, the Department of Interior adopted a policy of facilitating voluntary water transfers involving federal facilities as long as the transfers comply with federal and state law, have no adverse third-party impacts, and do not adversely affect facility operations.

- The Central Valley Project Improvement Act of 1992 authorized the transfer of federally supplied water outside the project service area. Although no off-project transfers have been approved yet, the act is potentially significant because the project is the largest water storage and delivery system in California and most of the project water is allocated to agriculture under highly subsidized terms.

- A proposed federal rule from the Department of Interior (*Federal Register*, December 31, 1997) is designed to encourage and facilitate voluntary transactions among the three Lower Colorado River Basin states by establishing a framework for approving and administering interstate agreements.

- In addition, the federal government as well as some states have been acquiring water for environmental purposes, such as the preservation of endangered species. These purchases help establish markets as viable mechanisms for allocating water.

More steps are of course needed. Uncertainties surrounding large but unquantified Native American water claims, for example, hinder the assignment of clearly defined, transferable property rights in water. Providing the tribes with rights that could be sold for uses off the reservations would foster water marketing as well as tribal welfare.

Finally, water scarcity and the potential benefits of water marketing are not limited to the West. In the East, riparian rights are gradually being replaced by or supplemented with permits. The advantages of using markets to allocate these permits will grow as the resource becomes increasingly scarce. Indeed, auctioning and trading permits are innovative approaches that might facilitate a more efficient allocation of water. It is unlikely, however, that markets resembling the ones we use to allocate most goods and services will ever become commonplace to transfer water. Finding expeditious ways to deal with the third-party effects that plague nearly all water rights transfers is critical if traditional market forces are ever to thrive. In the meantime, the enormous potential benefits of water marketing still wait to be tapped.

Suggested Reading

Committee on Western Water Management, Water Science and Technology Board, National Research Council. 1992. *Water Transfers in the West: Efficiency, Equity, and the Environment.* Washington, DC: National Academy Press.

MacDonnell, L.J., C.W. Howe, K.A. Miller, T.A. Rice, and S.F. Bates. 1994. "Water Banks in the West." Natural Resources Law Center Research Report No. 12. Boulder, CO: University of Colorado.

Wahl, Richard W. 1989. *Markets for Federal Water: Subsidies, Property Rights, and the Bureau of Reclamation.* Washington, DC: Resources for the Future.

23

Ecosystem Management
An Uncharted Path for Public Forests

Roger A. Sedjo

Since the 1960s, the U.S. Forest Service has relied on multiple-use management to balance demands for timber, recreation, and wildlife habitat. But new pressures, including the Endangered Species Act, have prompted the Forest Service to turn to ecosystem management, which seems to put the health of forests above all other considerations. Sedjo evaluates ecosystem management and suggests that it is a philosophy in need of refinement.

Should public forests be managed to reduce all traces of modern human activities or to produce goods and services? Recently, the U.S. Forest Service seemed to answer that question by saying that it would like to restore the forests of the northern Rockies to presettlement conditions—that is, to the way the forests were at the start of the nineteenth century. This is indicative of the Forest Service's new philosophy of ecosystem management and reflects its shift away from multiple-use management, which has been the practice on public forestlands since the 1960s.

The impetus for both approaches is the desire to sustain forests. Concern about the rapid rate of logging on public lands following World War II led to congressional legislation that called for multiple use management. This legislation explicitly recognized the worthiness of a range of goods or services provided by public forests—including market goods, such as timber, and nonmarket services, such as habitat for wildlife. Congress charged the Forest Service with managing forests to produce a mix of both within the context of sustainability.

In recent years, however, the leadership of the Forest Service has backed away from this goal as its attention has focused on forest ecology—the totality of relationships between forest organisms and their environment. This concern with forest ecology is embodied in the leadership's advocacy of ecosystem management. In accordance with this philosophy, the service has all but abandoned the notion of forests as primarily a vehicle for producing multiple goods (or "outputs") desired by society. Instead of practicing *multiple-use management*, which

Originally published in *Resources*, No. 121, Fall 1995.

emphasizes the sustainable production of myriad goods and services, the Forest Service has embraced *ecosystem management,* wherein the condition of forest ecosystems—the complex of forest organisms and their environment functioning as an ecological unit in nature—is considered to be the preeminent output.

Although an ecosystem-based approach has much to offer in the form of a broader, more integrated, and more comprehensive view of the forest—and thus contributes to the development of more effective management tools—its defect is its disregard for certain socially approved objectives. In essence, ecosystem management aims to restore forests to some biological condition that reflects fewer human impacts, but just *what* condition is a matter of arbitrary selection. Because ecosystem management has no real legislative mandate, decisions to seek any one of many possible conditions are being made by the Forest Service rather than by society at large, which makes its wishes known through the legislation of management objectives. More to the point from the perspective of taxpayers, these decisions are being driven almost exclusively by biological considerations, with little attention paid to economic and other concerns. In short, when identifying objectives, ecosystem management ignores the social consensus implicit in the congressionally legislated objective of producing multiple market and nonmarket forest outputs and, instead, attempts to achieve some arbitrary forest condition about which society has little say.

The comparison of ecosystem management and multiple-use management presented below highlights the pitfalls of the Forest Service's new philosophy. Despite these pitfalls, it would be unwise simply to dismiss ecosystem management. It has resulted in the development of some highly effective management tools and activities and reflects a concern for the health of ecosystems that traditional management may not sufficiently recognize. Management for multiple-use objectives should continue to be the practice on public lands, but perhaps with a view to incorporating some aspects of ecosystem-based management.

The Need for Clear Objectives

Management of public forestlands requires the identification of clear objectives and the development of a regime (procedures and tools) that will achieve the objectives without violating the constraints imposed by the availability of resources and the acceptability of actions and outcomes.

Forest management without objectives is meaningless. In the absence of stated goals, we cannot differentiate successful forestry activities from unsuccessful ones. And in the case of public forestlands, the ability to gauge the success of management efforts takes on added significance because these efforts are being financed by taxpayer dollars. Moreover, without specifying objectives, we cannot ensure that the preferences of society are being reflected in the way that our forests are managed. These preferences should inform goals as well as define the constraints within which a management regime will operate.

But where objectives dictate the management approach under multiple-use forestry, ends merge with means under ecosystem management. Indeed, in actual practice, the objective of ecosystem management is most often simply the application of an ecosystem, or ecosystem-based, approach that is concerned first and foremost with the state of the forest itself. Thus while the Forest Service has been embracing ecosystem management as its operating philosophy for several years, no clear vision of output goals, at least as traditionally understood, has emerged. What has emerged is a preoccupation with forest condition—that is, with biological attributes, such as a forest's structure (mixture of younger and older trees) and variety of tree species—rather than with the goods and services (particularly those consumed by humans) that forests provide.

Ecosystem Management versus Multiple-Use Management

Jack Ward Thomas, chief of the Forest Service, has said that ecosystem management means sustaining forest resources, from which will flow many goods and services. But our public forests have for decades been managed to sustain multiple uses. Is ecosystem management really different from multiple-use management?

The mandate for multiple-use forestry has been expressed by law since 1960, when Congress passed the Multiple-Use Sustained Yield Act. This act acknowledges that forests generate both market goods and nonmarket goods. The objective of multiple-use management is to produce the mix of these market and nonmarket goods that maximizes the value of forests to society.

If the objective of ecosystem management is simply the management of whole ecosystems for a variety of purposes, such management might be viewed as an expansion of the multiple-use approach. Under this expanded approach, the set of outputs under consideration would broaden to include the biological condition of the forest itself. In addition, the boundaries of the management unit would enlarge, because changes in forests affect the geographic area around forests. Finally, the potential uniqueness of each forest ecosystem would be recognized and new management techniques would be introduced. Conceptually, these considerations represent modest extensions of multiple-use management. The job of the public forest manager would continue to be producing the mix of outputs that would maximize the social value of the forest.

But proponents of ecosystem management are reluctant to treat such management as a mere extension of multiple-use forestry. Unlike multiple-use management, which focuses on distinct forest outputs, many of which are consumed directly by humans, ecosystem management focuses on forest condition as the dominant forest "output." In this context, timber, recreational opportunities, and other traditional forest goods are merely by-products of managing forests to achieve one of many possible forest conditions. Production of these other outputs is tolerable as long as it does not conflict with the primary objective of achieving one of these conditions. Thus, for example, timber harvests that improve the condition of a forest are

acceptable. But while under multiple-use management such harvests could be decreased in order to increase recreational opportunities, under ecosystem management such opportunities would not be augmented if they resulted in what was perceived as an undesirable change in forest condition. Under ecosystem management, forest condition—as the preeminent forest output—is not subject to trade-offs with other forest outputs, as it is under multiple-use management.

A clear statement of the objectives of ecosystem management appears in the Forest Service's proposed regulations dated April 13, 1995. In the proposed regulations, the management objective is stated as follows: "The principal goal of managing the National Forest System is *to maintain or restore the sustainability of ecosystems...*" (italics added). By this articulation, the goal of management is very similar to the constraints of other forest management sytems: sustainability. The proposed regulation goes on to suggest that the achievement of this goal will result in "...multiple benefits to present and future generations."

The Implications of Ecosystem Management

Given ecosystem management's focus on forest condition, the first question that arises is whether a given forest's current condition should be maintained or modified to some specified extent. Once such a decision is made, the vagueness of the management objective disappears. But, as I suggest below, the selection of desired or acceptable condition is essentially arbitrary. As a result, the objective chosen today may be sadly outdated in perhaps a few years.

Although not readily apparent, arbitrariness is reflected in the Forest Service's apparent preference for restoration, rather than maintenance, of forest condition. This restoration entails the return of forests to some state characterized by fewer human impacts—for example, the return of the forests of the northern Rockies to presettlement conditions. But why not aim for a forest condition that predates human activity?

On a philosophical level, such arbitrariness is perhaps easier to show if we compare the selection of desired condition for European forests with that for American forests. In the United States, landscape conditions before and after European settlement are readily distinguished, and the landscape conditions before European settlement often function as a model for desired forest condition. In Europe, however, the distinction between forests before and after human settlement is virtually impossible to make, and, as a result, determining desired forest condition is more difficult. Should forests there be returned to their pre-Celtic condition before about 1500 B.C., to their pre-Roman condition, to their condition in the Middle Ages, or what? This question inevitably raises more fundamental questions—namely, whether less human impact is always preferable to more human impact, and, if so, why. These questions do not have scientific answers.

At the same time, however, and despite assertions to the contrary, the perspective of ecosystem management is almost purely biological, with no serious attention given to social values and little real attempt made to relate forest outputs to human and social needs and desires. A critical question that is not being asked is whether achieving a particular forest condition is a sensible use of public funds. It is one thing to justify taxes to produce outputs, market or nonmarket, that are consumed directly by the public, but quite another for society to use its scarce tax dollars to achieve a biological objective that may or may not be valued by the majority of the taxpaying public.

Generating Benefits for Everyone

Public forests were established to generate benefits for all citizens, and in the past the objectives of forest management reflected a degree of political consensus. In recent decades, these objectives have been codified in congressional legislation: the Multiple-Use Sustained Yield Act of 1960, as well as the Resources Planning Act of 1974 and the National Forest Management Act of 1976. By contrast, forest management as practiced by the Forest Service in the mid-1990s has no clear political or social mandate. Indeed, ecosystem management marks a sharp shift away from legislatively supported multiple-use forestry—which recognizes many biological, social, and economic values—focusing instead on an arbitrary forest-condition objective that, in essence, is defined by biological considerations only.

While the Forest Service's adoption of ecosystem management may be inconsistent with legislation mandating multiple-use management, it is not inconsistent with the Endangered Species Act (ESA). In fact, recent court rulings that earlier Forest Service actions were contrary to the ESA do provide a rationale for the service's shift to ecosystem management. These rulings do not, however, provide sufficient justification for jettisoning the multiple-use objectives called for in existing legislation, at least until such time as a national consensus on new forest management objectives is codified by Congress.

The practice of ecosystem management, however, has arisen partly as a result of the difficulties inherent in multiple-use forestry. Achieving the optimal social mix of outputs is, obviously, no easy task. The selection of outputs has been complicated further by court interpretations of the ESA that constrained management decisions. In this context, the current administration and the new Forest Service chief have promoted the shift to an ecosystem management approach.

Changes in the administration or the ESA are likely to alter the way that ecosystem management is practiced, however, perhaps making the forest conditions managed for today undesirable tomorrow. And changes are likely. Administrations come and go, after all, and with them the leadership of the Forest Service. Moreover, the ESA is expected to be amended. In the absence of any kind of legislative mandate, then, ecosystem management could go by the wayside or it could constantly alter the goods that forests provide and do so without reference to public opinion.

If ecosystem management is to be practiced on public lands, the application of democratic principles suggests that such management be made law. In the absence of new congressional directives, however, management for multiple forest outputs should continue on public lands. But ecosystem-based management should not be dismissed altogether. Its tools and activities could and probably should be used by the Forest Service to achieve the objectives of multiple-use forestry. And if there appears to be some public support for returning forests to a specified condition of fewer human impacts, this condition could be added to the list of existing management objectives, such as producing timber and providing recreational opportunities.

The advantage of multiple-use management is that it tries to accommodate additional objectives and make trade-offs among them in order to increase social values. Such an approach, although sometimes flawed, is much more likely to benefit all members of society than ecosystem management, which makes one objective dominant and essentially impervious to trade-offs. In retrospect, we can see that multiple-use management's chief strength lies in its flexibility and in its responsiveness to changing social desires. By comparison, ecosystem management is rigid in identifying objectives and essentially arbitrary.

Suggested Reading

Sedjo, Roger A. 1996. Toward an Operational Approach to Public Forest Management. *Journal of Forestry* 94(8) August.

Thomas, Jack Ward. Forthcoming. "Challenges to Achieving Sustainable Forests: Is NFMA Up to the Task," in K. Norman Johnson and Margaret A. Shannon, eds., *The National Forest Management Act in a Changing Society 1976– 1996.*

24

The Global Environmental Effects of Local Logging Cutbacks

Roger A. Sedjo

The U.S. West has significantly cut back on its timber harvests as a result of logging restrictions. These restrictions, which are now being authorized in British Columbia as well, are intended to reduce regional environmental damage associated with logging activities. But the restrictions could simply relocate such damage because they are triggering increases in timber harvests elsewhere in the world. Ironically, the very environmental concerns that have led to decreased logging in the U.S. West could result in a net increase in global environmental damage.

According to a popular slogan, we should think globally and act locally—that is, regard the environment from a global perspective and act locally to protect it. In the context of land-use policy, however, acting locally often means that environmentally risky activities are curtailed in one locality only to be transferred to another. Depending on where these activities shift, a net increase in environmental damage could result. Such an increase might be the unforeseen consequence of restrictions on the volume of timber that can be harvested in western North America.

These harvest restrictions, which also are being authorized in British Columbia, already have been imposed on both federal lands and private lands in the U.S. West (especially Oregon and Washington). In the case of federal lands, the restrictions are the outgrowth of environmental concerns, most notably those over the spotted owl. In the case of private lands, they have resulted from a general tightening of various western states' forest practice acts.

Even as timber harvest restrictions help to allay some environmental concerns in western North America, they should arouse similar concerns in other parts of the world. This stems inevitably from the response of the world timber market to timber supply reductions: decreases in the timber harvests of one region spur increases in the harvests of other regions. By identifying the location of these increases, we can begin to determine whether global environmental damage associated with logging will be greater than before regional harvest restrictions were imposed.

Originally published in *Resources*, No. 117, Fall 1994.

Using a timber-supply model (TSM) developed at Resources for the Future in 1990, my colleagues and I have assessed where logging is likely to increase as a result of timber-harvest decreases in western North America. Below, I identify these regions and explain why the severity of environmental damage from logging depends significantly on where that activity occurs. In addition, I make some preliminary speculations about net changes in such damage in those regions where logging is potentially on the rise. Taken together, these regional damages can begin to indicate whether a net increase in global environmental damage will result from a regional restructuring of timber production. Finally, I make several suggestions regarding policies that address the environmental concerns associated with timber harvests.

Predicting Changes in the Timber Market

Because western North America is one of the world's largest producers and exporters of timber, major logging restrictions in that region could be expected to reduce significantly the volume of timber sold on the world market. However, the timber market typically adapts to such cutbacks. Consequently, reductions in the timber harvests of the U.S. West, which began in the early 1990s, now are being offset by increases in the harvests of other regions. To pave the way for an estimation of any net change in logging damage worldwide, we used our timber supply model to identify the regions where harvesting would increase.

The TSM projects timber production in response to changes in overall timber demand over the fifty-year period, 1990–2040. Its estimates of harvests are based on the assumption that timber-producing regions fall into one of two categories: those that are expected to be responsive to supply and demand forces and those that are not. The responsive regions are the U.S. South, the U.S. West, British Columbia, eastern Canada, the Nordic countries, the Asia-Pacific countries, and the emerging plantation region, which includes New Zealand, Chile, Brazil, and other major producers

of wood grown on plantations. The *nonresponsive* regions, which are assumed to be increasing their timber production slowly over time in accordance with historical trends, are the former Soviet Union, Europe (excluding the Nordic countries), and all other timber-producing regions of the world. Each of these two groups of regions accounted for about half the world's industrial wood production in the mid-1980s.

In the late 1980s, when we first ran our model to generate a fifty-year timber production forecast, the U.S. West had not yet reduced its timber harvests. In light of its subsequent harvest reductions and the reductions expected in British Columbia, we have revised our earlier forecast. To do so, we decreased the area and inventory of timber available for harvest in each region as originally specified in our model. Specifically, we decreased available inventory levels by 30 percent in the U.S. West and by 20 percent in British Columbia.

In our revised forecast, the level of timber harvests in the U.S. West and British Columbia is lower throughout the entire fifty-year forecast period than in our original base case scenario, and the average real (inflation-adjusted) price of timber is about 5 percent higher. During the first twenty years of this period (1990–2010), the principal focus of the analysis, our revised projections of the average annual volume of harvest in each of the seven responsive regions indicate that the decline in U.S. West timber harvests will be largely offset by harvest increases both in the United States and abroad.

Location of Increased Timber Harvests

Our revised projections suggest that the global timber-supply system can produce large volumes of wood in response to the incentive of higher prices brought about by harvest reductions. These higher prices are predicted to increase timber production in the Nordic region, the U.S. South, eastern Canada, the emerging plantation region, and other timber-producing regions. In turn, this increased production is predicted to replace about two-thirds of the harvest shortfalls created by harvest reductions in western North America. These forecasts are corroborated by recent experience.

Early in 1993, timber prices in the United States approximately doubled in a period of less than six months. During that period, rising wood prices around the world led to increases in timber harvests in the U.S. South and elsewhere. The upsurge in log prices was volatile, however, and fell rapidly after the initial rise, before rising once again. Today, timber prices have declined substantially from their peak levels, although they have yet to drop to their pre-1993 levels.

While prices were increasing in many other timber markets, they changed much less in the European market. The soft European prices, together with devaluations in the currencies of the Nordic countries, reduced the competitiveness of many North American timber producers, forcing them to curtail their activity in the European market. These producers redirected their production to the North American market. Thus, eastern Canada and the U.S. South, both of which had increased their timber production in response to rising wood prices, have been able to offset much of the reduction in timber harvests in the U.S. West. Likewise, the 50 percent decline in the U.S. West's wood exports, which resulted from the reduction in the U.S. West's timber harvests, has been offset by yet other regions. New Zealand, Chile, and Russia have filled most of the gap left by the decrease in U.S. West timber exports to the Pacific Basin.

This restructuring of the timber market indicates that the market has adapted well to the harvest reduction in the U.S. West. As British Columbia also reduces its timber harvests, the Nordic countries, eastern Canada, the U.S. South, and the currently major forest plantation regions will be joined by other regions in increasing their timber harvests. Notable among these other regions are Latin America, parts of Asia and Oceania, and Europe.

Latin America is likely to be a major wood supplier during the next century because it has established highly productive plantation forests. Brazil has assumed a major role in the production and export of wood pulp over the past decade or so. Argentina, Venezuela, and Chile are becoming important wood producers, as well as actual or potential wood exporters.

Plantation forests are not the only source of timber in Latin America. The vast timber resources of the Amazon are also potentially exploitable. Traditionally, wood exports from the Amazon have been modest, due in large part to the high degree of heterogeneity in the region's timber species and the inability of markets to utilize effectively lesser known species. These obstacles are being overcome, and tropical timbers are being used increasingly. Given limitations on supplies of tropical timbers from Asia, increased timber exports from the Amazon are anticipated. Nevertheless, environmental concerns might severely limit the volume of timber produced from the Amazonian native forest.

Like some countries in Latin America, several countries in Asia and Oceania may become bigger timber exporters in the near future. New Zealand, Vietnam, and Myanmar have increased their timber exports in recent years, a trend that is expected to continue. In Malaysia and Indonesia, timber from plantations and second-growth tropical forests could be for sale in major world markets within a decade.

Europe is already a major wood-producing region and is likely to remain so. Because its forests and wood production potential are expanding substantially, it could increase its timber harvests in the event that timber supplies become tight. The Nordic countries have already done so.

One European country with significant potential for increased wood production is Russia, whose timber exports have been declining since the mid-1980s. The question is whether Russia, the world's second largest producer of industrial wood, can recover as a major wood exporter. While opinions vary, the level of recent Russian wood exports to Japan offers evidence that it can. Russian wood exports rose 22 percent in 1993 and are anticipated to increase again this year. The future of these exports might be expected to depend in part on the advent of a reasonably orderly political process in that country. But given its vast timber inventories, Russia may not require democracy or even market capitalism for commercial exploitation of its timber resources. Ready markets, especially in the Far East, provide incentives for significantly expanded development of these resources under a variety of social systems.

Environmental Effects of Relocating Logging

As suggested above, in a world where wood products are heavily traded internationally, logging restrictions in one region will simply be offset by logging increases elsewhere. The issue, then, is not *whether* to log but *where* to log. Moreover, even if logging were to decline worldwide, the environmental consequences would not be altogether positive.

The issue of where to log is important because the environmental damage associated with logging may vary considerably from location to location. For example, damage that results from tree extraction (such as soil erosion) is greater on steep terrain than on flat terrain. Damage to old-growth and other unique forests, which are often highly prized for their preservation values, can be considered more serious than damage to either second-growth or plantation forests. Thus, the global environmental damage associated with logging can increase or decrease, depending on where the logging occurs.

Yet it would be a mistake to assume that net changes in environmental damage can be calculated simply by adding up damage in each locality where logging occurs. In assessing these changes, other factors must be taken into account, including the size of any particular type of forest being logged relative to the total area of forests of the same type. If the damage to a harvested forest is severe but the total area of that type of forest is large, the marginal damage to local and global biodiversity is likely to be modest. By contrast, if the damage to a harvested forest is modest but the total area of that type of forest is small, the marginal damage to local and global biodiversity could be large. As these considerations suggest, the damage associated with logging is not limited to the areas where timber is actually harvested.

Nor is logging damage necessarily the direct result of timber harvests. If timber production were reduced significantly worldwide, the consequent decline in timber availability would likely promote the substitution of other materials for wood. Although such substitution may appear to be environmentally desirable, it is not an unmixed blessing.

Most, if not all, alternative materials create their own serious environmental problems. For example, metals, cement, and other substitute materials are obtained through potentially environmentally damaging mining or quarrying activities. In addition, most substitute products require considerably more energy to produce than wood products. Increased use of fossil-fuel energy raises the level of carbon dioxide in the atmosphere, contributing to global climate change. Finally, few wood substitutes are renewable, recyclable, and biodegradable.

Environmental Effects of Timber Reductions in Western North America

The magnitude and nature of the global environmental effects of harvest reductions in western North America will depend significantly, but not solely, on the location of offsetting harvest increases. Assessing these global effects will require additional research, but the predictions of the TSM enable me to speculate about net changes in regional environmental damage. Such speculation is a starting point for determining whether the harvest reductions in western North America will lead to a net change in global environmental damage.

As noted above, the TSM predicts that the harvest reductions in western North America will trigger harvest increases in parts of Europe (notably the Nordic countries and probably Russia), parts of Asia and Latin America, and other parts of North America (notably the U.S. South and eastern Canada). Recent timber production and trade information suggests harvests have already increased in some of these regions. A consideration of the natural features of the forested area of three of the regions—the Nordic countries, the South American tropics, and eastern Russia—illustrates how increased logging could affect the severity of local logging-related environmental damage.

Increased harvests of the forests in the Nordic countries may generate only modest additional environmental damages. Logging in these forests does not cause serious erosion and water runoff problems because the forested terrain is generally flat. Since few of the forests contain old-growth timber, the loss of preservation value resulting from logging is negligible. Therefore, a sizable, but not huge, increase in harvest levels probably poses little additional risk to biodiversity.

Increased timber harvests in South America may involve either logging old-growth timber or expanding plantation forests. While the risk to biodiversity is great where old-growth habitat is destroyed, the risk to native habitat from plantation forests can be small. Contrary to popular impression, plantation forests are usually established on degraded agricultural lands, rather than on land cleared of native forests. Accordingly, the environmental effects of plantation expansion are usually negligible. Selection logging in tropical forests, in which only a few trees are harvested per hectare, could lessen damage, particularly if road building is minimized and if large areas of fairly inaccessible forest remain largely undisturbed. These precautions could be especially important in preventing erosion, although this problem is likely to be a small one in the Amazon, much of which is flat.

The environmental effects of increased logging are more difficult to assess in eastern Russia than in South America or the Nordic countries. Several natural features of the forests in eastern Russia suggest that damage resulting from logging is likely to be modest. The areas of native forest are vast, and much of the terrain is relatively flat. In addition, Russian forests, like other forests in cold climates, contain considerably less, yet more broadly distributed, biodiversity than tropical forests. However, other natural features of eastern Russia's forests suggest that logging could have serious environmental consequences. The relatively low volume of timber in many of the forests necessitates logging over

large areas. In addition, timber regeneration is difficult in many eastern Russian forests, especially in the more northerly regions. Land that remains without an adequate forest cover for a long period of time is at increased risk of susceptibility to environmental damage.

These speculations suggest the difficulty of making comparisons among different localities' logging-related environmental damage. In general, however, logging in plantation forests is likely to be the most environmentally benign, especially when these forests are established on former agricultural lands. Plantation sites are usually flat, and their volumes of old-growth timber and biodiversity are small. By contrast, logging in old-growth tropical forests is likely to be the most environmentally damaging, primarily because the biodiversity is greater in these forests than in any others.

Policy Implications

At the beginning of this essay, I referred to the slogan "think globally and act locally," and I suggested that acting locally to protect the environment sometimes could lead to a net increase in global environmental damage. This is certainly a possibility in the case of timber harvest restrictions in western North America. Because much of the damage associated with timber harvests is localized, many people presume that reducing the harvests in their own region will be environmentally beneficial. What they often do not consider is that much environmental damage is, in its essence, global. Thus the charge to think globally should be emphasized in planning any local action that affects the environment, even in a seemingly positive way.

At a minimum, policymakers should understand that a decision to protect the environment by reducing timber harvests in one region will not necessarily shield that region from the environmental effects of logging. Ultimately, new or increased timber harvests in other localities will affect the global environment. Whether the environmental effects of these harvests is positive or negative depends in large part on where the activities occur.

For this reason, national policies to address the environmental concerns associated with logging ought to follow the example of international policies to control climate change and to protect biodiversity. These international policies recognize that the most efficient way to deal with global environmental problems is to identify the regions of the world where the problems are most severe and to concentrate mitigation efforts there. With regard to logging-related damage, then, the most efficient strategy is to identify the areas where this damage is likely to be greatest and to devise incentives that discourage timber harvests in these areas. Such a strategy may even encourage timber harvesting in areas where that activity is likely to be most environmentally benign.

Suggested Reading

Sedjo, Roger A. 1995. Local Logging, Global Effects. *Journal of Forestry* 93(7) July.

Sedjo, Roger A., and Kenneth S. Lyon. 1990. *The Long-Term Adequacy of World Timber Supply.* Washington, DC: Resources for the Future.

25

Twenty Years after the Energy Crisis
What Lessons Were Learned?

Douglas R. Bohi and Joel Darmstadter

Last winter was the twentieth anniversary of the first of two oil price shocks, which had many Americans worrying about whether they could afford to fill up their cars and many analysts fearing severe consequences for the economy. At the time, the so-called energy "crisis" was blamed on the actions of the Organization of Petroleum Exporting Countries, and many experts believed that oil import independence was a crucial U.S. goal. It turns out that U.S. government policies facilitated and aggravated the crisis and that the independence goal is, for many reasons, unrealistic. These are but a few of the lessons that make up the legacy of the energy crisis for today's policymakers.

Twenty years ago, some motorists acquired the foresight to bring along reading matter while sweating out the inconvenience of long gasoline lines. Those Americans would have gotten the impression from their morning papers that they were the victims of a successful effort by the Organization of Petroleum Exporting Countries (OPEC) to dictate the price and supply of world oil. During the winter of 1973–74, a quadrupling of the world oil price encouraged the belief that a new OPEC-dominated era had dawned, with profound implications for oil, energy, and economic well-being. A renewed escalation of oil prices in the wake of the 1979 Iranian revolution reinforced that belief.

Although oil-related economic and geopolitical concerns can never be totally dismissed, a closer look at what occurred in the 1970s and at what has happened since serves to correct common misimpressions about the causes and consequences of the "energy crisis." As we mark the twentieth anniversary of the gasoline lines that symbolized that crisis, the world oil market seems calm. World oil prices, adjusted for overall inflation, are today but a fraction of what they were expected to be. Still, energy experts note that the proportion of imports in U.S. oil consumption is near its historic 50 percent share. Policy analysts are again asking how important it is for the United States to limit its dependence on oil imports, whether we can do more to use energy more efficiently, and how difficult it will be to manage environmental concerns associated with energy use.

Originally published in *Resources*, No. 116, Summer 1994.

To probe just how much has been learned from the energy crisis, a symposium at the University of Tennessee in April of this year considered the subject, "Twenty Years after the Energy Shock—How Far Have We Come? Where Are We Headed?" At the symposium, we presented a paper that addressed the topic, "The Energy Upheavals of the 1970s: Socioeconomic Watershed or Aberration?" Here we present our findings organized around five broad questions.

What Did the Energy Crisis Teach Us about the Strengths and Weaknesses of Government Intervention in Energy Markets?

The short answer: U.S. government policies facilitated and aggravated the energy upheavals of the 1970s. The effects of these policies were far greater than those of the Arab oil producers' limited 1973–74 production cutbacks and of their embargo. Among the government's counterproductive policies, three are especially worth recalling. First, price controls, which lessened incentives to find and produce natural gas, impeded a shift away from oil—this at a time when oil demand had been rising rapidly. Second, oil price and allocation controls, introduced by the Nixon administration in the early 1970s as part of a broader anti-inflation program of price and wage controls, had the effect of channeling U.S. oil demand into greater imports, rather than advancing the goal of reducing imports. This policy contributed, in due course, to abandonment of the mandatory oil import quota program, begun in 1959.

Probably the most misguided intervention during the oil price shock, however, was the entitlements program, under which refiners with access to cheap, price-controlled domestic oil in effect subsidized refiners dependent on costly imported oil. The resulting averaging of imported-oil and domestic-oil prices could not, of course, contain overall price increases as the oil import share rose, but it kept those prices below their unregulated level. The perverse result: Domestic consumption was encouraged, production discouraged. In the course of

one year, the United States switched from officially restricting to effectively subsidizing oil imports—an ironic twist to the then-popular view that we were in the grip of a cartel with a demonstrated capacity and will to wreak havoc on the international economy.

Although less directly interventionist, a whole series of government programs came into being. Some—notably an attempt to establish a synthetic fuels industry—were destined to collapse quickly, though not without some hefty bail-out from taxpayers. Other efforts continue to this day.

We note here only a few of those programmatic initiatives. In 1975, Congress enacted legislation mandating automotive CAFE (corporate average fuel efficiency) standards, continued oil price controls, and created the Strategic Petroleum Reserve. In 1978, it enacted the Public Utility Regulatory Policies Act (PURPA)—which sought to promote innovative resource and technology applications in electricity generation, as governed by avoided-cost criteria—and the Energy Conservation Policy Act (ECPA), which required utilities to provide conservation services and introduced mandatory equipment efficiency standards. The Natural Gas Policy Act of 1978 provided for phased decontrol of wellhead gas prices. Separate legislation committed the nation to decontrol of oil prices. In 1980, the Energy Security Act created the short-lived Synthetic Fuels Corporation (SFC). And so on . . .

With benefit of hindsight, it's easy to critique some of these efforts (like the hopelessly unrealistic synfuel production targets and poor management of the SFC). But other programs deserve a more tempered judgment. To this day, for example, there are respectable arguments over the respective influence of obligatory CAFE standards versus market forces in bringing about the impressive automotive fuel economy gains of the last twenty years. PURPA's directional nudge to much more competitive electricity generation was surely beneficial, notwithstanding some economic distortions occurring in the start-up years. And, of course, oil and gas price decontrol proved critically important.

Before long, the oil price shocks of the 1970s began to produce fundamentally changed views regarding the consequences of an interventionist government and, conversely, the value of unimpeded energy markets—a point reverted to below. While the initial response to the oil crisis, following a well-established tradition, was more government intrusion, within several years it became clear that government regulation would not only fail to extricate us from our problems; it was responsible for actually worsening the crisis. And so, by the late 1970s, even an interventionist-prone Congress began to realize that existing trends had to be reversed and that only by moving toward less regulated markets could the prevailing encouragement of economic inefficiency be reduced. That process continues to this day, notably in the case of electric and gas utilities.

What Insights Did the Energy Crisis Provide about the World Oil Market and, by Extension, Other Resource Markets?

Early in 1994, oil was trading at around $15 per barrel. In real price terms, then, oil is back to where it was prior to the first oil shock. At the same time, OPEC's share of world oil production is markedly down from its mid-1970s high, and energy sources other than oil make up a significantly increased proportion of the global energy mix. In the United States, per capita energy consumption is below its 1973 level. These facts reflect the substantial flexibility with which both supply and demand forces can adapt to changing market conditions.

Yet recall that, amid the concern sparked by the Arab oil embargo and quadrupling of oil prices in the winter of 1973–74, smart people saw OPEC as the likely harbinger of a string of successful commodity cartels around the world. In testimony before Congress in early 1974, Fred Bergsten, then at the Brookings Institution, said, "There can now be no doubt that a large number of primary producing countries will be making steady, determined, and often concerted efforts to raise substantially their returns from a wide range of

commodities which they produce. . . through the formation of new OPECs. . . [and] many of them are in an excellent position to do so." And in what seemed to signal a reversal of this country's general aversion to international commodity price agreements, in 1975 U.S. Secretary of State Henry Kissinger indicated a willingness to contemplate such agreements, at least on a case-by-case basis.

As it turned out, over the last two decades, adjustments in oil specifically and in energy generally have conformed to what our understanding of energy markets should have told us would have happened. As already noted, the energy mix shifted away from oil, particularly outside the United States. Demand slowed, and energy began to be used more economically. Incentives created by the new price realities favored exploration in and new supplies from non–OPEC oil sources (see Figure 1). And, in the face of these pressures from both the demand and competitive supply sides, OPEC's ability to make its members respect allotted market shares within a shrinking pie began to weaken and then falter.

Amid all this, the introduction of market instruments long utilized in other commodity markets (for example, futures and spot markets) and new business strategies (for example, dual-fuel capability by utilities) would contribute to both forestalling disruptions and cushioning the effect of those that occurred.

On a somewhat more subtle level, the past twenty years have also made clear the futility of our trying to insulate the United States from the instability of the world oil market. Back in 1974, President Nixon's Project Independence envisaged complete self-sufficiency as a viable American objective. We have since learned that the domestic economy cannot be shielded from events in the world oil market, regardless of how much oil we import. Domestic oil prices are determined by world oil prices.

We have also learned that the United States cannot influence the world oil market without taking into account the actions of the rest of the world. An increase in oil demand or oil supply, whether it originates in the United States or elsewhere, has the

Figure 1. Major oil-producing regions' percentage shares of world output (1960–92).

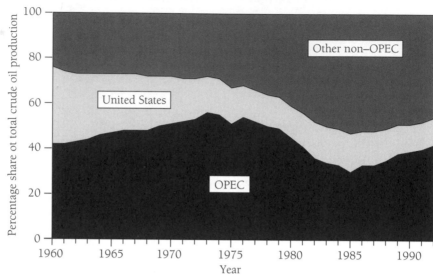

Note: Crude oil production excludes natural-gas plant liquids.
Source: U.S. Department of Energy (1993).

same effect on that market. This interdependent feature of the world oil market also means that policies implemented by the United States alone will have limited effects on the world oil market and could be offset by the actions of other countries. What has long been true for other commodities is now recognized as true for oil.

How Worried Should We Be about Oil Import Dependence?

With falling world oil prices since the mid-1980s raising the imported share of U.S. oil consumption, energy dependence—and the bearing which such dependence may have on energy insecurity and economic vulnerability—remains, for many policymakers and special interests, a charged issue. (Note the successful recent petition to the U.S. Department of Commerce by the Independent Petroleum Association of America for an examination of the security implications of oil import dependence. A finding that a threat to national security exists could be used to invoke protectionist measures.)

At issue are, first, the likelihood of major oil price shocks to oil-importing countries—whether purposefully or accidentally triggered—and, second, the economic consequences of such disruptions. In the spirit of camaraderie that prevailed among oil-importing countries for a few years (partly through the existence of the International Energy Agency, which was created in 1974 to coordinate the energy policies of Western countries), there was hope of being able to stave off such impacts through the use of strategic stockpiles and coordinated demand restraint measures. In fact, that joint strategy never blossomed—in part because the OPEC threat is seen as having receded. Among the initiatives of individual countries, the most visible defensive measure continues to be the U.S. Strategic Petroleum Reserve (SPR), now amounting to nearly 600 million barrels. The evolution of the SPR as an important, but limited, thrust of U.S. energy policy was the early recognition that only at an intolerably high cost could the self-sufficiency preached by Nixon be approached, much less achieved.

Regarding the economic consequences of major oil price shocks, it is worth underscoring two points elaborated by Douglas Bohi in *Energy Price Shocks and Macroeconomic Performance* (RFF, 1989). First, empirically, the economic damage through lost national output and inflation accompanying the second oil price shock in 1979–80 was uncorrelated with the degree of oil import dependence. And second, conceptually, if what matters is the price of oil—the domestic price of which is determined by the world price—then reducing imports would not alone improve energy security. That recognition shifts the burden of oil import policy to stabilizing the world price of oil during crisis situations, and, to this end, the SPR can be said to offer a sort of backstop strategy, although it begs the question of what stockpile magnitude is justified on cost-benefit terms.

A more basic and unresolved question is whether the vulnerability of the economy to energy price shocks is really as great as some interpretations of the events of the 1970s would have us believe. What is less debatable is that we cannot go wrong by doing what we can to increase the elasticity of energy supply and demand. For example, we can encourage technologies that extend the range of energy options. In addition, through environmentally justified policies (such as, arguably, a higher gasoline tax), we can lower the energy intensity of the economy.

But emotions and myths continue to influence the issue of how best to protect the country from energy shocks. One example: oblivious to the price controls and allocation schemes responsible for the gasoline lines in the 1970s and to our limited capacity to influence the world price of oil, Senator Patty Murray (of Washington) recently voiced her support for an extension of the ban on Prudhoe Bay oil exports, observing that "Alaska North Slope oil provides an insurance policy to consumers on the West Coast that the giant gasoline lines of yesterday will not reappear because of the irrational acts of some Middle East despot or a group of crazed religious zealots" (Environmental and Energy Study Institute's *Weekly Bulletin*, March 14, 1994).

What Did the Energy Crisis Reveal about Our Ability to Analyze, Model, and Project Energy Developments?

Again with the benefit of hindsight, it's easy to point to misjudgments made fifteen or twenty years ago, and not just in the policymaking arena. Academics and business planners—in whose judgments policymakers presumably placed some confidence—also turned in a pretty spotty performance. An example in the electric utility sector was the costly failure to perceive the dramatic reduction in electricity demand brought about by higher prices (see Figure 2). Construction programs based on historic growth rates, and often sanctioned by regulatory commissions, soured as excess capacity, higher interest rates, and (especially in the case of nuclear plants) cost overruns all took their financial toll, at which point many regulators blamed the utilities for imprudent planning.

A key analytical problem was that econometric studies of electricity or, say, gasoline demand provided little historical empirical basis for judging elasticities: the demand—or, for that matter, supply—response to sharply higher prices. In an interesting case of asymmetry, ten-year projections made by the U.S. Department of Energy in 1975 wound up overstating energy consumption by about 25 percent, not because of faulty economic growth assumptions, but because of flawed elasticity measures. At the same time, the projection for the U.S. oil supply led to a 35 percent overstatement. But then economists have always had trouble forecasting oil supply, which after all is predicated not merely on firms' strategic behavior but on technological and geological success as well.

Forecasting failures and scorekeeping aside, the general experience of the last twenty years reminds us that society does respond rationally to economic incentives. People alter the way they consume energy; firms invest in new technology; and new institutions arise that inject greater efficiency into world energy transactions. In other words, what is taught in Economics 101 tends largely to be true. While these lessons are no guarantee against

Figure 2. Actual and projected U.S. electricity consumption (1960–92).

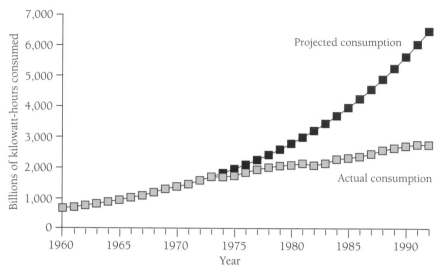

Note: Projected electricity consumption for the period 1973–90 was extrapolated from actual consumption during the period 1960–72.
Source: U.S. Department of Energy (1993).

future energy shocks, the more reliable analytical insights and stronger empirical base that we now have should help us, at the very least, to avoid doing harm and, at best, define more judicious policy choices than those we embraced in the past.

How Have Oil and Related Energy Shocks Altered the Way We Think about the Environmental and Other Social Consequences of Energy?

It would be wrong to ascribe to the energy upheavals of the 1970s all the credit for the way in which our consciousness has been raised on the broader social impacts and ramifications of energy—namely, environmental and public health threats, resource scarcity, and sustainability.

Those issues had drawn visible attention prior to 1973 in the academic, ideological, and policy arenas. While some expressions of concern had an alarmist edge, there were also more restrained efforts to consider the possible dilemma of having to trade environmental integrity for economic growth and resource demands. The decade preceding the first oil shock also saw enactment of important statutes directed at health, safety, and environment—for example, the Occupational Safety and Health Act and the Clean Air Act. And plenty of examples in the economic literature can be cited that argued for socially efficient means to force polluters to bear the cost of their assaults on common-property resources.

But clearly the energy shocks of the 1970s had the catalytic effect of elevating these issues to a much more prominent plane, in part because alternatives to oil seemed especially vulnerable on environmental grounds. Mining coal was dangerous, and the use of coal released unwelcome combustion products. Synfuels posed a threat of major land disturbance and water contamination. Nuclear power, chronically confronted with questions of radioactive waste management and proliferation, also had to contend with safety concerns.

These unpalatable alternatives lent substantial credence to those who saw—and continue to this day to see—aggressive attention to conservation as the principal route out of the quandary. Many individuals and groups have sparred over that point, sometimes with only casual fidelity to underlying facts. For example, differences among countries in energy/gross domestic product ratios tended to be almost reflexively equated with differences in successful conservation practices (and therefore capable of being emulated by the wastrels), rather than being seen, at least to a considerable extent, as

reflections of differences in industrial structure, housing patterns, and many other factors responsible for variations in aggregate energy intensity among countries.

Closely related to the conservation debate was the more legitimate question about the extent to which market imperfections led to price signals inherently favoring supply options rather than conservation options and, moreover, supply options that (in the view of some) were tilted toward unwieldy "hard-path" facilities favored by technological traditionalists. In that view, planners who are conditioned to equate electricity expansion needs with large fossil-fueled power plants would not be alert, say, to "soft-path" solar energy options.

Debate on these issues, though not stilled, has surely become less contentious. Analysts tend to the view that energy systems do pose social impacts that betray market failures, but that efficiency, flexibility, and the overall interests of the community are best served by using market-like or economic instruments to achieve desirable outcomes. These outcomes might be limits on sulfur dioxide releases, which actually have been put into effect, or, for example, congestion pricing of rush-hour automotive commuting, which has not.

Update

Since our observations in 1994, the diminished ability of OPEC to allocate and control oil production demonstrably has weakened still further. During the last four years, world oil prices have declined another 60 percent in real terms—reducing them to a level significantly below where they stood *prior* to the 1973–74 oil disruption. Not surprisingly, this development has meant rising U.S. oil-import dependence—a state of affairs whose implications, as argued in the article, need not be nearly as threatening as it is sometimes judged to be. For example, some people view our considerable military presence in the Mediterranean and Persian Gulf as an economic burden necessitated largely by oil-import dependence. They forget that much of this country's defense posture is governed by the pursuit of wide-ranging strategic goals—irrespective of whether we import 30 or 60 percent of our oil requirements.

In general, the momentum driving energy market reforms and their attendant economic efficiencies has continued since 1994. But there is an enduring challenge in dealing with the unpriced environmental costs accompanying current and increased future energy use.

Conclusion

Did the oil price shocks of the 1970s constitute some kind of watershed or defining moment in our understanding of and ability to deal more rationally with energy upheavals, and more broadly, with the larger resource and environmental issues of which they are a part? Clearly, characterizations like "watershed" or "defining moment" are overly theatrical labels for the 1970s, which did not usher in an era in which cartels manipulated petroleum and everything else from Brazil nuts to bauxite. In a couple of respects, the 1970s did represent a significant benchmark: a sobering lesson on the misplaced confidence in the effectiveness of government intervention and, conversely, an appreciation (or maybe rediscovery) that markets work and that energy is not wholly different from other economic necessities bought and sold in the marketplace. At the same time, notwithstanding the sometimes diversionary and hyperbolic preoccupation with doomsday scenarios, we have developed a heightened consciousness about the prevailing and long-term social impacts of energy that we must continue facing up to. And that is welcome.

Suggested Reading

Feldman, David L., ed. 1996. *The Energy Crisis: Unresolved Issues and Enduring Legacies* Baltimore: Johns Hopkins University Press. (That volume contains a longer version of this chapter as well as contributions with diverse views on the significance and long-term effects of the energy market upheavals on the 1970s.)

Gordon, Richard L. 1997. Book review. *The Energy Journal* 18(3): 142–46. (This review is a critical evaluation of the Feldman book.)

Part 6

Biodiversity

26 Preserving Biodiversity as a Resource

Roger A. Sedjo

Wild plants and animals can provide natural chemicals and compounds for producing drugs and other products, information and ideas for developing synthetic chemicals and compounds, and genes for engineering plants and animals with desirable sets of traits. Despite their value, wild species are threatened by destruction of natural habitats. Because there are no property rights to wild species or the genetic resources embodied in them, habitat protection tends to be undervalued, particularly in developing countries. However, contractual arrangements that allow these countries to trade the right to collection of their wild genetic resources in return for compensation could foster habitat protection in the absence of such property rights.

The rationale for the preservation of the world's biodiversity runs from the highly spiritual to the pragmatic. On the spiritual side is the growing feeling among some groups that wholesale disturbances of natural systems are somehow unethical or immoral. On the pragmatic side, it is well recognized that the genetic constituents of plants and animals have substantial social and economic value from which all members of the global community may potentially benefit. Genetic information provides direct and indirect inputs for plant breeding programs, development of natural products (including pharmaceuticals and drugs), and increasingly sophisticated applications of biotechnology. The substantial increase in world agricultural output since the early 1970s has been due primarily to the ability of plant breeders to develop high-yielding varieties of the various food and feed grains by utilizing genes drawn from often overlooked plant species. More recently, recognition of the potential of wild genetic resources in development of drugs has led the National Cancer Institute to initiate a massive plant collection project that seeks to identify plants with chemical constituents effective against a variety of cancers. In recent years a number of widely used drugs have been developed from plants, including two important anti-cancer drugs derived from the now well-known rosy periwinkle found in tropical Madagascar.

Originally published in *Resources*, No. 106, Winter 1992.

Making Use of Wild Species

The benefits of using wild plants (or animals) as a resource may be obtained in three general ways. First, a species—or its phenotype, the individual plant or animal—can be consumed directly or it can be a direct source of natural chemicals and compounds used in the production of "natural" drugs and other natural products. Second, a species' natural chemicals can provide information and ideas—a blueprint—indicating unique ways to develop useful synthetic chemicals and compounds. For example, aspirin, an early synthesized drug, is a modification of the natural chemical salicylic acid (found in plants), which is too strong to be taken orally. And third, a wild species can be the source of a gene or set of genes with desired genetic traits that can be utilized in breeding or in newly developed biotechnological techniques. For example, germplasm from wild species is used to maintain the vitality of many important food crops. The latter two utilizations are essentially nonconsumptive, employing the genotype—the characteristics embodied in the genetic constituents of plant and animal species—as a source of information.

One recently publicized example of a useful natural chemical is taxol—a promising anti-cancer compound occurring naturally in the Pacific yew tree found in western North America. In 1985 taxol was found to shrink tumors in many ovarian cancer patients. In addition, its unique anti-tumor properties have been demonstrated in about 50 percent of advanced breast cancer patients treated with the drug. In two recent studies taxol has proved successful in treating tumors that had not responded to conventional treatments such as chemotherapy. It is the first and, to date, only member of a new class of anti-tumor compounds whose unique mechanism of action is distinct from the action of any currently used cytotoxic agent.

The current process for extracting taxol—peeling the bark of the yew—destroys the trees involved. It is anticipated that naturally occurring yews will provide most of the taxol through the mid-1990s, after which other sources will gradually be developed. These could include the conversion of compounds similar to taxol into taxol, the generation of taxol from plant tissue cultures, and biosynthesis. Synthetic production of taxol may also be possible, although this could be difficult due to the complexity of the compound.

With recent breakthroughs in biotechnology, the potential for development of useful products from wild plant and animal species would appear to be limitless. Species that have no current commercial application, contain no useful natural chemicals, or are as yet undiscovered, nevertheless may have substantial value as repositories of genetic information that may someday be discovered and exploited. The ability of modem biotechnology to transfer genes to unrelated natural organisms opens the possibility for the development of a wide variety of engineered plants and animals with hitherto unattainable sets of traits. As biotechnology develops, the scope for utilization of genetic information embodied in wild plants and animals will almost surely increase. Moreover, the ability to utilize the information from different organisms is likely to increase as genetic engineering expertise grows. The benefits of sustaining a rich and diverse biosystem are likely to be large since technology and natural genetic information may well complement each other in economic activity.

Loss of Genetic Resources

Despite the acknowledged social value of sustaining wild plants and animals, destruction of natural habitats in which they are found is widespread, posing a serious threat to genetic resources. Species with potentially useful characteristics for biotechnological innovations may be lost through tropical deforestation, for example. It has been estimated that 70 percent of the 3,000 plant species known to have anti-cancer properties are found in tropical forests. Considerable criticism has been directed at Third World countries with large areas of tropical forest for not protecting and properly appreciating the values of their native forests, particularly the values of biological diversity.

If preservation were without cost, then all genetic resources would be preserved. However, as the pressures on natural habitats rise due to alternative uses for the land, such as cropping or grazing, the costs of protection and preservation also rise. In earlier periods of human existence preservation of genetic resources was essentially costless. Recently, in situ and ex situ approaches have been used to protect the acknowledged values of genetic resources. The in situ approach involves protection of species in their natural habitats, whereas the ex situ approach involves protecting plants and animals in permanent collections such as zoos and botanical gardens, and preserving seeds and other genetic material in controlled environments such as germplasm banks. Although the ex situ approach has the advantage of lower costs, it is feasible for only a small fraction of species. This approach obviously cannot be used for species as yet unknown. Furthermore, the ex situ approach preserves selected species, not ecosystems, and thus risks the longer-term loss of species that are reliant upon the symbiotic relationships within ecosystems.

Although the destruction of a unique genetic resource base can occur from the consumptive use of a particular plant or animal itself, in practice a much more ominous threat comes from the process of land-use change. Land-use changes that destroy existing habitat and individual phenotypes can inadvertently drive to extinction potentially valuable genotypes, many as yet undiscovered, that are endemic to certain ecological niches.

Sustaining and Preserving Wild Genetic Resources

One way to view conceptually the problem of sustaining wild genetic resources is to think of these resources as a lottery containing a vast number of genetic "tickets," each with a different potential payoff. The timing and size of their economic returns vary greatly. Some of these tickets are currently generating payoffs. Others could or might generate future payoffs if the habitat is preserved long enough to allow their discovery and develop-

ment. Still others would have to await further biotechnological developments before their potential returns could be realized. Although most of the lottery tickets will ultimately provide no payoff in terms of new chemicals, compounds, or transferable genes, a few will eventually result in substantial payoffs—jackpots—in the sense that these genetic resources will eventually generate large social benefits. However, it is difficult to differentiate in advance between those with significant potential future value and those with none.

Today, no ownership of the genetic lottery tickets exists. Individuals and countries, having no unique claim to the returns of the genetic information embodied in the wild plants or animals on the land they are developing, will tend to ignore the potential economic value of the existing habitat. The destruction of genetic resources thus becomes an unintended consequence, an external effect, of land-use changes that destroy natural habitats.

Although the costs of investing in habitat protection and preservation can become substantial, the industrial world has argued that such investment is needed because wild genetic resources are global resources from which the development of better lines of food grains, new medicinal products, and other advances generate global benefits that accrue to inhabitants of all countries. Nevertheless, a landowner—public or private—whose land provides the habitat for a unique genetic resource has no unique claim to its benefits.

The paradox is not hard to comprehend. Most public goods lend themselves readily to investments by the national state. The state perceives itself as readily capturing the returns to goods such as defense and lighthouses. However, it is much more difficult for the state to capture the returns to a global public good such as genetic resources. There are two reasons for this. First, international law does not recognize property rights to wild species or wild genetic resource genotypes, and hence any rents associated with valuable natural genetic resources typically cannot be captured simply through domestic management of the resource, even by a national authority. Second, the tradition that natural genetic resources are the common heritage of mankind and thus should be available without restriction provides an obstacle to the introduction of barriers to the unrestricted flow of wild genetic resources out of a country.

Protecting Public Goods

One result of the lack of private or national property rights to wild genetic resources is that, to date, most efforts to preserve and protect these resources have been altruistic. Most proposals for protecting them have involved actions by governments and the international community to preserve habitat. The usual approach is for environmental groups and the governments of industrial countries to try to persuade governments of developing countries to protect habitats rich in biodiversity, such as tropical rain forests. Some progress is being made—for example, in maintaining plant genetic resources used for breeding food and feed crops. An international system of germplasm preservation, commonly called seed banks or germplasm collections, has been developed. The collections are in both public and private ownership, with the private collections often being held by plant breeders who capture returns through the development of improved stocks to which some forms of exclusive rights exist. However, the system of collections is much less well developed for genetic resources that might have potential for drugs and pharmaceuticals than is the system for plant genetic resources used in crop breeding. In either case, collections can preserve only a small fraction of the total genetic resource base.

Progress in preserving this base is being made as individual countries, often in concert with international organizations, protect unique lands and habitats, including tropical forests, wetlands, and coral reefs. The world total of protected land doubled between 1970 and 1980 and increased another 50 percent in the first half of the 1980s. By the mid-1980s there were more than 400 million hectares of protected land (1 billion acres or 7 percent of land worldwide, excluding Antarctica), up from about 100 million hectares in 1960.

Altruism has motivated greater protection of unique lands and habitats in developed countries than in developing countries, many of which have been indifferent to seriously protecting habitat preserves and have pursued protection haphazardly at best. This situation is beginning to change as the "common property" difficulty is recognized and various attempts are made to address it. For example, the Keystone International Dialogue Series on Plant Genetic Resources (talks among a high-level group of scientists and researchers from around the world) has identified as a "gap" the failure to develop an institutional framework for dealing with issues of plant genetic resource conservation related to ownership and intellectual property right (IPR) systems for plant genetic resources. In a June 1991 workshop on property rights, biotechnology, and genetic resources, held in Nairobi as part of the preparation for the United Nations Conference on Environment and Development (UNCED), the participants reached consensus on two key points. First, it was found that, as presently practiced, the treatment of biodiversity and genetic resources as a common heritage of humankind may have the unintended effect of ultimately undermining steps to conserve the resource. Second, it was agreed that any international negotiation on intellectual property rights should ensure that countries are free to decide whether or not to adopt IPR protection for genetic resources. Given this degree of interest, it is virtually certain that property rights for plant genetic resources will be an important item on the UNCED agenda.

A Coasian Solution

Perhaps the most exciting developments in the search for vehicles to facilitate protection of genetic resources and to ensure that some portion of the benefits accrue to developing countries are changes in legal arrangements, driven in part by market forces. It was first recognized by Ronald Coase, a Nobel laureate in economics, that external social benefits can often be "internalized" or captured through the simple legal instrument of the contract if transaction costs are small. In the last few years, contractual arrangements have begun to appear that allow developing countries to capture some of the rewards associated with the development of commercial drugs and other products that utilize genetic constituents of wild genetic resources found in their countries. These contractual arrangements require no new property rights. Rather, they utilize the ordinary legal instrument of a contract to, in effect, trade the right to collection in return for a guarantee of some portion of the revenues generated by the commercial development of a product that utilizes a genetic constituent from a unique wild genetic resource collected within the country. The judicious use of contract arrangements can allow for the capture of at least some benefits without de jure property rights to the individual natural genetic resources.

Organizations are also modifying their practices to allow them to enter into contractual arrangements with tropical countries to transfer the development rights to unique wild genetic resources to institutes in developed countries. For example, the National Cancer Institute in the United States is developing transfer agreements with tropical countries that have provisions for compensation, or revenue sharing, or both.

In addition, private collector firms are beginning to enter into contractual arrangements with tropical countries to offer royalties from revenues generated by future product developments in exchange for collection rights to wild plants. The most advanced activity of this type is occurring in Costa Rica, which recently created the National Biodiversity Institute to identify all of the wild plant species in the country, undertake preliminary screening of the various natural plants, and make agreements with pharmaceutical companies for further utilization of promising plants and natural chemicals. In 1991 the institute signed an agreement with the Merck pharmaceutical firm, whereby Merck would provide $1 million over the next two years to help the institute build its plant collection operations. In return, Merck would acquire exclusive rights to screen the collection for useful plant chemicals and extracts. Indonesia is currently investigating the possibility of establishing a similar system that would allow for the capture of some portion of product benefits derived from its biological resources.

Whatever emerges from UNCED, those concerned with biodiversity will confront an extremely complex and rapidly evolving resource issue. In addition to the traditional approaches to protecting areas where biodiversity is high, innovative approaches are evolving that give promise of providing financial incentives for protecting habitat where biodiversity can be preserved and for returning some of the proceeds of the successful development of a natural-based product to the country that provided the genetic constituents. The challenge for UNCED will be to serve as a catalyst for facilitating further development of these innovations, while being careful not to advance procedures and controls that inhibit, rather than promote, such constructive processes.

Suggested Reading

Barton, John H. 1991. Patenting Life. *Scientific American* 264: 40–46.

Sedjo, Roger A. 1992. Property Rights, Genetic Resources, and Biotechnological Change. *Journal of Law and Economics* XXXV, April.

27 Biodiversity Prospecting
Shopping the Wilds Is Not the Key to Conservation

R. David Simpson

Preserving biodiversity may have little bearing on whether the next miracle drug is found. Better arguments should be stressed in developing conservation policies.

Some people say "biodiversity prospecting" offers a compelling reason to save as much as possible of the world's immense variety of genes, species, and ecosystems. Sifting among genetic and biochemical resources for something of commercial value, as biodiversity prospectors do, could lead to the discovery of a wild plant or animal that contains the key for curing AIDS, cancer—or some other disease the world has yet to identify. The desire to capitalize on new and better products for industrial, agricultural, and especially pharmaceutical applications provides strong incentives for conserving nature, so the argument goes. What's more, by acting on advances in biotechnology, researchers are better equipped than ever to investigate organisms at the genetic level, providing fresh financial reasons to conserve as many product leads as possible.

But several RFF studies show that losses in biological diversity may have little bearing on whether the next miracle drug is found. That's because there are so many wild plants and animals that can be used by researchers engaged in biodiversity prospecting. With millions and millions of species, sources of useful products are either so common as to be redundant or so rare as to make discovery unlikely. Either way, the sheer numbers involved weaken the argument that biodiversity prospecting generates any appreciable economic value.

That is not to say natural leads are not important in the development of new products. Natural organisms have evolved a staggering variety of chemical compounds to escape predators, capture prey, enhance reproductive success, and fight infection.

Originally published in *Resources*, No. 126, Winter 1997.

Some of these chemical compounds have proved to be of great value when adapted for industrial, agricultural, and pharmaceutical uses. In the United States, for instance, nearly 25 percent of prescription medicines contain active ingredients derived from plants, while many other drugs are synthesized to replicate or improve naturally produced molecules. Today we treat leukemia with medicines derived from the rosy periwinkle of Madagascar, and the bark of the Pacific yew tree is the source of a promising treatment for ovarian cancer.

It is not surprising, then, that natural scientists, legal scholars, and even economists often cite nature's contribution to new product research and development as one of the most important considerations in formulating biodiversity conservation policy. Given the passions that biodiversity and its protection arouse and the varied backgrounds of the people making proposals, however, it is also not surprising that many of the arguments are less than watertight.

Thinking clearly about the values that surround biodiversity is important, since destruction of a habitat is often irreversible. Over and above its potential as a source of new products, a plant or animal may be valuable for any number of commercial, ecological, esthetic, and ethical reasons. Given this fact, we should be cautious in making related policy decisions and choose to devote scarce funding to only the most effective strategies. Placing too much emphasis on biodiversity prospecting may divert attention—and funds—from potentially more effective conservation strategies.

The Value of "Marginal Species"

As a source of leads in new product research, natural organisms would be very difficult to replace. There is simply no substitute for biodiversity as a whole. Economically speaking, however, biodiversity is valuable to the extent that it makes sense to save a little bit more, and not to the extent of its admittedly astronomical value overall. That is because we typically are not concerned with actions that would wholly eradicate biodiversity, but rather

with the costs and benefits of actions that would result in incremental reductions. In thinking about the role that biodiversity plays in new product development, therefore, it is important to consider the contribution of biodiversity on the margin.

In economics, the worth of something is its "marginal" value; in other words, the incremental benefit that a little bit more of the thing provides. In the case of biodiversity prospecting, the value of the "marginal" species is the contribution an additional species makes to the probability that researchers find what they are looking for. Put in another way, things are valuable to the extent that there are few substitutes for them. Within the immense set of living organisms, many species are likely to be adequate substitutes for one another as leads in the development of commercial products.

This may not be apparent, however. After all, aren't different species identified as such precisely because each is genetically unique, and therefore not a perfect substitute for any other? Biologically, yes; but let's consider the economic interpretation of this fact. Each species represents a research opportunity and is a substitute for another in the sense that time and costs incurred in pursuing one research opportunity could be devoted to another. The question of economic interest is "how valuable is an additional research opportunity?" This, in turn, really boils down to the question, "how much does having additional species to test increase the probability that a new product will be found?"

Let's consider a couple of extreme cases. While species are genetically different, different species can produce the same chemical compound (caffeine, for example, is found in both tea and coffee). Suppose, then, that it is relatively likely that several species produce the same chemical compound. How helpful will it be to maintain additional biodiversity for use in the search for new products? The answer is "not very," as species are very likely to prove redundant when there are large numbers from which to choose for testing.

Now consider the opposite extreme. Suppose that it is very unlikely that two or more species among the millions in existence will prove to contain a chemical useful in the treatment of AIDS, cancer, or some other condition. But if it is very unlikely that two or more species among millions will prove redundant as sources of new product leads, it must mean that useful leads are so rare as to make it very unlikely that any species will contain the key for a cure.

Researchers at RFF have shown that, regardless of the probability that any one species chosen at random will yield a particular commercially valuable product, the value of the marginal species is negligible when there are large numbers of species available for testing. The value of the "marginal" species is equal to the expected payoff from testing it times the probability that all other species fail to provide the product that researchers seek. This figure is necessarily small when there are lots more species to choose from.

Arriving at a numerical estimate of the value of a marginal species depends on how many species are available for testing, how many new products are being sought, the financial rewards earned from developing new products, and the relative value placed on future, as opposed to current, earnings. Even if these conditions are fairly favorable, however, the estimated economic value of biodiversity for use in new product research is modest.

Incentives for Habitat Conservation

The greatest threat to biodiversity probably comes from the conversion of natural habitats, particularly those in tropical rain forests, to agricultural or residential use. Such conversions take place because those undertaking them expect to gain some benefits. Making the economic case to preserve biodiversity means showing that the benefits to be had from preservation are as good or better than those to be had by converting the habitat for other purposes.

But such a case seems hard to make. RFF research shows that pharmaceutical researchers are not willing to pay much to preserve natural habitats even in some regions that are highly imperiled and rich in biodiversity. (See Table 1.)

RFF researchers arrived at the estimated prices shown in Table 1 first by using a formula often employed by biologists to predict how species extinctions are related to habitat loss to estimate the effect of a "marginal hectare" in preserving species. That is, we considered the effect of such a plot of land in supporting and sustaining endangered species. Taking the estimated effect of an extra hectare of land, we then multiplied it by our estimate of the value of the marginal species to be sustained. This procedure yielded an estimate of only a couple of dollars per hectare for the preservation of even some of the hottest of global biodiversity "hot spots."

Even in some relatively biodiverse regions of the globe, these figures might amount to only pennies per hectare. Values on the scale shown in Table 1 might have some small impact on biodiversity preservation incentives in some of the more isolated parts of the regions indicated. Many of these habitats are, however, imperiled by metropolitan expansion; the incentives provided by biodiversity prospecting are negligible in proportion to the pressures for land conversion in these areas.

Estimates of Value and Conservation Strategies

Estimates of value on the margin are important for devising workable conservation strategies. In particular, they raise some serious doubts concerning the efficacy of two popular strategies intended to encourage the conservation of biodiversity. One of these strategies involves expanding biodiversity prospecting activities with the idea that conservation will follow. The fact that a resource may have a relatively low value on the margin does not imply that the activity in which it is used is not worthwhile: water tends to be relatively cheap, but it is essential to businesses that use it in producing important products and services. By the same token, however, investment in businesses that use water does not necessarily have much effect on the value assigned to this plentiful resource.

Similarly, increasing investment in biodiversity prospecting activities may have some socially desir-

Table 1. Pharmaceutical Company Willingness To Pay To Preserve a Hectare of Land for Biodiversity Prospecting in Eighteen "Hot Spots."

Hot spot	Value in dollars per hectare
Western Ecuador	$2.29
Southwestern Sri Lanka	1.87
New Caledonia	1.38
Madagascar	.76
Western Ghats of India	.53
Philippines	.52
Atlantic Coast Brazil	.49
Uplands of western Amazonia	.29
Tanzania	.20
Cape Floristic Province of South Africa	.18
Peninsular Malaysia	.16
Southwestern Australia	.14
Ivory Coast	.13
Northern Borneo	.11
Eastern Himalayas	.11
Colombian Choco	.08
Central Chile	.08
California Floristic Province	.02

Sources: R. David Simpson and Roger A. Sedjo, 1996, "Valuation of Biodiversity for Use in New Product Research in a Model of Sequential Search," RFF Discussion Paper 96-27; R. David Simpson, Roger A. Sedjo, and John W. Reid, 1996, "Valuing Biodiversity for Use in Pharmaceutical Research," *Journal of Political Economy* 104, 163–185; Myers, Norman, 1990, "The Biodiversity Challenge: Expanded Hot-Spots Analysis." *The Environmentalist* 10, 243–256; Myers, Norman, 1988, "Threatened Biotas: 'Hot Spots' in Tropical Forests." *The Environmentalist* 8, 187–208.

able effects—the number of new products discovered may, for example, increase—but it is unlikely to increase the value assigned to the marginal species by much. It may be a little too simplistic simply to say that "you can't get something from nothing," but attempts to increase the value added in biodiversity prospecting operations are unlikely to have any appreciable effect on conservation incentives. Biodiversity will remain plentiful with respect to the needs of new product research.

A second focus of conservation efforts has been on the establishment of property rights in biological diversity. The argument here is that when people own a resource with commercial value, they take effective measures to conserve it. Increasing incentives for preservation will be irrelevant, however, if the values generated by protection do not outweigh the costs of forgoing alternative uses of the land. The RFF estimates of value suggest that incentives for conservation would still be negligible even if property rights were perfectly well defined and the owners of a hectare of land were entitled to all benefits arising from the biodiversity prospecting conducted on it.

Prospecting and Preservation

The point of recent RFF research on biodiversity prospecting is not that diversity in nature is without value. In fact, the point is almost the opposite.

Biodiversity may be important for any number of commercial, ecological, esthetic, ethical, or even spiritual reasons. However, when it comes to commercial prospecting among natural sources for new products, the value of biodiversity is not as high as some conservationists might suppose. Since that is likely to remain the case, it is important that other, more workable, incentives for conservation be developed.

Suggested Reading

Simpson, R. David, Roger A. Sedjo, and John W. Reid. 1996. Valuing Biodiversity for Use in Pharmaceutical Research. *Journal of Political Economy* 104(February): 163–185.

Weitzman, Martin L. 1992. On Diversity. *Quarterly Journal of Economics* CVII: 363–406.

28 Contracts for Transferring Rights to Indigenous Genetic Resources

R. David Simpson and Roger A. Sedjo

Pharmaceutical companies and other organizations are prospecting for potentially valuable chemicals derived from natural organisms in tropical rain forests. Such prospecting would increase protection of these forests if the countries in which they are located were paid for the use of their genetic resources. Complex contracts may be needed for the transfer of these resources to ensure that neither buyers nor sellers will be exploited. Although most of the tasks required to commercialize genetic resources are performed by buyers, many sellers wish to conduct their own research on these resources. Their reasons for doing so must be examined carefully. Unwise investments in research capacity may lead to excessive costs, inefficient contracts, and reduced incentives to preserve irreplaceable ecosystems.

The chemicals produced by natural organisms to resist infections or repel pests might be valuable in agricultural, industrial, and, especially, pharmaceutical applications. Since of all ecosystems tropical rain forests may have the greatest variety of life, these ecosystems may yield the greatest number of chemicals that could be used in the development of new products such as pesticides and drugs. Payments for the use of genetic resources—the natural organisms from which the chemicals are taken—could aid in the development of the poor countries in which most of these forests are found. Such payments would also provide greater incentives for poor countries to preserve their rain forests. Given that these forests are disappearing at an alarming rate, this is an important consideration.

Genetic resources are unusual in one respect. As nonrival goods—that is, goods that can be used or consumed by one person without affecting the ability of another person to use or consume them—they can be exploited by any number of people. This may affect the ability of the countries in which they are first found to obtain payment for their use.

Chemicals to be used in commercial products must be manufactured in large quantities. Once it has been established that an organism is the source of a valuable chemical, it is generally more efficient to produce the chemical by some means other than harvesting the organism in its original environment. For example, the organism may be cultivated on farms outside its original habitat. Its genes might also be transplanted into other organisms, which would then produce the desired chemi-

Originally published in *Resources,* No. 109, Fall 1992.

cal. In addition, the molecular structure of the chemical can sometimes be used as a model for developing a similar synthetic chemical. Under each of these production alternatives, a person who sells the chemical or a product containing the chemical would not need to rely on the original source of the organism to acquire the chemical. Thus, if the country from whose plants and animals commercial products are developed is to reap any benefits, it must have some way of controlling access to these organisms.

Historically, genetic resources have been commercialized without any payments to the countries or other parties that originally provided them. For example, Europeans found plants such as quinine, rubber, and potatoes in the New World, but they never made payments to the peoples on whose ancestral lands these plants were grown or in whose cultures their uses were first discovered. Because the plants were regarded as products of nature, no person could claim to have created them, and hence no person could claim to deserve payment for them.

This attitude is now changing. Perhaps no one can claim to be the creator of plants or animals that will later be found to be the source of valuable chemicals, but certain people do have the power to preserve or destroy these resources. Population growth and development are threatening to ravage habitats and extinguish species at catastrophic rates. If those who have the power to destroy ecosystems rich in genetic diversity are not paid for the products that may be derived from them, they will have less incentive to preserve them.

This realization motivates in part the Biodiversity Convention offered for signature at the United Nations Conference on Environment and Development (UNCED) held in Rio de Janeiro. Although the United States has refused to sign the convention, it is likely that some of its provisions will come to be generally accepted. Among these are declarations that countries have sovereign rights in their indigenous genetic resources and that such resources cannot be used by others without the prior informed consent of the country. In essence, the Biodiversity Convention establishes that countries have property rights in their genetic resources. This is an important first step in creating economic incentives to use these resources efficiently and to preserve the areas in which they are found. However, countries wishing to commercialize their genetic resources must either develop ways in which to transfer them to foreign firms that have greater expertise in research, development, and marketing, or they must acquire such expertise themselves.

The Necessity of Contracts

Simple arrangements for the transfer of genetic resources are unlikely to work; these resources cannot simply be sold in a single, once-and-for-all transaction. This is because large amounts of raw materials from which genetic resources are obtained may be needed to conduct research to develop new products. In the development of pharmaceuticals, for example, initial tests of chemicals may require a few kilograms of sample materials, but if the tests show promise, several hundred kilograms of the materials may be required for the next round of tests. If the latter tests show promise, thousands of kilograms of the material may be needed for clinical trials. Production of commercial quantities of drugs may require millions of kilograms. Even if the drugs are to be produced from organisms cultivated on farms outside their original habitats or are eventually to be synthesized from inorganic materials, several stages of testing and large quantities of the organisms are likely to be required.

It would be impractical to collect very large quantities of organisms before any tests are conducted, however. Experts estimate that only about 1 in 10,000 natural materials sampled yields a commercial product. It would be grossly inefficient to collect many samples of materials to be tested when the probability that any one of these materials will be useful is so low. The practical implication is that a researcher testing natural materials will need to have continuing access to the source of the materials.

The need for continuing access may raise several problems that, in turn, explain the creation of complex contracts between buyers and sellers of genetic resources. The first problem is that the buyer may fear exploitation if he or she requests more materials of the type originally purchased. If the buyer makes such a request, the seller may infer that the buyer has found something useful. The seller would then want to charge the buyer more for the next batch of samples. If the buyer anticipates that the seller will behave in this way, he or she would have little incentive to begin research in the first place: if a discovery is made, the buyer knows that the seller will try to deprive him or her of the profits by increasing the price of samples. A contract in which the price of subsequent samples is specified in advance will relieve such worries.

The second problem that necessitates contracts is that destruction of tropical forests may limit or curtail the continuing availability of sample materials. Rain forests are disappearing because people in the countries where they are located perceive it to be more lucrative to chop them down than to maintain them. As long as rain forests represent a potential payoff, however, they will be preserved. This suggests that a once-and-for-all payment for the right to prospect for genetic resources is unwise. Once such a payment is made, there is no further incentive for conservation.

Of course, a contract might require a seller to take specific steps to maintain the ecosystem from which the buyer takes samples. Promises to do so may not be credible, however. It is often difficult for a buyer to discern how much effort the seller is putting into ongoing conservation activities. The buyer may not be able to tell why some prospecting activities are unsuccessful. Were the resources the buyer had hoped to find lost due to the seller's negligent conservation efforts, or did they not exist in the first place?

Given that poor performance cannot be observed directly and thus cannot be punished, a buyer would want to provide an incentive for conservation efforts rather than rely on promises. Such an incentive would be a guarantee that the seller would be rewarded if a valuable chemical is discovered. Contract terms that call for royalty payments contingent on discovery would give the seller a continuing incentive to make discoveries more probable by conserving ecosystems.

Not all of the problems that motivate contracts arise from the buyer's concerns about the seller's performance. Once the buyer has amassed enough material that he or she no longer needs to depend on the seller, the seller may worry about whether or not the buyer will fulfill his or her obligations. If the seller has accepted a contract in which he or she will be paid royalties, for example, he or she may want the contract to contain provisions for auditing the buyer to be sure he or she is not being cheated.

Vertical Integration and Contracting

Contracts are means, often imperfect means, of committing one party to perform in a way that another party desires it to perform. The problems that necessitate complex contracts would not arise if the same party were responsible for all stages of the commercialization of genetic resources. To avoid these problems, one party may attempt to vertically integrate these stages. The degree of vertical integration is the extent to which the same organization engages in the collection of wild species, the classification of these species, the testing of the chemicals they contain, the development of products containing the chemicals or synthetic variants, and, ultimately, the marketing of the products.

Complete vertical integration in the commercialization of genetic resources is unlikely. A major pharmaceutical company is not likely to incur the expense of a purchase of vast tracts of tropical forest, even if it could overcome objections to such a purchase on the grounds of national sovereignty. Nor are many developing countries where tropical forests are found likely to have the financial resources to buy a major pharmaceutical company or the technical know-how to establish one. However, countries rich in tropical rain forests are interested in partial vertical integration. They have expressed a wish to acquire the capability to undertake domestically at least some of the tasks required to produce pharmaceuticals derived from their genetic resources. Such tasks might include collection and classification of natural organisms, extraction of chemicals from the organisms, and some testing of the chemicals.

There are several reasons why a seller of genetic resources might wish to undertake part of the commercialization process. One is cost advantage. A seller may have greater knowledge about the location of raw materials and thus a better vantage point from which to direct collection activities than the buyer. In addition, he or she may have greater knowledge about which organisms may be valuable or about the uses to which the organisms may be put. It should be noted that if the seller can realize a cost advantage in performing certain tasks, it is to the advantage of both the seller and the buyer to let the seller do so. The more efficient the commercialization process is, the more profits both parties may realize. Thus, in the absence of other considerations, the party that can perform a task most efficiently should be entrusted with the task.

Another reason why a seller might wish to perform collection or other commercialization activities is to lower the cost of monitoring the performance of the buyer. Although buyers have an incentive to discover any valuable chemicals produced by the organisms with which they are supplied, they do not necessarily have an incentive to be honest about their profits from sales of these chemicals. Sellers may be compelled to monitor buyers to ensure that they receive their fair share of these profits. However, if a seller knows that one of the resources he or she sold is a promising antibiotic, for example, he or she could simply monitor the buyer's sales of antibiotics rather than monitor all of the buyer's revenues. Thus, by conducting some amount of research and testing, a seller might reduce the cost of ensuring that he or she is not cheated in royalty payments.

Yet another reason why sellers may prefer to perform commercialization tasks themselves is to improve their bargaining position. In general, sellers make more attractive deals when there is a lot of competition among buyers for their genetic resources. In the absence of such competition, a seller may offset the advantage enjoyed by a single powerful buyer by developing capabilities similar to those of the buyer.

A number of large and sophisticated pharmaceutical and chemical companies might bid for access to a particular seller's genetic resources. None of these companies is likely to have an appreciable advantage in terms of technology and general research expertise. It is possible, however, that one of the companies might have greater experience in working with natural organisms or with the types of organisms offered by the seller. In this situation, a less-well-informed company knows that if it receives the contract to commercialize the seller's genetic resources it will be because it has offered more than its better-informed rival, who, presumably, has a better idea of what the resources are worth. Thus less-well-informed bidders will bid less aggressively for contracts. The better-informed company will take this into account, and the seller can expect to receive less than he or she would have if all potential buyers had the same information. In this scenario, sellers may find it advantageous to establish their own research capacity in order to increase their knowledge about the value of their genetic resources and pass this knowledge on to buyers. A similar, albeit more complex, argument suggests that the seller would like to provide information to bidders when all have different, but not objectively better, information.

However, a seller may encounter several problems in providing this information. On one hand, buyers would anticipate the seller's incentive to make self-serving announcements; they would not believe unsubstantiated claims. On the other hand, verifiable claims—a statement, for example, that a particular plant contains a compound that will cure cancer—might be an invitation for unauthorized appropriation. Large quantities of sample materials may be required to identify a useful compound when research on the materials is starting from scratch, but much smaller lots might suffice to develop a product of proven value.

These problems, to the extent that they are in fact problems, are not insurmountable. Legal institutions may evolve to prevent unauthorized appropriation of valuable products. Contracts that emphasize royalties rather than up-front payments may make it unnecessary for sellers to provide information about the value of their products. A seller who is confident in the value of the product he or she provides should be willing to rely on royalties, and this willingness may reveal the value of the product.

Although several developing countries with tropical forests have expressed interest in acquiring relatively advanced research capabilities, their rationales for wishing to vertically integrate research activities should be examined carefully. Cost advantage may not explain this interest. If developing countries have a comparative advantage in pharmaceutical research, why have they not already been chosen to host research facilities? The argument that sellers could reduce the expense of monitoring the activities of buyers if they conducted their own research might be more relevant. While foreign research organizations are likely to have appropriate incentives to work hard in making discoveries, it may be difficult for a country that provides genetic material to collect the payments it is due. The argument that the seller would be able to strike a better bargain with would-be buyers if they conducted their own research can make sense. When one buyer dominates the market due to superior information, it may be in the seller's interest to generate competition by conducting its own research and passing the results on to all potential buyers. However, the decision to do so should be made carefully, as it may further complicate contracting.

The Evolution of Contracts

In recent years a number of organizations have entered into contracts for the commercialization of genetic resources. The National Cancer Institute (NCI) of the United States has negotiated contracts for access to genetic resources in Zimbabwe, Madagascar, Tanzania, and the Philippines. Biotics—

Update

While interest in the commercialization of genetic resources continues, enthusiasm has been tempered by more careful assessment of the values involved. In fact, R. David Simpson went on to investigate these values (see chapter 27 of this book) and arrived at very pessimistic findings, as summarized in this chapter.

---

a British firm that matches sellers of genetic resources with buyers and provides some extraction and processing services—has negotiated contracts with suppliers in Ghana, Malaysia, and New Zealand. Perhaps the most sophisticated agreement is that recently signed by Merck and Company, a leading U.S. pharmaceutical firm, and the Instituto Nacional de Biodiversidad (INBio), a quasi-governmental organization charged with oversight of Costa Rica's biological diversity.

All these contracts require that the parties promise to perform continuing or contingent obligations. The standard contract forms employed by NCI and Biotics provide for royalties to be paid in the event of discovery. While the Merck/INBio contract calls for a one million dollar up-front payment, there are also provisions for potentially substantial royalties. Reliance on royalties might seem somewhat strange, since sellers might be expected to prefer the certainty of receiving a smaller sum of money in the present to the remote possibility of receiving a larger sum of money in the future. As noted above, however, royalties are a way of creating incentives for the preservation of ecosystems.

In addition to royalties and up-front payments, some contracts specify that buyers will provide assistance to sellers who wish to increase their research capability. Biotics is helping some source countries increase such capability under its agreements with these countries. INBio's agreement with Merck calls for Merck to provide equipment to be used by Costa Rica for pharmaceutical research. Many countries are likely to follow Costa Rica's lead in establishing institutions like INBio, which has undertaken a massive project to catalogue Costa Rica's entire biological inventory in order to develop domestic collection and research capabilities.

Existing arrangements for the commercialization of genetic resources contain many different provisions for distributing risks, motivating conservation of biologically diverse ecosystems, revealing information about the potential value of genetic resources, and assisting in the development of the sellers' research capability. To some extent, the substantial variation among the terms of the contracts negotiated between buyers and sellers of genetic resources reflects the different circumstances of sellers. The fact that INBio has entered into the most sophisticated of such contracts, and the only one in which substantial up-front payments have been made, is probably related to the fact that Costa Rica enjoys greater political stability than many developing countries in the tropics. It is unlikely that the different circumstances of sellers explain all the variation in contract forms, however.

As parties learn from trial and error, they may adopt different contract forms and different divisions of the tasks required to commercialize genetic resources. However, it may be unwise to simply wait for the most efficient contract forms to evolve.

A lack of credible contracts may translate into a lack of incentives to preserve irreplaceable ecosystems.

It would be both unfair and inaccurate to describe existing arrangements as arising from random experimentation. Contracts are often structured in accordance with expert advice from attorneys and natural scientists. There is, however, an extensive economics literature on risk sharing, incentives, vertical integration, and related issues from which insights should be drawn. Researchers at Resources for the Future are applying and extending the methods of economic analysis to issues arising in the commercialization of genetic resources. Some of the implications of this study have been sketched above. A more detailed treatment of these implications is likely to be of great value in drafting contracts, making investment decisions for new research capability, and, by extension, promoting the conservation of endangered ecosystems.

Suggested Reading

Reid, Walter V., Sarah A. Laird, Carrie A. Meyer, Rodrigo Gamez, Ana Sittenfeld, Daniel H. Janzen, Michael A. Gollin, and Calestous Juma. 1993. *Biodiversity Prospecting: Using Genetic Resources for Sustainable Development.* Washington, DC: World Resources Institute.

Part 7

Environmental Justice

29 Unpopular Neighbors
Are Dumps and Landfills Sited Equitably?

Vicki Been

Incinerators, landfills, and other "locally undesirable land uses" (LULUs) are not popular neighbors, however essential they may be to the community at large. The fact that many of them are located in poor and minority communities may look at first glance like a clear case of discrimination in siting. A closer look reveals, however, that the problem may not be solely with the siting of LULUs, but also with the housing-market dynamics that come into play after a LULU has been established. If so, attempts to achieve a more equitable distribution of LULUs would have to extend beyond changes in the siting process.

Studies show that communities hosting waste management facilities and other locally undesirable land uses (LULUs) have, on average, higher percentages of racial minorities and the poor than other communities. Advocates of environmental justice contend that this is unfair, arguing that environmental risks should be distributed more equitably among races and socioeconomic classes. They assert that the disproportionate burden LULUs impose on poor and minority communities is the result of racism and classism in the siting process. As evidence, they point to studies that reveal a correlation between the racial and class characteristics of communities and the presence of LULUs in those communities.

If the siting process does discriminate against the poor and racial minorities, it should be reformed. However, there may be other reasons why poor and minority communities host a disproportionate number of LULUs; one may be that housing-market dynamics lead the areas surrounding LULUs to become disproportionately poor or minority *after* LULUs have been sited.

If this is the case, we must look beyond the siting process if we are to remedy inequities in the distribution of LULUs. Indeed, if the free market is in part the cause of these inequities, even a siting system that ensured a perfectly fair initial distribution of LULUs would not result in any long-term benefit to the poor or people of color.

Originally published in *Resources*, No. 115, Spring 1994.

The GAO and Bullard Studies

More than a dozen studies document the fact that poor and minority communities now host a disproportionate number of the nation's LULUs. Two of the most notable studies are "Siting of Hazardous Waste Landfills and Their Correlation with Racial and Economic Status of Surrounding Communities," which was conducted by the General Accounting Office (GAO) in 1983, and "Solid Waste Sites and the Black Houston Community," which was published by Dr. Robert Bullard in the same year.

The GAO study examined the racial and class characteristics of communities surrounding four hazardous waste landfills located in three southeastern states. GAO found that, in 1980, African Americans made up between 52 and 90 percent of the population of three of the four communities where the landfills were sited, but only between 22 and 30 percent of the host states' populations. The study also found that between 26 and 42 percent of the population of the host communities was living below the poverty level, but that the host states' poverty levels ranged only from 14 to 19 percent.

Bullard's study sought to determine whether the siting of waste facilities in Houston, Texas, discriminated against African Americans. It found that, in 1980, six of Houston's eight incinerators and mini-incinerators, as well as fifteen of its seventeen landfills, were located in predominantly African American neighborhoods. At that time, African Americans made up only 28 percent of the Houston population.

Both the Bullard and GAO studies are cited as proof that the current distribution of LULUs is the result of discrimination in the siting process. However, neither establishes that the siting process caused the disproportionate distribution. Each study considered only the *current* demographics of host and nonhost communities, ignoring the demographics of communities at the time siting decisions were made. This failure begs an obvious question—namely, whether host communities were poor and minority communities at the time they were selected as LULU sites or only became so in subsequent years.

If neighborhoods were minority neighborhoods at the time they were selected to host LULUs, the choice of sites may have been racially discriminatory. If so, then the siting decisions would have been unfair. But if the neighborhoods were *not* minority neighborhoods when they became LULU hosts, some factor other than discrimination must account for the fact that they now are disproportionately populated by minorities. That factor may be the dynamics of the housing market.

The Role of Housing-Market Dynamics

Each year, between 17 and 20 percent of the U.S. population moves to a new home—often to a different neighborhood in the same city or to a different city. The decision to move is based in part upon individuals' dissatisfaction with the quality of their current neighborhoods. A new neighborhood is selected in part because of its characteristics and cost of housing. These two factors are interrelated because the quality of the neighborhood affects the price of housing.

Accordingly, the siting of a LULU can influence the characteristics of a neighborhood in two ways. First, an undesirable land use may cause those who can afford to move from the neighborhood to do so. Second, it may decrease property values in the neighborhood, making housing available to low-income households and unattractive to high-income households. As a result of both influences, the neighborhood is likely to become poorer than it was before it hosted the LULU.

The neighborhood also is likely to become home to an increasing number of people of color, whenever racial discrimination in the sale and rental of housing relegates them to less desirable neighborhoods than are available to whites. Once a neighborhood becomes a community of color, racial discrimination in the promulgation and enforcement of zoning and environmental protection laws, the provision of municipal services, and the lending practices of banks may cause neighborhood quality to decline further. That further decline will induce those who can leave the neighborhood—namely, the least poor and those least subject to discrimination—to do so.

The dynamics of the housing market, therefore, are likely to force the poor and people of color to move to or remain in the neighborhoods in which LULUs are located, regardless of the demographics of the communities when the LULUs were first sited. Indeed, as long as the market depends upon existing wealth to allocate goods and services, it would be surprising if, over the long run, LULUs did not impose an undue burden upon the poor. And as long as the market discriminates on the basis of race, it also would be remarkable if LULUs did not impose an undue burden upon people of color.

Extending the Studies

To determine whether the current distribution of LULUs is the result, at least in part, of market dynamics, I extended the GAO and Bullard studies. While those studies documented only the then-current demographics of the communities in question, I documented demographics roughly concurrent with the years in which siting decisions were made. I then traced subsequent demographic changes through 1990.

The GAO study examined the racial and class characteristics of communities surrounding four large hazardous waste landfills in the southeast. Sites for three of these landfills were probably chosen in the early or mid-1970s; the site for the fourth landfill was chosen in the late 1970s. Therefore I examined the 1970 demographic data for the first three sites and the 1980 demographic data for the remaining site.

My analysis of these data reveals that all four host communities studied by the GAO were predominantly African American at the time they were selected as LULU sites. The percentage of African Americans in the host communities' populations at the time the LULUs were sited ranged from 1.6 to 3.3 times that of the host states' populations. In the

GAO's analysis, however, only three of the communities were predominantly African American in 1980.

Accordingly, demographic data from the time of the sitings, rather than from the 1980 census, strengthen the inference that siting choices had a disproportionate impact upon African Americans. This does not necessarily mean, however, that the process was discriminatory. Siting decisions may be based upon land prices, proximity to sources of waste, transportation networks, or other factors unrelated to race or poverty that nevertheless have an incidental, disproportionate effect upon people of color or the poor.

At the same time, the data provide no support for the theory that market dynamics cause host neighborhoods to become increasingly populated by African Americans. In all the communities the GAO studied, the landfill sitings were followed by *decreases* in the percentage of African Americans populating the communities. Between 1970 and 1990, the decreases in two host communities were 32.3 and 35.8 percent, even though the decrease in African Americans making up the total population of South Carolina, where the communities are located, was only 2.3 percent.

Demographic data for the time when the landfills were sited also provide no support for the theory that market dynamics cause host neighborhoods to become increasingly populated by the poor. If this theory were correct, the data should show decreases in relative median family income and relative median housing values, as well as increases in relative poverty subsequent to the sitings. According to my analysis, the relative poverty and relative median family income of the host counties changed only marginally between 1970 and 1990. During the same period, the relative median housing value also changed only slightly, and in two of the four host communities it actually increased.

My extension of the Bullard study offers somewhat different results. As noted above, Bullard examined demographic data for Houston communities hosting solid waste management facilities. In redoing his study, I eliminated the community demographics for communities where facilities had ceased to operate by the 1970s, since these sites were selected long ago and meaningful demographic data were not available. Consequently, my analysis was confined to the Houston communities that host the three mini-incinerators and seven landfills cited in the original study. While the original study used "neighborhoods" as its unit of analysis, I examined census tract data. I changed the unit of analysis because I had no information about how Bullard defined neighborhoods and therefore could not replicate his analysis.

In my extension of Bullard's study, then, I examined the 1970 census data for seven communities, because all three of the mini-incinerators and four of the landfills were sited in the early 1970s. For the community hosting the two landfills sited in the early and mid-1950s, I examined both 1950 and 1960 census data. (Because the 1950 census tract containing the landfills was so large, the 1950 data are not particularly meaningful.) The remaining landfill was sited in 1978, so I examined the 1980 census data for its host community.

My analysis of the census data reveals that three of the seven landfills and two of the three mini-incinerators in question were located in areas where the percentage of African Americans was significantly greater than that of Houston as a whole at the time the facilities were sited. Even though about 25 percent of Houston's population was African American, five of the ten facilities were sited in areas where African Americans made up 60 percent of the population. This indicates that the siting process had a disproportionate effect upon African Americans.

Yet analysis of the host communities' demographics in the decades *after* the LULUs were sited reveals that the siting process was not the sole cause of the undue burden Houston's African Americans now bear. Between 1970 and 1980, the percentage of African Americans in the neighborhoods surrounding the landfills increased by as much as 223 percent, while the percentage of African Americans citywide increased by only 7 percent. And while the number of African Americans as a percentage of the

Update

The results of the study described in this article are published in Been and Gupta (1997) and Been (1995). In brief, the study found no evidence that communities selected to host commercial hazardous waste facilities after 1970 were disproportionately African-American at the time of the facility was sited. There was significant evidence, however, that the communities were disproportionately Hispanic at the time they were selected as hosts. The study found no substantial evidence that, between 1970 and 1990, market dynamics following the introduction of a facility into a neighborhood led host communities to become increasingly populated by racial or ethnic minorities. By the 1990 census, host communities were, on average, disproportionately Black and Hispanic, but that disproportion was largely accounted for by facilities sited prior to 1970. Because of limitations on the census data prior to 1970, it was impossible to determine whether those pre-1970 facilities were sited in disproportionately Black and Hispanic communities, or whether market dynamics (not found after 1970, but possibly existing before then) led them to become disproportionately Black or Hispanic after the facility was opened.

total Houston population changed little in the following decade, the number of African Americans as a percentage of host communities continued to increase in all but one of the communities. By the 1990 census, all of the communities hosting landfills had become home to a disproportionate percentage of African Americans.

Analysis of the host neighborhoods' economic characteristics reveals a similar pattern. Only three of the ten communities studied had poverty rates significantly higher than Harris County, where the communities are located, at the time the facilities were sited. Between 1970 and 1980, the poverty rate of all but two of the host communities (as measured by the percentage of the communities' population with incomes under the poverty level) increased, while that of Harris County dropped. Between 1980 and 1990, most of the communities hosting landfills experienced significantly higher increases in their poverty rates than did Harris County. By the 1990 census, five of these communities and two of the three communities hosting mini-incinerators had become significantly poorer than the county.

Similarly, median family incomes in all but one of the communities hosting landfills decreased relative to those of Harris County between 1970 and 1990. In addition, all but one of the communities in which landfills were sited suffered marked declines in their housing values relative to Harris County in the decades following the sitings.

According to my analysis of data from the census closest to the date of the siting decisions in question, the siting process had a disproportionate effect upon African Americans. But it also provides considerable support for the theory that market dynamics contribute to the burden LULUs impose upon people of color and the poor. The data I examined lend force to the argument that LULUs change a community's demographics by driving down property values. True to that argument's prediction, the homes surrounding the landfill sites in most of the host communities became less valuable properties relative to homes in other areas of Harris County after the landfills were sited. The host communities then became increasingly populated by African Americans and the poor.

Implications

Using demographic data from the census nearest in time to siting decisions (rather than data from the most recent census) and then tracing changes in demographics significantly changes the implications of the GAO and Bullard studies. My analysis of the sites in the GAO study indicates a correlation between neighborhood demographics and siting decisions, but suggests no evidence that market dynamics are forcing the poor or people of color to "come to the nuisance." My analysis of the sites in Bullard's study, on the other hand, indicates that market dynamics may play a significant role in the distribution of the burdens LULUs impose. This finding suggests that even if siting processes can be improved, market forces would be likely to create a pattern in which LULUs become surrounded by people of color or the poor.

My research shows that we can make no easy generalizations about the cause or causes of the current inequity in the distribution of LULUs. More data and analysis are needed to prove either that discrimination in the siting process is the sole cause of this inequity or that both siting decisions and housing-market dynamics—demographic changes caused by a LULU's effect on property values and by discrimination in the sale and rental of housing—are to blame.

If further study of LULU sitings confirms the findings of my analysis of the sites in Bullard's study, however, the debate about the fairness of the distribution of environmental risks would have to shift gears. It would become a debate, not just about the process of siting LULUs, but also about the free market and poverty and racial segregation in residential areas. Moreover, discussions about remedies would have to extend beyond the siting process, because changes in this process would be unlikely to achieve real, long-term improvement.

Suggested Reading

The results of the study described in this article are published in:

Been, Vicki. 1995. Analyzing Evidence of Environmental Justice. *Journal of Land Use and Environmental Law* 1.

Been, Vicki, and Francis Gupta. 1997. Coming to the Nuisance or Going to the Barrios? A Longitudinal Analysis of Environmental Justice Claims. *Ecology Law Quarterly* 24(1).

For recent studies of the demographics of communities hosting undesirable land uses, see:

Anderton, Douglas, and others. 1994. Environmental Equity: Hazardous Waste Facilities: "Environmental Equity" Issues in Metropolitan Areas. *Evaluation Review* 18: 123.

Goldman, Benjamin A., and Laura Fitton. 1994. *Toxic Wastes and Race Revisited.* Washington D.C.: Center for Policy Alternatives. Executive summary is available online at http://www.cfpa.org/publications/environment/ej-twrr.html.

Lambert, Thomas, and Christopher Boerner. 1997. Environmental Inequality: Economic Causes, Economic Solutions. *Yale Journal on Regulation* 14: 195.

Oakes, John Michael, and others. 1996. A Longitudinal Analysis of Environmental Equity in Communities with Hazardous Waste Facilities. *Social Science Research* 25: 125.

See www.nyu.edu/pages/elc/ej/or for an up-to-date listing of empirical studies, many of them available on the World Wide Web and hyperlinked from this site.

30 Measuring Environmental Equity with Geographical Information Systems

Theodore S. Glickman

Concern that racial minorities and the poor are shouldering a disproportionate share of the burden of environmental hazards has prompted interest in ways to redress existing environmental inequities. Many efforts have been made to identify these inequities, but not in terms of the actual risks associated with environmental hazards. Researchers at Resources for the Future are now combining risk assessment techniques with geographical information systems (GIS) software to do just that. They are analyzing environmental equity with respect to the risks from industrial hazards in Allegheny County, Pennsylvania. This test case of the use of GIS to analyze environmental equity has suggested that those most exposed to environmental risks are not always nonwhites and the poor.

Environmental justice, or the equitable distribution of environmental hazards, is currently attracting more attention than perhaps any other environmental issue. In 1993, the White House issued an executive order that requires federal agencies to consider the impacts of their decisions on environmental equity, and the U.S. Environmental Protection Agency has created a special office to facilitate such analyses. Both actions were motivated by concern that racial minorities and the poor may be shouldering a disproportionate share of the impacts of environmental hazards.

Indeed, racial minorities and the poor, who in many cases are one and the same people, typically *do* have greater exposure to environmental hazards than those who are more economically advantaged. The poor often live in areas that are likely to have more environmentally undesirable facilities—for example, factories, power plants, waste incinerators, and so on—than the areas where other groups live. And, unfortunately, these subpopulations may include a disproportionate number of young children or elderly people, two groups that are generally believed to be especially susceptible to the health effects of pollution.

Before cases of existing environmental inequity can be remedied, they must first be identified. Fortunately, a new information technology has emerged that can be used to provide real data about environmental equity impacts in any selected location. This technology—known as geographical information systems, or GIS—is a type of software that was originally devel-

Originally published in *Resources*, No. 116, Summer 1994.

oped for combining different types of spatial data, such as information about a region's topographical features and its distribution of natural resources. Using GIS requires appropriate data and expertise, but most of the data are already available, and some systems are friendly enough to be learned easily.

GIS is becoming an increasingly important technology for analyzing environmental equity. The information provided injects an essential degree of objectivity into environmental justice deliberations. In turn, this objectivity helps decision makers to establish priorities based on information about which hazards create the greatest disparities in impacts and which groups of people are most affected.

In the Center for Risk Management at Resources for the Future, we are conducting a study that demonstrates the potential of GIS to shed light on the distribution of environmental burdens. An important part of the study is a comparison of proximity-based measurements and risk-based measurements of environmental equity. Below we describe how we have used GIS to generate both kinds of measurements in an analysis of environmental equity with respect to industrial hazards in Pittsburgh and surrounding Allegheny County, Pennsylvania, circa 1990.

This analysis differs in three important respects from related analyses of environmental equity that have been done elsewhere. First, it considers not only *chronic* hazards in the form of air pollution from industrial facilities but also *acute* hazards in the form of potential exposure to accidents involving the airborne release of toxic chemicals from facilities where the chemicals are stored. Second—and more important—our analysis of equity is based not only on *proximity* to hazards but also on the actual health and safety *risks* associated with each kind of hazard, separated and combined. These dimensions of equity are absent in most related studies, which measure equity based only on people's proximity to hazardous facilities. Third, for selected facilities, our analysis will trace changes in the distribution of environmental hazards using historical data on hazards, land use, property val-

ues, demographics, and other agents or indicators of change, whether legal, political, or economic.

Our study using GIS reveals the need to look beyond aggregate results when analyzing equity and to be cautious when using worst-case assumptions. Also, somewhat surprisingly, it shows that, in the face of hazards that have the potential to affect large areas (such as major accidental chemical releases), most of those who would be exposed do not belong to the most disadvantaged groups of people—that is, racial minorities and the poor.

When our study is completed this fall, we expect it to benefit a wide audience, ranging from community groups to professional peers, by demonstrating how to assemble data germane to environmental equity, how to analyze the data using moderately priced GIS software, and how to interpret the results. We believe our approach will be particularly useful in showing how to measure "outcome inequity," that is, in determining whether one socioeconomic group bears more of the burden of a particular environmental hazard than another. In addition, the part of our study that deals with the use of historical data to examine the evolution of environmental inequities should prove useful in showing how the distribution of burdens changes with time and—perhaps—in helping understand why.

Proximity-Based Measurements

One way to measure environmental equity is based on people's proximity to facilities that pose environmental hazards. In our analysis of environmental equity with respect to industrial hazards in Allegheny County in 1990, we used a GIS to avoid some of the pitfalls that attend more simplistic approaches to proximity-based equity measurements.

Previous approaches hinged on a comparison of the percentage of minorities or poor people in the census areas that contain environmentally hazardous facilities with the percentage of minorities or poor people in nearby census areas that do not contain such facilities. This kind of approach is problematic for several reasons. First, it draws no distinction between areas that are home to only one facility and those that host two or more facilities. Second, it does not account for the possibility that the hazardous facility or facilities may be so close to the edge of the host area that a neighboring area is affected as much, if not more. Third, and perhaps most important, this approach does not consider that census tracts and counties do not generally represent either the affected neighborhood or the range of the hazard associated with a facility. A more sensible way to represent both is to construct an imaginary circle centered at each facility, although the question of how large the radius of the circle should be is open to question. In the case of facilities in urban areas, a radius of one or two miles seems reasonable, since neighborhoods do not usually extend any further than that.

Our environmental equity analysis accounts for all three of the above shortcomings. In this analysis, we divided Allegheny County's industrial facilities into two types: those that may pose chronic hazards and those that may pose acute hazards. We refer to the former facilities as TRI facilities after the Toxic Release Inventory (a national database of reports of industrial air pollution), from which we obtained the location of these facilities and information about emissions from each. We refer to the latter facilities as EHS facilities because they store "extremely hazardous substances"; the risks associated with these facilities arise in the event of an accidental chemical release rather than from continual, routine chemical releases. We were able to identify these facilities using the federally required reports that indicate where EHSs are stored in quantities above a certain threshold.

Next, we constructed circles with radii of one-half mile, one mile, and two miles around each TRI and EHS facility. Then, for each radius, we divided Allegheny County into two parts, one being the area formed by the circles and their overlapping portions, and the other being the rest of the county (the areas outside of all the circles). We made this division for the EHS facilities and then, separately, for the TRI facilities. In each case, the combined area within the circles, which may not all be contiguous, is what we call the "close-proximity region"—the region where people live in close proximity to the facilities. We assumed that, for a given choice of radius, the close-proximity region is homogeneous with regard to proximity effects—that is, the hazard burden is the same no matter which facility you are close to or how close you are to it, as long as you live within the region.

Using a GIS, we then calculated the proportion of nonwhite residents and poor residents inside and outside the close-proximity region of the sixty-two facilities in the county that stored large quantities of EHSs in 1990. We found that *nonwhite* residents made up 16 percent of the population inside the close-proximity region but only 11 percent of the population outside this region in 1990 (see Figure 1). Similarly, *poor* residents made up 16 percent of the population inside the close-proximity region but only 10 percent of the population outside that region in 1990 (see Figure 2). Thus the percentage of nonwhites and the percentage of the poor among people who live close to the EHS facilities are slightly higher than those elsewhere in the county.

When we calculated the percentage of nonwhite residents and poor residents inside and outside the close-proximity regions of the county's TRI facilities, we obtained similar results: the percentages of nonwhites and the poor living inside the close-proximity regions were greater than the percentages of these people living outside those regions.

Risk-Based Measurements

Risk-based equity measurements are superior to proximity-based equity measurements because they take into account other major factors on which risk depends—factors that can actually change the picture of environmental equity given by proximity-based measurements. These other factors include the probability of an accidental release of chemicals; the size of the area affected by such a release (which depends, in turn, on the substance released, the quantity released, the nature of the release, the

Figure 1. EHS facilities and areas with large nonwhite populations.

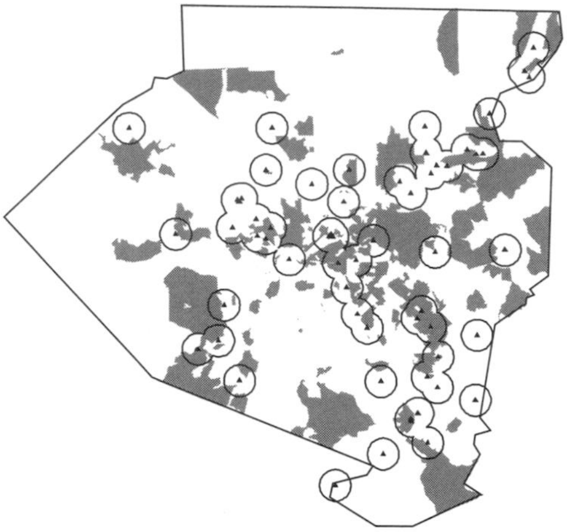

The triangles indicate the locations of the sixty-two facilities in Allegheny County that stored more than a minimal quantity of EHSs in 1990. Circles indicate areas within a one-mile radius of the facilities. The shading indicates the areas formed by the 25 percent of census block groups that had the highest number of nonwhite residents in 1990.

Miles

0   5   10

release rate, and the weather at the time of the release); and the wind direction at the time of release. Risk also depends on the toxicity of the chemical released and on the level of exposure of the population of concern.

As yet, the results of our risk-based environmental equity analysis for Allegheny County are confined to acute hazards associated with EHS facilities in 1990. We defined the risk posed by these hazards as the expected annual number of persons exposed to accidental chemical releases, and we developed an exposure assessment procedure that takes all the above-noted factors into account. We did so using a formula that multiplies the probability of an accidental chemical release by the size of the impact area and the population density in that area. This procedure allows for the possibility that any person might be exposed to several such accidents in any given year, thereby contributing several "person-exposures" to the annual total.

Because population exposure varies by time of day, so does risk. Analysts commonly calculate only nighttime risks, because doing so requires only residential census data. However, because nighttime and daytime risks can differ significantly, it is important to account for each separately. Therefore we used residential census statistics and "journey-to-work" data, which reflect the weekday comings and goings of commuters, to calculate both the nighttime and the daytime risks that each EHS facility poses for nonwhites and for the poor (see Figure 3). Then we calculated the total risk to nonwhites and to the poor for each facility alone and for all EHS facilities taken together.

We defined the average risk that EHS facilities pose as the weighted combination of the nighttime risks, which only take the residents of each impact area into account, and the daytime risks, which take into account the working population and the nonworking residential population in each impact area. We counted twice any risk in the overlap between two impact areas, which is appropriate since the total risk to any person is essentially the sum of the two risks. Based on these measurements, equity for nonwhites (or the poor) is said to exist if their percentage of the total risk is the same as that of nonwhites (or the poor) among the entire county population.

According to our calculations, which are based on the most hazardous chemical stored at each facility, the percentages of nonwhites and poor people at risk from accidental chemical releases are 9 percent and 8 percent, respectively. The percentages of nonwhites and poor people in the county are 13

Figure 2. EHS facilities and areas with large poor populations.

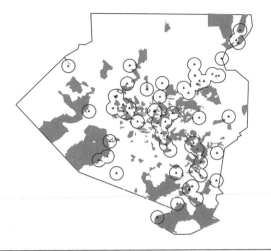

The triangles indicate the locations of the sixty-two facilities in Allegheny County that stored more than a minimal quantity of EHSs in 1990. Circles indicate areas within a one-mile radius of the facilities. The shading indicates the areas formed by the 25 percent of census block groups that had the highest number of residents living below the poverty line in 1990.

percent and 12 percent, respectively. In other words, nonwhites and poor people actually bear proportionately slightly *less* of the risk than they would if equity existed.

At first, this finding comes as a surprise because environmental inequities are generally expected to work in favor of the white, more affluent majority, as demonstrated by the above-noted proximity-based equity measurements. Upon reflection, however, the reasons for the outcome of our exposure risk-based equity measurements are clear.

First of all, this outcome is an aggregate result obtained by combining the results for all the facilities where EHSs are stored. On a facility-by-facility basis, the direction of the inequity varies, sometimes working in favor of whites who are not poor and sometimes against them. The aggregate result shows that, on balance, it worked against them more than it worked for them.

Second, nonpoor whites are often at greater risk from hazards that affect a large area, such as major accidental chemical releases, than from hazards that affect only a small area. This becomes apparent when we consider that the radius of the area affected by a major chemical release accident often exceeds one mile and that nonwhites and

poor people tend to live closer to EHS facilities than whites and nonpoor people. Thus, nonpoor whites will be affected at larger radii.

This phenomenon calls for caution in risk assessment. Although it is commonly suggested that worst-case assumptions be used in assessing risks, doing so introduces a bias when disadvantaged people live closer to hazardous facilities than other people. Why? Because the impact of hazards on nonpoor white individuals increases as the hazard area increases. As a result, risk assessments that use worst-case scenarios may show disproportionate risks in nonpoor, white communities because these communities' share of the hazard burden is larger than it would be under average-case assumptions—for instance, if the "plume" of a toxic vapor cloud were assumed to dissipate quickly, rather than more slowly and over a larger area.

We are still in the process of generating risk-based measurements of equity for the chronic hazards associated with air pollution from the TRI facilities in Allegheny County. This is a more time-consuming process, because it requires that concentration contours representing countywide pollution patterns first be modeled for common pollutants such as particulates, as well as for less ubiquitous air toxics. When these concentration

Figure 3. Highest risk aeas by night and day.

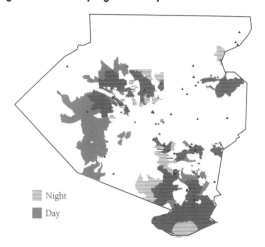

Night

Day

Crosshatching indicates the distribution of nighttime and daytime risks posed by accidental chemical releases in Allegheny County in 1990. These shaded areas represent the 5 percent of census block groups that had the highest risks in each of the two time periods during that year. The shaded areas *appear* to represent more than 5 percent of the county because census block groups in rural parts of the county usually are larger in area than those in more densely settled areas.

Miles

0   5   10

contours are grafted into the GIS as a "data layer," they will be combined with the aforementioned estimates of population exposures in order to assess the associated risks to nonwhites and the poor. We will use the resulting risk estimates, which will be expressed not just as person-exposures but as predicted cases of cancer or disease, to assess distributional equity. We will also analyze equity on the basis of the *combined* risks of accidental chemical releases and air pollution, which means that the acute impacts of accidental injury or fatality and the chronic health effects of pollution exposure will have to be measured on a common scale, such as the total expected reduction in life expectancy.

Future Developments

While the use of GIS to measure environmental equity is still in its infancy, we feel safe in making certain observations about this practice. Given the widespread availability of census and TRI data, as well as the increasing availability of user-friendly GIS packages, the capability to produce proximity-based estimates of industrial air pollution hazards is within the reach of many interested parties. Naturally, such estimates should not be considered the "last word" on environmental equity, since TRI

data and EHS storage data are self-reported, TRI facilities are but one source of air pollution, and proximity is not a surrogate for risk. Other pollution sources and any environmental hazards that are of a nonpolluting nature or that are unrelated to health effects also can be readily subjected to a proximity-based analysis, provided that the data are available, complete, and "clean." GIS may be a new technology, but the oldest maxim in computing—"garbage in, garbage out"—still applies.

Risk-based analysis of environmental equity is another matter entirely. Such analysis is still the province of specialists, requires much more data than proximity-based equity analysis, and yields results that are more difficult to interpret. However, these obstacles will become less formidable as more research of the kind we are conducting is done, as better risk assessment software becomes available, and as risk education and communication improve in general. In the meantime, much more research is needed on how to combine risks, especially those that are difficult to measure in common units, such as carcinogenic and noncarcinogenic risks (or even health and ecological risks), and those that do not merely sum when accumulated, namely health risks that are exacerbated in the presence of certain other health risks.

In the near future, two principal benefits are likely to emerge from the use of GIS to measure environmental equity. One is the capability of concerned parties, such as public interest groups or government agencies, to use GIS as a screening tool to evaluate a region and determine which facility or facilities are contributing to the inequitable distribution of risks in the region. The other benefit is the contribution that GIS can make to the process of facility siting. Ideally, all the stakeholders in this process—whether they be industries, government agencies, or community groups—would participate in the process of identifying and evaluating the candidate locations for an undesirable facility, with the assistance of a GIS.

If, at some point in the future, an inventory of risk estimates could be developed for each region—whether it be a city or county—the facilities considered in the screening or siting process could be evaluated not only in terms of the absolute risk they pose to each population group of concern but also according to their relative contributions to the overall risk burden of each group.

One overarching policy issue should be confronted in the not-too-distant future: Is it ultimately better for all parties concerned to spread out a region's environmental hazards in order to achieve short-term equity? This would be the outcome of making immediate, piecemeal improvements in the status quo. Or is it better to concentrate these hazards in one or more "hazard zones" and effect long-term equity by reducing the associated risks and putting programs in place to enable affected residents to relocate over time? This issue goes well beyond the use of GIS to measure environmental equity, although GIS could help in such a policy analysis.

31 Protest, Property Rights, and Hazardous Waste

Robert Cameron Mitchell and Richard T. Carson

In 1980, the U.S. Environmental Protection Agency estimated that despite the need for as many as 125 new sites for hazardous waste facilities in the near future, local opposition would make finding these sites an "exceptionally difficult task." That pessimism was understated. Not one major facility has been sited anywhere in the United States during the last six years, and, according to the 1985 state-by-state review published in *Hazardous Waste Consultant*, the outlook for the future is "even more bleak," due in large part to a worsening of the "emotional atmosphere" surrounding siting efforts. This situation has come about regardless of assurances by government and company officials that new facilities would pose negligible risks to local residents. Attempts have been made to break local deadlocks by instituting procedures for extensive public participation, establishing state siting boards with the power to overrule local decisionmakers, and requiring facility owners to compensate local governments for safety services, all to little avail.

The ambiguity of existing property rights that govern the siting of hazardous waste facilities is an important cause of the stalemate. What is called for is a new approach to siting—we suggest a political market, via a referendum mechanism—that recognizes the de facto property rights assumed by local communities. The referendum, supervised by the state, would be held at the request of the firm wishing to site the facility. The developer, in effect, would offer a comprehensive package of incentives to the community in exchange for a yes vote.

Originally published in *Resources*, No. 85, Fall 1986. An earlier version appeared in *AEA Papers and Proceedings*, vol. 76, no. 2, May 1986.

From Theory to (Almost) Practice

In May 1986 AEA Papers and Proceedings *published a proposal by RFF's Robert Cameron Mitchell and Richard T. Carson of the University of California, San Diego, for mandatory public referenda on the siting of hazardous waste facilities. Almost simultaneously, and quite by coincidence, the* New York Times *(May 6, 1986) reported that something very like the Mitchell–Carson proposal was in the works in rural Lisbon, Connecticut.*

According to the Times, *Philip Armetta—who is proud to be called Mr. Garbage—had proposed to locate in 3,400-resident Lisbon a modern incinerator that would generate both energy from waste and $1 million in tax revenues. Despite the financial incentive and assurances that the incinerator would be equipped with the latest antipollution devices, garbage still was garbage in the eyes of the town: Armetta was rebuffed. Indeed, the issue so galvanized the electorate that in November 1985 forces opposed to the incinerator captured control of Lisbon's Planning and Zoning Commission. In January 1986 the commission majority delivered on its campaign promise and voted to prohibit waste plants.*

At this point Armetta put a brilliant new spin on his proposal. In place of saying it would bring the town $1 million a year in new tax revenues—a solid-enough figure but one lacking appeal to individual voters—he promised to pay the 1986 property taxes of every landowner in Lisbon and to continue paying the same amount for the next twenty-five years. At an average of $900 per homeowner, Armetta had shrewdly calculated,

his promise came to a rough annual total of $1 million.

The dollars involved may have remained constant, but garbage in Lisbon suddenly took on a more attractive aroma. Political winds shifted, minds changed, the local newspaper modified its editorial stance, and a referendum was scheduled. In a surprising turnaround from the November election, Lisbon voted 680 to 590 to rezone the town to allow incinerators.

Had the Mitchell–Carson proposal been law, Mr. Garbage would have his incinerator and Lisbon's property owners would have their $900 per year for twenty-five years. But Connecticut does not permit binding referenda in such matters and the vote thus was only advisory. In a meeting on August 25, the town Planning and Zoning Commission cast a 5 to 4 vote against the incinerator. Is the issue then dead? "Nothing's dead when a substantial number of people still want it," said Lisbon First Selectman Jeremiah Shea in a post-vote telephone interview with Resources.

Lisbon's citizens are no more greedy than any others, and it may be that Armetta's catchy property-tax offer only provoked closer examination of a development that had some appeal regardless of how it was presented. But what Lisbon's experience does show is that the Mitchell–Carson hypothesis has considerable practical merit. Offered a package of incentives by a developer and empowered to vote their property rights, citizens can be trusted to act in their own best interests.

—Kent A. Price

To understand the rationale for our approach, it first is necessary to examine the evolving nature of the property rights in question, an evolution driven by changing perceptions of the risks associated with toxic waste disposal and a social movement of considerable power that has raised the cry of "not in my backyard." Of course, citizens as individuals have much to gain by opposing hazardous waste facilities near them. But their resistance imposes large costs on society as a whole. After all, blocking new waste facilities does not make the waste itself

disappear. Quite the contrary: growing quantities of toxic chemicals held in temporary and deteriorating storage conditions as they await destruction or a permanent home create strong incentives for illegal "midnight dumping."

Protest Is Effective

Until recently, waste disposal was not considered a social problem. Dumps containing hazardous materials were treated by the public and planners as minor extensions of garbage dumps and sanitary landfills; and opposition, if any, was based on the dumps' nuisance characteristics, not on their perceived safety risks. As for property rights, the developer's entitlement to engage in waste handling was preeminent as long as the facility was located in an industrial area.

Passage of the Resource Conservation and Recovery Act (RCRA) in 1976 marked official recognition that these wastes, many of them disposed of improperly in the past, posed a potentially serious threat to health. Three years later, the Superfund legislation targeted existing toxic waste dumps for cleanup. In between, the issue exploded into public awareness when the problems at New York's Love Canal reached the national news media. Subsequently the entire town of Times Beach, Missouri, was abandoned after authorities found dioxin contamination there in 1982, and news reports of contaminated drinking water wells now are commonplace.

Proposed hazardous waste facilities quickly became the subject of widespread and effective protest, despite stringent federal design and operation safety standards imposed by RCRA and augmented by state regulations. For example, four years of work and $1.5 million were spent on a comprehensive treatment and land disposal facility in Los Angeles County before its corporate owner withdrew in the face of seemingly insurmountable public opposition. In Texas, a regional authority proposed a high-temperature incinerator for toxic wastes from the area (a solution favored by environmentalists). Notwithstanding a well-demonstrated need for such a facility and initial support from local governments, citizen opposition caused the developer to give up after a three-year battle when it became apparent that political approval was not forthcoming.

Hazardous waste facilities are a prime example of a regulated entity whose costs are highly concentrated geographically while its benefits are distributed far beyond the local area. The principal costs of hazardous substances are the health risks believed to be posed by groundwater and soil con-

tamination in the case of landfills, and pollution of the air by cancer-causing substances in the case of incinerators. What local citizens see is an abrupt threat that involves a visible source (the site) for which clear responsibility can be ascribed (the developer)—characteristics that heighten public awareness of the perceived risk. Moreover, in contrast to nuclear power plants or industrial plants, which usually have a local constituency, a hazardous waste facility provides few offsetting benefits in the form of jobs or tax revenues, and residents may fear a decline in local property values.

"Not-in-my-backyard" aptly captures the views of those who resist facility siting. The syndrome itself is not new—homeowners long have resisted undesirable facilities in their neighborhoods. What is new is the scale and intensity of protests provoked by facilities perceived to be risky. Figure 1 shows the percentage of the public in a national survey willing to accept (without protesting or moving) each of five hypothetical facilities.

Three distinct "siting aversion profiles" emerge, with corresponding "backyards" and protest constituencies. Reactions to a ten-story office building—more than half would accept one if it were at least a mile from their homes—represent a useful baseline. Majority acceptance of an industrial plant or a coal-fired electric power plant, both likely to be perceived as dirty and potentially obnoxious neighbors, occurs at about nine miles. High contrast is provided by the two facilities posing potentially catastrophic but extremely low probability risks. Both a nuclear power plant and a new, well-regulated disposal site for hazardous wastes reach majority acceptance only at the 50-mile mark, a "distance premium" of 49 miles from the office building baseline. This suggests a crucial difference between an ordinary industrial facility and one involving hazardous wastes: the neighbors affected by the latter involve entire communities. Another difference is the number of people who feel strongly about the issue. Whereas only 9 percent express the extreme view that they did not want the two industrial facilities as neighbors "at any distance," 29 percent took this stance about the two "risky" facilities.

Figure 1. Cumulative percentage of people willing to accept new facilities at various distances from their homes.

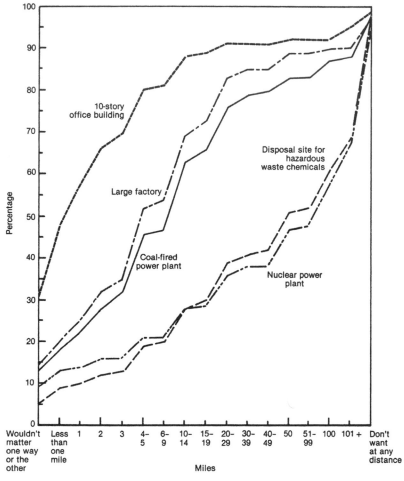

Source: Based on data from R.C. Mitchell, *Public Opinion on Environmental Issues: Results of a National Public Opinion Survey* (Washington, D.C., Council on Environmental Quality, 1980).

Property Rights in Flux

Property rights specify how persons may benefit or be harmed and, therefore, who must pay whom to modify the actions taken by affected parties. In a now-famous article, "The Problem of Social Costs," Ronald H. Coase argued in 1960 that the assignment of property rights to one party or another does not, in the absence of transaction costs, affect economic efficiency, although it does affect the distribution of wealth. Coase's insight was deep: resources are put to their most efficient use regardless of how the political system initially chooses to allocate property rights. The problem with the hazardous waste situation is that currently no one really has clear title to site a hazardous waste facility—not the firm, not the community, and not community residents as individuals.

Update

Our work on siting undesirable land uses contributed to three distinct lines of research and policy initiatives. The first of these is directly related to the siting issues raised in the paper. Several states and many localities have effectively adopted measures that require local approval of a locally unwanted land use. This has often been implemented in the form of requiring voter approval and, in some instances, has even been extended to development projects that require very large changes in current zoning. The result of these measures has been what we would have predicted: a shift in power to local officials or the local public and bargaining by the developer to provide the lowest-cost compensation package that will win approval. The role of trust in the government to enforce agreement provisions also has been shown to influence the amount and nature of the compensation demand by the local community in order to voluntarily accept the undesirable land use.

The second line of research emphasized the importance of assigning the property rights in terms of the strategic aspects of the decision, as well as in terms of the compensation required to get voluntary acceptance of the land use by the local community. At the time of the article, it was widely accepted from a theoretical perspective that maximum willingness to pay (WTP) and minimum willingness to accept compensation (WTA) measures should be close together, although the empirical evidence from contingent valuation surveys suggested this was not true. Michael Hanemann later showed that for imposed quantity changes, such as a locally unwanted land use, WTP and WTA measures could be very far apart and hence who was assigned the property rights to the siting decision could have a very large influence on outcomes.

The third line of research emphasized the difference between individual and collectively held property rights. One of the persistent problems in implementing contingent valuation surveys that ask WTA questions is that respondents often refuse to answer such questions because they contend that they do not have the individual right to give up public goods. This has lead to reformulating such survey questions (and actual government policies) to provide for the compensation in either the form of other public goods or in the form of a reduction of a widely paid tax or utility bill.

As noted, firms wishing to site a hazardous waste facility have lost their unfettered right to locate where they wished as the public and government officials became alarmed over the possible risks posed by the technology. Local residents increasingly have been able to delay (and thus effectively block) siting efforts in administrative and judicial hearings, and communities have taken a leading role in stopping the construction of new hazardous waste facilities through the use of their extensive police powers to regulate zoning and safety. With a few exceptions, however, communities do not have the legal right to ask for sizable payments in exchange for issuing the necessary licenses and permits.

The recent establishment of state siting boards with the power to preempt local governments is an attempt to reassert the former property right regime. The concurrent establishment of schemes for compensating communities for the presence of a hazardous waste facility represents a movement in the opposite direction—toward giving the property right to the community. The innovative Massachusetts siting law has both features, going further in the direction of bargaining for compensation and less in the direction of preemption (calling for binding arbitration only in the case of irreconcilable differences) than any law in the country. No facilities yet have been sited under this law, suggesting that compensation without ultimate local veto power over a facility may not be a successful strategy.

Community Rights

If local residents were individually to hold the property right, developers could not bargain efficiently with the large number of potentially affected residents and one holdout could block a well-conceived project. We suggest, therefore, that a collective property right be established by having states pass a law specifying the use of referenda to determine local approval or rejection of a proposed facility. Such a law would require the relevant political authorities to hold a referendum when requested by a qualified developer meeting state requirements. Specific plans for the facility and for compensation to the community for its perceived drawbacks would be proposed by the developer and incorporated into the ballot proposal. A number of possible compensatory measures have been suggested in recent years, and the contents of a developer's particular package probably would vary according to the nature of the facility, the characteristics of the site, and the community's concerns. The types of measures that might be offered include guarantees against declines in property value, incentive payments to the community (which could be earmarked to reduce property taxes or for other purposes), outside monitoring, accident insurance, credible guarantees of nonabandonment, donation of land for use as parks, and in-kind services like free waste disposal for community residents and businesses.

Should the decision rule be a simple majority, or something larger, such as the often-used two-thirds majority? Although a two-thirds majority requires a more expensive package, we believe it is more likely to produce a situation of maximum benefit and minimum cost and therefore greater community harmony.

Who would administer and enforce the contract established by the referendum? This undoubtedly would fall first to the local political authorities and ultimately to the state. This must be made clear beforehand, because doubts about enforcement would increase the payments required to pass the referendum. In addition, there must be sufficient administrative flexibility to respond to new EPA regulations and to technological change. The boundaries defining who should be allowed to vote on the proposal is a difficult political question that the state legislature would have to decide.

The advantages of a referendum law are several. The developer and the state have strong incentives to address the issues of most concern to the community. The community's incentive to be intransigent is minimized because it has the power to say no, and it is protected from unwittingly accepting too great a risk because the facility would have to meet strict federal and state safety regulations. Moreover, the debate occasioned by the referendum should ensure close scrutiny of the developer's proposal. Paying for the compensation package transforms the costs hitherto concentrated on the local community into more equitably shared burdens that are borne by the ultimate beneficiaries of the facility. Finally, to the extent that this plan increases the costs of handling hazardous wastes, those who produce the wastes will have an incentive to engage in in-plant waste-stream modifications and resource recovery.

Suggested Reading

Hanemann, W. Michael. 1998. The Economic Theory of WTP and WTA. In *Valuing the Environmental Preferences: Theory and Practice of the Contingent Valuation Method in the US, EC and Developing Countries,* Ian Bateman and Ken Willis (eds.). Oxford: Oxford University Press.

Mitchell, Robert Cameron, and Richard T. Carson. 1989. *Using Surveys to Value Public Goods: The Contingent Valuation Method.* Washington, DC: Johns Hopkins University Press for Resources for the Future.

Muton, Don (ed.). 1996. *Siting by Choice: Waste Facilities, NIMBY, and Volunteer Communities.* Washington, DC: Georgetown University Press.

Rabe, B.G. 1994. Beyond NIMBY: Hazardous Waste Siting in Canada and the United States. Washington, DC: Brookings Institution.

Part 8

Global Climate Change

32

A "Crash Course" in Climate Change

J. W. Anderson

Climate change is, by a wide margin, the most complex environmental challenge to which governments have ever attempted to construct a cooperative international response. It requires changing the ways in which the world uses fossil fuels, the most common source of energy in nearly every country on earth. Any possible agreement would affect the commercial rivalries among nations and might well affect their standards of living. The kind of agreement now under discussion would require supranational institutions that do not yet exist to undertake the work of verifying emissions and mediating disputes.

The central political issue currently is the relationship between the rich countries and the poor ones. The United States, fearing an emigration of industry to unregulated sites, says that it will not accept a commitment to reduce its emissions unless the major developing countries also accept some kind of commitment. But the developing countries see global warming as a threat that has been created by the industrial countries; in their view, the industrial countries are obliged at least to take the lead in meeting it. It is not easy to work out an emissions-reduction formula that all sides will regard as fair when the United States emits 5.4 tons of carbon per capita each year, China emits 0.7 tons, and India 0.2 tons. The negotiators have not yet thought through the questions of fairness, and no consensus has emerged. One possible solution involves a flow of financial aid to the poor countries to help them improve efficiency in their use of energy. But there is little indication so far that the rich countries are prepared to increase their aid.

One key question for policy is whether the negotiations up to this point provide a firm foundation for further agreements in the future. Or will it be necessary to abandon some of the agreements already drafted—for example, the Kyoto Protocol—and begin over again, in a different direction?

Rio de Janeiro and Kyoto

When they met in Rio de Janeiro, Brazil, in 1992, the world's governments agreed to try to stabilize emissions of the gases that can change the climate. But it was an agreement only to make a voluntary effort, with no binding commitments and no penalties for countries that failed to meet their goals. When the same governments met again in Kyoto, Japan, in December 1997, they were attempting a much more difficult feat. They undertook to negotiate a binding agreement to firm and specific limits on greenhouse gas emissions in at least the world's major economies.

As the Kyoto Protocol to the Rio Treaty emerged on December 11, 1997, it was a success in the sense that it represented a solid consensus of the 168 governments represented at the conference. But it was only a partial success, for it achieved that consensus by avoiding, or deferring to future conferences, a number of issues that would be crucial to the treaty's operation generally and to its ratification by the country producing the greatest volume of emissions, the United States.

Since greenhouse gases are generated mostly by burning fossil fuel in power plants, factories, and automobiles, it is not easy to reduce emissions in growing economies. With the Kyoto conference, the concept of limiting emissions has become a political issue in many of the industrial countries. In the United States, for instance, ratification of the protocol will require a two-thirds vote in the Senate. On one side are fears that emissions limits will interfere with economic growth and the rise in standards of living. On the other are fears, expressed by President Bill Clinton and Vice President Al Gore among others, that global warming will mean bizarre weather with disastrous flooding, droughts, and epidemics.

Climate change is an unusually difficult issue for the people who make the decisions in democratic governments. First of all, the science is uncertain at a time when governments have to make firm policy decisions—if only the decision to do nothing—long before these uncertainties can be resolved. Political leaders are already beginning to overstate the clarity of the science in order to attract public support. A lot of money is now going into climate research, and new findings with varying political implications will continue to appear.

Any serious attempt to cut emissions will have clear and immediate costs, but the benefits may not appear for a long time. To the extent that the benefits may be disasters that don't happen, they may never be obvious. But the costs will be. As the debate develops, much of it is being cast in terms of the restraint that the present generation owes to future generations.

The Negotiating Record up to 1996

Climate change is a newcomer to the international political agenda, having emerged as a major policy issue only in the late 1980s. But scientists have been working on the subject for decades. They have known since the nineteenth century that carbon dioxide (CO_2) in the atmosphere retains heat from the sun and that temperatures on the Earth's surface are affected by it. Carbon dioxide is generated by burning any fuel containing carbon, from coal and oil to wood and cow dung. With the industrial revolution, the world's consumption of these fuels increased hugely.

For many years, scientists ignored the issue because, they assumed, the CO_2 increases were being absorbed by the oceans. But in the International Geophysical Year 1957 to 1958, they decided to test that assumption with a series of measurements from the top of Mauna Loa, the Hawaiian volcano, chosen as a site far removed from the influence of any local smokestacks. The Mauna Loa readings quickly demonstrated that CO_2 concentrations in the atmosphere were rising steadily.

That finding led to a series of scientific meetings, most of them organized by various agencies of the United Nations. Concern about the global climate was reinforced by a number of events, particularly the devastating drought along the southern rim of the Sahara Desert. As these meetings continued, scientists began to realize that in addition to carbon dioxide, other common gases could also affect climate. Methane was one example, nitrous oxide another.

The year 1988 is the point at which climate change and global warming emerged as a major political issue throughout the industrial countries. In the United States, Senator Timothy Wirth (D–Colorado) had been deeply exasperated by his inability to draw public attention to the subject. When summer arrived, he waited for a day forecast to be spectacularly hot and called a hearing at which several experts testified. With the temperature at 98 °F and anxiety rising about the drought gripping the Midwest and South, one of the experts, James E. Hansen, told the senators that the world was warmer than at any time in this century. It was 99 percent certain, he continued, that the cause was human-made gases and not natural variation. "It is time to stop waffling so much and say the evidence is pretty strong that the greenhouse effect is here," he told a reporter for the *New York Times*, which put the story at the top of page one. Hansen's testimony had unusual force because he was director of the National Aeronautics and Space Administration's Institute for Space Studies and the first scientist of that stature to declare flatly that the rising temperatures were related to burning fuel.

Four days later, a conference opened in Toronto, attended by several hundred scientists, politicians, and officials from forty-eight countries and the United Nations. It started the push for action by calling for a 20 percent reduction in CO_2 emissions by the year 2005. Political leaders in several countries picked up the issue. One of them was Margaret Thatcher, the prime minister of the United Kingdom, who had been trained in chemistry as an undergraduate. In December 1988, the U.N. General Assembly approved the establishment of an Intergovernmental Panel on Climate Change (IPCC) to review the science.

The following year at their annual summit meeting, the heads of state of the seven big industrial democracies called for a treaty—a "framework convention" as it became known—to limit the world's CO_2 production. Negotiations shortly got under way.

But strains between the United States and most of the Western European countries soon became visible. The Bush administration was uneasy about the scientific base for policy and wanted more time for research. It was also hearing from industries threatened by the prospect of limitations on fuel consumption. The Europeans wanted to begin to move immediately. In early 1990, President George Bush held a White House conference, attended by most of the industrial countries, where this dispute was ventilated but not resolved.

Later in 1990, the first IPCC reports appeared, demonstrating a broad consensus among scientists in the field that the possibility of global warming at least had to be taken seriously. If warming had not yet started, the IPCC said, continuing the increases in concentrations of greenhouse gases would certainly lead to it sooner or later. The Europeans cited the warnings in the report, while the Bush administration pointed to the uncertainties.

Throughout a long series of negotiating sessions, the United States flatly opposed any firm targets for CO_2 reduction in the emerging Framework Convention on Climate Change. The Europeans pushed and pulled vigorously, but got nowhere. In its final form, the text gave them half a loaf by acknowledging the desirability of reductions and setting the voluntary goal of cutting CO_2 emissions back to the 1990 level by the year 2000. Although that goal was merely an aspiration with no firm commitment or means of enforcement, it represented substantial movement from the original inclination of the United States to do no more than study the situation. The text of the Framework Convention was completed in time to be signed with great ceremony by nearly every country on Earth at the huge and colorful Conference on

Environment and Development held by the United Nations in Rio de Janeiro in June 1992.

The 1992 elections brought to Washington an administration more sympathetic to action on environmental issues than its predecessor. The Clinton administration immediately signaled its intention of moving away from the Bush administration's position, and in February 1993 President Clinton proposed a broad tax on all energy consumption. It became known as the "Btu tax," since it was to be based on each fuel's energy content measured in British thermal units. In April 1993, to celebrate Earth Day, Clinton announced that he would reverse the government's previous position and adopt a program to stabilize greenhouse gas emissions at the 1990 level by 2000, as the Framework Convention urged.

That was the high point of environmental concern in the first Clinton administration. The president was soon distracted by the great struggle over his budget. Congress, hostile to the idea of an energy tax from the beginning, whittled the original Btu proposal down to an increase of 4.3 cents a gallon in the gasoline tax—too little to have any significant impact on consumption. When the president's Climate Change Action Plan appeared later in the year, it turned out to be entirely voluntary. The details of the plan included the promotion of products and technologies that use energy efficiently, encouragement of industry to commercialize efficient technologies to bring down their prices through mass production, and review of regulatory rules affecting energy production and use. Many environmental advocacy organizations denounced the action plan as inadequate.

They turned out to be right. By 1996, greenhouse emissions in the United States were 8.3 percent above 1990 and continuing to rise steadily. But the Bush administration also turned out to have been right about the Europeans when it accused them of making promises that they could not keep. By 1996 it was clear that, of the world's major industrial powers, only three would have their emissions under the 1990 level in the year 2000—and none of them for reasons arising from environ-

mental policy. Russia would make it because of the tremendous drop in industrial production there since the collapse of the Soviet Union. Germany would succeed because it had been closing down the grossly inefficient plants, mostly fueled by brown coal, that it had inherited from the defunct Communist regime in the former East Germany. Great Britain would also succeed, because the government was cutting off its subsidies to the obsolescent coal industry. But nowhere, in any of the large economies, was there any sign of a serious and purposeful effort to reduce CO_2 emissions for environmental reasons.

The Science of Global Warming

At that point, the IPCC published its second survey of the subject in three fat volumes, the first of which was a review of the science that was markedly more decisive in its tone than its predecessor five years earlier. The statistical evidence, it concluded in a widely quoted line, "now points towards a discernible human influence on global climate." But it followed that sentence with a warning about the limitations on present knowledge: "Our ability to quantify the magnitude of this effect is currently limited by uncertainties in key factors, including the magnitude and patterns of longer-term natural variability and the time-evolving patterns of forcing by (and response to) greenhouse gases and aerosols."

This report is, in effect, a textbook representing mainstream opinion among the specialists in climate change and the many sciences that it touches. It does not represent unanimous opinion. Some researchers believe that the case for human influence on the climate is still unproved. But the IPCC report is now at the center of the debate over the science of global warming, which at least in Washington goes forward largely in terms of what the report said, or did not say, or should have said. This report is not as sharply conclusive as many politicians would have liked. It strongly emphasizes the many questions not yet answered and the uncertainties that make judgment difficult. But the appearance of this report marked the point at which it became clear that most of the scientists involved had concluded that, to one degree or another, human activity appears to be playing a part in global warming.

The concentration of carbon dioxide in the Earth's atmosphere, the report said, was about 280 parts per million (ppm) in 1750, before the industrial revolution began. By 1994 it was 358 ppm and rising about 1.5 ppm per year. If emissions continue at the 1994 rate, the concentration will be around 500 ppm, nearly double the pre-industrial level, by the end of the twenty-first century.

Other greenhouse gases, such as methane and nitrous oxide, have also been increasing. The effect is that the atmosphere retains more of the sun's heat, warming the Earth's surface. Not all human-made additions to the atmosphere increase warming. Aerosols—tiny particles of solid or liquid suspended in the air—tend to reflect heat and diminish warming. But aerosols are mostly short-lived while the carbon dioxide thrown into the atmosphere will stay there for decades.

Although the pattern of future warming is very much open to debate, it is indisputable that the surface of the Earth has warmed, on average, 0.3 to 0.6 °C since the later nineteenth century, when reliable temperature measurements began. Recent decades appear to be the warmest since at least 1400, according to the fragmentary evidence available. (Too little is known about the world's climate before 1400 to allow generalizations.)

In early 1998, American scientists at the National Climatic Data Center reported that the worldwide average surface temperature in 1997 was the highest ever recorded and the 1990s the warmest decade. The continuing trend, they said, tended to strengthen the evidence that greenhouse gases are a cause.

The warming has not been uniform across the Earth's surface. It has been greatest in the mid-latitudes in winter and spring. Night temperatures have increased more than daytime highs. Patterns of precipitation have also changed, with greater rain and snow fall in the high latitudes of the Northern Hemisphere and less in the subtropics from Africa to Indonesia. Global sea level has risen 10 to 25 centimeters over the past century.

The IPCC panel estimated that, on a world-wide average, temperatures will rise by 1 to 3.5 °C by the year 2100 and sea level will rise another 15 to 95 centimeters. Warmer temperatures could mean more droughts and floods. But the IPCC report stated that present data are inadequate to judge whether climate is becoming more unstable or whether storms are becoming bigger and more destructive.

Within the span of recorded history, the report observed, there is no time when the climate has been changing as rapidly as it now appears to be doing. Historical experience is consequently not a reliable guide. That's why it is necessary to rely on the models, although they still show systemic errors. There are also uncertainties about natural variation, since accurate records go back less than 150 years.

The IPCC report clearly warns that the current models are unlikely to include all of the feedback mechanisms in the planet's highly complex climate system. This point is highly important and cuts both ways in the political debate. It means that there may be triggers hidden in the system that could change the warming trend line suddenly and sharply. To some people, it seems intolerably dangerous to run the risk of a feedback that might accelerate the warming process or set off other disastrous effects. But the possibility that feedbacks might work in the other direction, to stabilize the climate, is the basis for much of the criticism of the work of the IPCC panel.

In its general outline and tone, the IPCC report gives great prominence to the uncertainties and gaps in the present scientific understanding of the climate. As it notes, politicians will have to make decisions on climate policy without waiting for conclusive scientific evidence on the key questions.

If the physical science of this subject is uncertain, the economics is even more so. The third volume of the IPCC report is devoted to the economic and social consequences of warming. The right approach to policy, the report proposed, is to ask what actions to take over the next decade or two to position the world to act on new findings as they appear. "Climate change demands a decision process that is sequential and can incorporate new information."

In the present state of knowledge, it is impossible to quantify the costs of doing nothing. A large factor is the risk, which is unknown, of catastrophic consequences of warming. Another factor, which usually goes unmentioned in the current debates, is the possibility of offsetting benefits of warming. One example is the prospect of longer growing seasons in northern countries like Canada and Russia.

A good deal of work has been done on the costs of mitigation, but it has resulted in a range of estimates too wide to be of much use in making policy. Many models have been constructed, but their results reflect the assumptions on which they rest. One of the strongest reasons for getting started on at least a small scale with some mandatory mitigation programs is to begin to accumulate reliable cost data based on actual experience.

Recent Negotiations

In early 1995, the United Nations had held in Berlin a "conference of the parties" to the Rio de Janiero meeting—in the jargon of the negotiators, COP-I. The idea was to assess progress on the grand promises made at Rio. The 120 governments represented at COP-I agreed to a plan, known as the Berlin Mandate, to pursue over the following two years an attempt to set specific and binding targets and timetables to reduce greenhouse gas emissions. These targets and timetables were to apply to the industrial countries, but not the developing countries. Although that provision seemed innocuous at the time, it became increasingly controversial as time passed.

The negotiators met again for COP-II in Geneva in the summer of 1996. At that session Timothy Wirth (the former senator, now undersecretary of state for global affairs and the chief American spokesman in this process) announced that the United States would support legally binding limits on emissions if other countries also did so. That was a clear and important reversal of long-standing American policy.

But there was still friction between the Americans and the Europeans. In December 1996, at an interim meeting, Wirth emphasized that the United States favored great flexibility in setting targets. Here he was following the advice of economists, who argue that flexibility means greater efficiency and lower costs in reaching targets. In the international talks, though, as in American domestic environmental politics, there is sharp debate between economists seeking efficiency and regulators who suspect that flexibility is merely a synonym for loopholes.

The Europeans were vigorously trying to turn up the pressure for action. On March 3, 1997, the fifteen-nation European Union called for a reduction of emissions by all industrial countries of 15 percent below the 1990 level by the year 2010. In the United Kingdom, where an election campaign was under way, the Labour Party pledged in its manifesto to put the U.K. at the front of the world environmental movement by supporting a 20 percent reduction in carbon dioxide by 2010. Labour's huge win in May gave further momentum to its demand.

The reasons for the tension between the Americans and the Europeans are rooted in deep differences between the economies and politics of the two continents. Only a few of the European countries have substantial energy industries under private ownership. Both the United States and Canada have hugely important energy industries with great political influence, not only through corporations but through the labor unions. By 1997, a number of American labor unions, led by the United Mine Workers, were fiercely hostile to any treaty on global warming. Europeans have a stronger conservation tradition and already have energy prices much higher than those in the United States. Some Europeans would also point out that they live much closer to the horrific pollution that the former communist governments of Eastern Europe created in their desperate efforts to wring faster growth out of their badly run industries. In some European countries, there are vigorous Green parties nipping at the heels of their governments.

In addition to the disputes between Americans and Europeans, the U.N. session also displayed a broader range of differences between North and South—the rich countries and the poor. At Rio the rich had promised new aid, amounting to about $6 billion, to the poor to help them toward more efficient use of energy. Very little of that money had actually been forthcoming. Some of the oil-exporting countries were trying to claim compensation for any worldwide attempt to drive down energy consumption and with it their incomes. A group of small countries inhabiting low-lying islands plaintively observed that continued rises in sea level threatened to sweep them and their people off the map altogether. More important, many people in the big countries that are well-embarked on industrialization, especially China and India, suspected that the whole idea of global warming was a device invented by the rich countries to hold down growth among their newest competitors. In general, the Third World regarded the threat of global warming, if it actually existed, to be a creation of the rich countries and felt that it was up to the rich countries to deal with it.

While the United States was committed in principle to binding limits, the Clinton administration was having trouble deciding exactly how much to reduce and over what period of time. Those questions were being hotly debated in endless meetings in Washington. The debate was being forced not only by criticism from abroad but by a deadline. A third meeting of the conference of parties, COP-III, was scheduled to be held in Kyoto in December 1997, and the negotiators hoped to sign a treaty there that would set the promised legal limits.

In this atmosphere President Clinton went to New York in June 1997 to address a special session of the United Nations. "The science is clear and compelling," he said. "We humans are changing the global climate." He spoke of new technologies and economic strategies such as emissions trading that would, he argued, allow reductions in greenhouse gases without damaging economic growth.

Clinton clearly felt that, before he proceeded to any serious action, he would have to get more Americans focused on the subject. On July 24, he held a White House conference in which he and Vice President Gore, the administration's ranking environmentalist, opened a campaign for greater public awareness. "We see the train coming," the president said, "but most Americans in their daily lives can't hear the whistle blowing."

On the following day, as though in response, the Senate passed, 95 to 0, a resolution telling the president not to agree at Kyoto to any treaty putting limits on the developed countries' emissions unless it also committed the rest of the world to take action. While lobbyists for some of the energy companies had been pushing this idea, it spoke to much wider concerns in Congress. Specifically, the Senate feared that the treaty might hamper American industrial expansion to the benefit of China—with which the United States is already running a very large trade deficit—and other newly industrializing economies.

The President's October 22 Program

President Clinton defined his position on Oct. 22, 1997, at the National Geographic Society in Washington—the proper setting, he thought, for an address on global climate change. He said that the United States would commit itself at Kyoto to "the binding and realistic target of returning to emissions of 1990 levels between 2008 and 2012. And we should not stop there. We should commit to reduce emissions below 1990 levels in the five-year period thereafter, and we must work toward further reductions in the years ahead."

Regarding the developing countries, he declared that, "both industrialized and developing countries must participate in meeting the challenge of climate change. The industrialized countries must lead, but developing countries also must be engaged. The United States will not assume binding obligations unless key developing nations meaningfully participate in this effort." The administration had decided not to fight with a unanimous Senate over the role of China and the other poor but rapidly growing countries.

As the administration conceived it, the Clinton program was to fall into four five-year phases. In the first, from now until 2002, the inducements would be all carrots and no sticks. The federal government would provide $5 billion in incentives, meaning subsidies and tax preferences, to encourage the development and installation of new technologies to produce and use energy more efficiently. To industrial plants that reduce their greenhouse gases now, the government would offer special credits to be cashed in when the mandatory restraints on emissions took hold later. As Gene Sperling, the president's assistant for economic policy, put it in a press briefing, this first phase would be devoted to "the things America can do without waiting—without, we think, conflict." The concept was to exploit first the technologies that already exist.

The second phase would be devoted to evaluation of what has been accomplished and taking account of new developments in science. Nothing envisioned in these first two stages, other than minor tax cuts and other incentives, was to require congressional action. The method would be, again, carrots alone.

The binding and mandatory limits were to be applied in the third phase, from 2008 through 2012, with the intention of getting the country's greenhouse gas emissions back down to the volumes of 1990. By the administration's calculation, if there was no change in policy, emissions would be 34 percent above 1990 by 2010. As the president described it, voluntary reductions might well bring emissions close to the necessary 1990 target even before the mandatory limits went into effect. The enforcement system, he suggested, might follow the model of the present market for sulfur dioxide rights to control acid rain. That mechanism applies only to large industrial power plants, most of them run by electric utilities.

The suggestion here was that the future emissions reduction regime would operate mainly, if not exclusively, through the large industrial sources. Through two decades of experience, environmental

policymakers have discovered that the American public has a fairly high tolerance for costs of protection as long as they are filtered gradually through utility bills. But it has very little tolerance for directly imposed costs, as in higher taxes for gasoline. There was nothing in the president's program that remotely hinted at restraints on individual citizens' activities, such as driving.

Nor was there any reference to nuclear power or hydroelectric power, at present the only technologies that generate electricity on a large scale without producing carbon dioxide. Any indication of an expansion of either would incense much of the environmental movement, illustrating one of the many hazards in climate politics.

In the fourth phase of this program, the years after 2012, the president said that the country would push its emissions down below the 1990 level to a target not yet specified.

Another notable omission in the president's plan was any discussion of preparations for life in a somewhat warmer world. If the president was correct in believing that CO_2 emissions were raising temperatures, some measure of warming is going to be inevitable in the twenty-first century. No one in political office, in this country or any other, is talking about the kind of draconian reductions in current emissions that would be required to hold the CO_2 concentrations in the atmosphere at their current levels. The question is not whether the world will get warmer, but how fast it will happen. Adaptation to a changing climate will be unavoidable. It is a subject that carries a heavy ideological freight: many people in the environmental movement suspect that any discussion of adaptation can only distract attention from the efforts to cut emissions.

President Clinton's purpose in his October 22 address was less to lay out a detailed program than to get Americans thinking about climate change and taking the prospect seriously. Before anything else could be done, he had evidently concluded, he needed to build a political base for action. Throughout his address, the president sought to elevate the issue of global warming without getting entangled in the difficult and controversial choices that a serious policy will, sooner or later, require. The time for vigorous administrative and regulatory blueprints would come only when, to use Clinton's metaphor, Americans begin to hear the train's whistle.

There is another side to this subject. However accurate the administration's political calculations might have been, certain economic realities also applied. In particular, the administration was caught in two dilemmas.

The first involved prices. Far from talking about increased energy prices, the administration was saying that under its plans they would stay low or perhaps even fall. But lower prices generally mean higher consumption and higher CO_2 emissions. The administration did not deal with this contradiction, merely deferring it—with much else—to the future. That led to the second dilemma.

The president's program depended heavily on the credibility of the prospect that, a decade from now, mandatory reduction in emissions would be enforced by federal law. If people do not believe that—especially the people making decisions on the design of long-lived industrial equipment and consumer goods—they are unlikely to begin making the substantial investments that a low-emissions regime requires. The administration's offer of $5 billion in subsidies and tax breaks would not begin to cover the enormous costs of shifting to that regime. To have any significant effect without greater outlays of federal money, it would be necessary that both industry and, ultimately, consumers consider it highly probable that a time will soon come when cooperation is no longer voluntary. The sketchiness of the Clinton program, its evasiveness on pricing, and its postponement of mandatory action past the next two presidential elections all militated against its credibility.

The Kyoto Conference

Less than six weeks after President Clinton announced his program, the international conference to draft a treaty opened at Kyoto, Japan. It was a huge affair, with some 10,000 officials from nearly every national government on Earth plus a large following of lobbyists and observers from a great variety of nongovernmental organizations representing environmental interests and industries. Like many hard-fought negotiations, it picked up speed only in the final hours when the meetings had run into overtime.

On December 8, Vice President Gore flew in from Washington for a one-day stop to push the process forward. Three days later, after an all-night session, the conference approved by consensus the text of the Kyoto Protocol.

Where the U.S. negotiators had wanted a target of returning to 1990 emissions levels and the European Union had wanted one 15 percent below it, they compromised on a target for the United States 7 percent below the 1990 level and for the European Union 8 percent below it. Within the European Union, the text allows wide country-by-country variations. Japan would cut emissions to 6 percent below 1990 levels. Several smaller economies would be allowed to increase emissions: Iceland 10 percent above 1990, Australia 8 percent above and Norway 1 percent above. Several would be required only to return to 1990 levels: Russia, Ukraine, and New Zealand. The others would reduce emissions by varying amounts in the range of 5 to 8 percent below 1990. These limits would apply to thirty-eight countries—those conventionally called developed, including Eastern Europe and some of the former Soviet Union. In the language of the protocol, these are the Annex B countries. These cuts are calculated to amount to a reduction in the industrial countries' total emissions of 5.2 percent below 1990.

The conference accepted the U.S. idea applying the targets to an average over the five-year target period 2008 through 2012, rather than to a single year. Flexibility was the theme of the U.S. position. The conference also accepted the U.S. proposal to include all six of the major greenhouse gases. There had been some sentiment in favor of covering, at this stage, only the three most important: carbon dioxide, methane, and nitrous oxide. As completed, the text also covered hydrofluorocarbons, perfluoro-carbons and sulfur hexafluoride. Emissions were to be measured by methodologies established by the IPCC in its role as scientific advisor to the process.

On a less conspicuous but equally important point, the governments were finally able to agree to count net, not gross, emissions. A forest is a sink for carbon dioxide; that is, it absorbs the gas out of the atmosphere. Most of the developed countries strongly favored allowing for sinks, on grounds that it would encourage planting trees and discourage burning them. Some developing countries, notably Brazil, were uneasy about this language, apparently fearing that it would turn into a condemnation of their land-clearing practices. But the developed countries won that one, and the final text (Article 3, paragraph 3) permits them to count net changes in sinks "resulting from direct human-induced land use change and forestry activities" since 1990.

The Kyoto Protocol was to be open for signature from March 16, 1998, to March 15, 1999. It was to come into effect ninety days after the fifty-fifth government ratifies it, assuming that those fifty-five countries account for at least 55 percent of the CO_2 emissions of the developed countries in 1990. That percentage was chosen to ensure that most of the developed countries' emissions would be covered, but not so high a proportion that any one country—meaning the United States—could veto the protocol by refusing to approve it. This provision was not, of course, likely to have much practical effect since few of the major economies seemed likely to ratify this agreement unless the United States did.

The Issues Left Unresolved at Kyoto

The Kyoto conference was a landmark, for it came to agreement on a document that covers an extremely difficult subject touching on many interests. It represented substantial progress toward a worldwide agreement on greenhouse gases and, by implication, the production and use of energy. But the negotiators arrived at that achievement only by explicitly postponing a series of crucial questions to later meetings. They also used notably vague language to patch over several troublesome disputes. There are five major points on which the Clinton administration needed further progress or clarification before it could safely send the Kyoto protocol to the Senate for the vote on ratification..

First and most obvious, the United States got nowhere at Kyoto with its demand that the largest of the developing countries commit themselves to some kind of limitation on emissions. In the weeks following the Kyoto conference, the Clinton administration flatly promised Congress that the protocol would not be submitted to the Senate until it had been revised to bring those countries into some kind of significant participation in the control regime.

Second, the United States has also insisted on international trading in emissions rights and on joint implementation, and it wants the protocol revised to provide for them more clearly. Trading means that a country that is under its limit can sell emissions rights to a country, or to a corporation in a country, that is over its limit. Joint implementation means that a company in, say, the United States, wishing to expand and increase its emissions, can earn the right to do it by making investments that reduce emissions elsewhere in the world. But, to make these markets work, the countries on both sides of the deal have to be under emissions limitations. Otherwise there is no baseline against which to score the increases and decreases. The greatest opportunities to reduce emissions cheaply are in the Third World. That adds to the American interest in bringing developing countries into the system.

Trading and joint implementation have run into substantial political opposition. Some Third World governments fear that the developed countries will use their great financial power to buy their way out of emissions restrictions and transfer those limits to poorer countries, where they will interfere with industrial development. Third World officials have spoken of joint implementation as environmental colonialism. Meanwhile, in Washington, it has not escaped congressional notice that trading and joint implementation would result in a substantial flow of investment capital from the United States to other countries. Representative Bill Archer (R-Texas), the chairman of the House Ways and Means Committee, called them "another form of foreign aid."

Third, the protocol provides for explicit financial aid in the form of funds given by rich countries to poor ones to help them set up emissions accounting procedures and to promote the transfer of technology. But it is vague about the amounts, speaking only of "new and additional financial resources" (Article 11, section 2 (a)). These financial issues have received a great deal of attention among Third World governments. Their disappointment with the performance of the rich countries under the Rio Treaty and the promises made in that treaty has contributed heavily to the suspicion and resentment with which many of these governments reacted to the U.S. proposals at Kyoto. While the financial aid provisions have hardly been mentioned so far in the United States, many developing countries consider them crucial to any further cooperation on their part.

Fourth, the whole question of sanctions and what to do about countries that fail to meet their commitments was deferred to a future meeting of the Conference of Parties (Article 17). The question was, in effect, too hard to resolve at Kyoto. Diplomats point out that the history of international agreements shows that most countries make a good-faith effort to live up to them most of the time. But that assurance is probably not going to be accepted by the opponents of the Kyoto Protocol, who observe that in this case violators would gain a substantial commercial advantage over their competitors.

Fifth, no government has yet offered any clear outline of the actual policies and methods by which it would meet its commitments under this Protocol. In this respect, the Clinton administration is typical. President Clinton's last word on the subject was his October 22 program calling for little more than voluntary efforts until 2008. But the protocol would commit it to an emissions reduction of 30 percent from the level that, by the administration's calculations, it would otherwise reach by 2010. In a period of only twelve years, a 30 percent reduction is a massive change with deep implications for industrial practices and consumers habits alike. If the administration has no intention of using taxes and higher energy prices to push that change, it is difficult to see how it can begin to meet this very ambitious target.

In the United States and every other democracy, emissions reductions on this scale will require public debate and public engagement with the issues of energy use of an intensity not yet visible. Some governments have recently begun to talk about action to meet the Kyoto requirements. But the governments at Kyoto set their goals with no clear strategy for getting there or clear estimates of the costs of doing it.

A Final Comment

The Kyoto conference was the beginning, not the culmination, of the first serious international attempt to address greenhouse gas emissions and the prospect of climate change. Kyoto was not the last word but rather only a stage in working toward genuinely binding international agreements. That has disappointed those people who had hoped for dramatic action. But it was consistent with the advice that scientists are still far from a reliable grasp of the planet's climatology, just as its diplomats and politicians are far from a consensus on dealing with it.

The global climate is not the only factor that will affect the quality of life on Earth in the twenty-first century. The population of the world is also rising rapidly. While its rate of growth has slowed over the last several decades, population is still rising several times as fast as the CO_2 concentrations in the atmosphere. The world's economic output in recent years has been rising half again as fast as its population and there are few people—certainly not President Clinton—who are prepared to slow down productivity in order to protect the climate. Global warming, population increase, and economic expansion are all related to each other.

"It is our solemn obligation to move forward with courage and foresight to pass our home on to our children and future generations," Clinton said at the National Geographic Society in October 1997. But because of all that a restless and creative humanity does from day to day, our children and future generations are going to live in a world very different from the present one. A changing climate will be only one of those differences.

Suggested Readings

Bettelli, Paola (ed.). 1997. *Earth Negotiations Bulletin* 12(76; December 13). Available on the Internet at http://www.iisd.ca/linkages/vol12/enb1276e.html. This gives a detailed and useful account of the Kyoto proceedings.

Intergovernmental Panel on Climate Change (IPCC). 1996. *Climate Change 1995:The Second Assessment Report*. In three volumes from Cambridge University Press: *The Science of Climate Change* (vol. 1); *Impacts, Adaptations and Mitigation of Climate Change* (vol. 2); *Economic and Social Dimensions of Climate Change* (vol. 3). The IPCC Web site, http://www.ipcc.ch, is a useful source about these reports, subsequent technical reports, and the ongoing Third IPCC Assessment, the report of which is due out around 2000.

Mintzer, Irving M. and J. A. Leonard (eds.). 1994. *Negotiating Climate Change: The Inside Story of the Rio Convention*. Cambridge: Cambridge University Press.

Paterson, Matthew. 1996. *Global Warming and Global Politics*. London and New York: Routledge.

33 Climate Change and Its Consequences

Michael A. Toman, John Firor, and Joel Darmstadter

Are humans changing the climate? In its latest assessment, scientists on the Intergovernmental Panel on Climate Change say we probably are, and the consequences could be serious. But uncertainties about risks and response costs make it difficult to formulate a specific long-term action plan. The potential risks the panel identifies, however, are sufficient to warrant additional actions beyond those now under way in the United States and other countries.

The Intergovernmental Panel on Climate Change, an international group of scientists, economists, and decision theorists convened by the United Nations, recently completed its second assessment of the current state of knowledge regarding human-induced changes in the Earth's climate and possible consequences. The goal of stabilizing atmospheric greenhouse gas concentrations—which the nations of the world agreed to under the 1992 Framework Convention on Climate Change and which the IPCC is charged with helping to effect—is difficult, touching as it does on national interests in varying ways. Thus controversy, including the allegation that politics has tainted the science, has arisen over the IPCC's latest findings (just as it did after the panel released its first report in 1990).

In this article, however, we focus on the substance of the reports that the IPCC's three working groups most recently produced, with particular emphasis on issues related to the use of fossil fuel and emissions of carbon dioxide (CO_2)—the greenhouse gas that contributes most to climate change. The three groups assessed the available information on (1) effects of human activity on climate conditions through modifications of the atmosphere; (2) potential impacts of this climate change, along with the technical potential for mitigating and adapting to it; and (3) socioeconomic consequences of climate change, including human responses to potential impacts.

Originally published in *Resources*, No. 124, Summer 1996.

Human Impacts on Climate

A striking feature of the new IPCC assessment is the conclusion by Working Group I that a human cause for the climate change now observed is likely, not just possible—a much stronger conclusion than the one reached in the first assessment. In 1990, the IPCC stated that although all signs pointed to human-induced climate change, crucial evidence for cause and effect was not available. The evidence then available indicated that atmospheric greenhouse gas concentrations had increased in the previous 130 years and that the global climate had warmed; however, when complex computer simulations of climate processes were applied retrospectively, they predicted a larger warming than had actually occurred and did not adequately represent climate changes in different regions and at different altitudes.

The latest generation of models can now replicate the past with greater realism. In particular, new models include analysis of the cooling effect of aerosols—tiny particles—in the air formed from sulfur emitted during the burning of fossil fuels.

By including in their analyses the cooling effects of aerosols and stratospheric ozone depletion, most of the latest studies have detected a significant climate change and, in the conclusion reached by Working Group I, show that the observed warming trend is "unlikely to be entirely natural in origin." The balance of evidence suggests a "discernible human influence on global climate."

Despite recent improvements, climate models are still unable to project the details of climate change on a regional scale, complicating assessment of potential impacts and response options. A further complication is the possibility that future climate change will be neither gradual nor continuous, but abrupt and surprising, as Working Groups I and II caution repeatedly.

Potential Natural Impacts

What constitutes a "dangerous" level of interference with the climate is a complicated question. In its latest assessment, the IPCC addresses many impacts of climate change, including the effects on agriculture, forestry, terrestrial and marine ecosystems, hydrology and water resource management, human health, human infrastructure, and financial services. While the potential impacts of climate change are broad, some aspects of human society are more sensitive than others. In particular, more highly managed systems like agriculture, where skills and resources for investing in adaptation are available, may be less sensitive than less managed systems like wilderness areas. However, some of the adverse effects of climate change may fall disproportionately on poorer, less-adaptive parts of the world.

The IPCC puts greater emphasis than it has in the past on the potential adverse effects of climate change on human health. Periods of sustained higher temperatures not only could increase mortality but also foster the spread of disease through greater water contamination and a wider geographic dispersion of disease-carrying organisms such as mosquitoes.

The ability to quantify future damage and adaptation potential varies greatly across sectors. The physical consequences of a given magnitude of sea level rise or the impacts of climate change on agricultural yields and forest conditions can be projected with higher confidence than, say, impacts on wetlands and fisheries. Yet even when confidence is high that a certain effect will occur if climate changes, its magnitude cannot be predicted precisely.

Working Group II also points out that damage to ecosystems and human structures arising from such other causes as population growth, industrial expansion, and changes in land use could combine with effects of climate change to push already stressed systems "over the edge." Particularly if climate change were very rapid, damage could be severe and long-lived, perhaps irreversible. However, such rapid change may be unlikely and is difficult to predict.

Socioeconomic Consequences

Decisionmakers contemplating public policies to deal with climate change need to understand the socioeconomic consequences that might follow from the physical and biological impacts of climate change. Uncertainties about these consequences are compounded not only by remaining scientific questions but also by diverse views about how socioeconomic consequences should be defined and measured.

The latest IPCC assessment notes the practical limits of conventional benefit-cost analysis as applied to climate change issues. Climate change involves risks of natural impacts that would be very long term, spanning multiple generations. Moreover, these impacts could be very large in scale and not so readily offset by substituting other capital investments. Such risks are not easily incorporated into conventional benefit-cost analysis.

Nevertheless, Working Group III rightly asserts that an economic benefit-cost assessment can help guide decisionmaking when coupled with an assessment of other factors. These include impacts that are not easily monetized and the distributional effects of climate change within and across generations.

In reviewing the available evidence on the economic impacts of climate change, Working Group III looked at a number of potential effects. These include impacts on agriculture and forestry, effects on water supplies, damages from sea-level rise to coastal areas and expenditures to protect them, increased mortality risks, effects on fisheries and wetlands, and effects of changes in conventional air and water pollutants.

But the group's assessment was not exhaustive. For lack of data, several important impacts of possible climate change were either only partly addressed or not addressed at all. These include broad ecosystem damages and the consequences of increased nonfatal illnesses. Moreover, the estimates reflect individual damage components, without fully accounting for the effects of multiple stresses brought on by forces other than climate change. The estimates also are based primarily on a single scenario in which the global climate has reached equilibrium after a doubled atmospheric CO_2 concentration. This formulation does not capture the cost of adjusting to climate change or the possible consequences of even higher greenhouse gas concentrations.

Aggregate damage estimates under these assumptions, expressed as a percentage of GDP to provide a sense of scale, tend to cluster around 1 to 1.5 percent for advanced industrial economies and 2 to 9 percent for developing countries. For some individual countries—say, small island states subject to flooding from sea-level rise—substantially higher costs could be incurred. Clearly, a number of thorny issues related to adaptive capacity and equity lie beyond estimates of aggregate damages.

The range of estimates for individual types of damage is wide, and the assessment recognizes the possibility of benefits, such as a longer growing season in some locations (leaving aside the costs of adjusting to climate change). Moreover, all damage figures are point estimates, lacking probability ranges or confidence intervals, and in many cases the estimates are simply educated guesses.

Effectiveness and Cost of Response Options

Emissions of CO_2—almost all generated by burning fossil fuel—account for about two-thirds of all enhanced heat trapping by greenhouse gases. Greater efficiency in the conversion and use of energy would obviously slow emissions, but no meeting of the minds exists on what it would cost to increase energy efficiency. Indeed longstanding differences of opinion about that cost enter into the IPCC's debate over the cost of reducing greenhouse gas emissions.

Some analyses reviewed by the IPCC indicate that decreases in energy use of 10 to 30 percent can be achieved at low or even negative cost by widespread adoption of technologies that people do not use now because of such market barriers as lack of information, uncertainty about product performance or lifetime costs, high up-front costs, the distorting effect of energy subsidies, and the "chicken and egg" problem created by low initial purchase volume and high initial price. By reducing these barriers, the argument goes, government policies could reduce greenhouse gas emissions very cheaply.

Economists accept the idea that energy and other markets do not always work effectively, which certainly is the case in many countries. But many economists remain skeptical that the apparent lack of interest in more energy-efficient products necessarily is a market failure, citing other explanatory factors, such as customer dissatisfaction with some product attribute that overwhelms consideration of its energy efficiency.

Accordingly, some of the analysts whose studies Working Group III surveyed do not believe that barriers to widespread adoption of technologies are a major problem, at least in advanced economies where markets generally work, or that their elimination offers a truly cost-effective way of lowering mitigation costs. Thus their estimates of these costs are often higher than the technological state of the art would imply. For example, these latter analyses suggest that the cost to OECD* countries of stabilizing carbon dioxide emissions at 1990 levels over the next several decades could range from –0.5 percent (a small net increase in gross domestic product) to as much as 2 percent of GDP. (In evaluating these estimates, it is important to keep in mind the fact that GDP is not an accurate measure of social well-being.)

Moreover, these estimates tend to assume use of the most cost-effective emissions control policies, such as carbon taxes or emissions trading. If the policies put in place were actually less cost-effective, the estimated economic burden would increase.

Regardless of one's position in this debate, an important conclusion to arise from Working Group III's review is that the total costs of meeting a long-term target for reducing greenhouse gas concentrations in the atmosphere may be reduced substantially by stretching out the time period of emission reductions and providing emission sources with flexibility in the timing of reductions. Such flexibil-

---

*OECD is the Organisation for Economic Co-operation and Development.

ity could cushion abatement costs by reducing premature obsolescence of existing capital and permitting greater development and deployment of new, efficient technologies.

With regard to the eventual necessity for global participation in curbing greenhouse gas emissions, the IPCC notes the tension between that ideal and the need to respect the economic development priorities of the world's lower-income countries. Yet by meeting those priorities, poorer countries might expand their capacity to cope with climate change stresses, in addition to raising their living standards.

If fairness and implementation issues can be resolved, the IPCC points out that significant opportunities exist for international cooperation to lower the costs of emission reductions. These opportunities include "joint implementation" projects in which richer countries make investments in reducing emissions in poorer countries. Properly structured, such projects can convey tangible economic and environmental benefits to recipient countries while lowering the total costs of greenhouse gas emission reductions.

As for permitting countries some flexibility as to when and where emissions are reduced, while it is true in principle that such flexibility can increase economic efficiency, in practice the ability to do so may be constrained by political considerations. Permitting delays in emissions reductions may lack credibility because of skepticism that governments will honor previously made commitments to pursue aggressive reductions. Developing countries also have expressed suspicion about the motivations for joint implementation and a desire for more concrete action by developed countries themselves.

To overcome these concerns may require developed countries to carry out greater and more immediate emissions reductions than a simple analysis of economic efficiency would indicate. The IPCC assessment also considers the way in which adaptation measures—for example, improved water management—can contribute to both economic efficiency and increased resiliency to weather fluctuations and climate change. Indeed, given the IPCC's conclusion that some climate changes have already been set in motion, some adaptation is already essential as well as desirable in order to avoid some of the costs of mitigation. More detailed attention to adaptation is needed in future assessments, however.

Final Comments

Based on the insights, information, and findings of the IPCC's second assessment, it is now much more difficult to argue that human activities are not changing the climate. It is also now easier to argue that the impacts of climate change may be substantial, surprising, and unfair.

Unfortunately, the continuing uncertainties about the scale and nature of climate change, its consequences, and the costs of response make it difficult to specify a long-term plan of action at this time. Legitimate debate continues about what constitutes—and how best to avoid—a "dangerous" interference with the climate system.

For our part, we believe that the latest IPCC assessment justifies some degree of policy intervention that goes beyond actions to improve economic efficiency *without* reference to climate change, although neither the United States nor the other industrialized nations have yet to exhaust all opportunities for these "no-regrets" actions. While the potential risks are difficult to quantify, the IPCC assessment strongly suggests that they are not zero. Given that society is not impervious to risks, some anticipatory efforts to reduce threats as well as efforts to improve the understanding of their magnitude are called for.

The task is not easy. The second assessment underscores the challenge of understanding and responding to the ecological and socioeconomic aspects of climate change and other closely intertwined global problems, as well as the need for further understanding of how the climate is affected by human activities. Climate scientists need to focus on the regional manifestations of climate change and the variability of these changes; impact studies must become more quantitative and effective adaptations need to be better identified; and economists

must extend and supplement their tools for assessing the consequences of global change and the costs of policy responses.

The opportunities, as well as the needs, for new approaches in these fields are substantial. To reap these opportunities, governments and other sources of research funding should maintain or increase their budgets for climate change analysis, and a greater share of future research budgets should be allocated to ecological and socioeconomic research.

Suggested Reading

Firor, John. 1994. Resource Letter: GW-1: Global Warming. *American Journal of Physics* 62: 490–495. (This contains a list of readings about global warming and its consequences prepared for the American Association of Physics Teachers.)

Nordhaus, William D. ed. 1998. *Economics and Policy Issues in Climate Change.* Washington, DC: Resources for the Future.

Repetto, Robert, and Duncan Austin. 1997. *The Costs of Climate Protection: A Guide for the Perplexed.* Washington, DC: World Resources Institute.

Toman, Michael, and Jay Shogren. Forthcoming. "Climate Change Policy." In *Public Policies for Environmental Protection,* 2nd ed., Robert Stavins and Paul R. Portney, Eds. Washington, D.C.: Resources for the Future.

The Energy Modeling Forum at Stanford University has carried out a comprehensive review of the economic cost of controlling greenhouse gas emissions (the EMF12 Study) and is in the midst (as of 1998) of a follow-on study (EMF16) concerned with mitigation costs and climate change damages. The EMF Web site, www.stanford.edu/group/EMF/Publish.html, provides information on these materials. A summary of the EMF12 findings also can be found in: Darius Gaskins and John Weyant, 1993, Model Comparisons of the Costs of Reducing CO_2 Emissions, *American Economic Review* 83(May): 318–323.

The three-volume report on the Second Assessment by the Intergovernmental Panel on Climate Change, published in 1996 by Cambridge University Press under the heading *Climate Change 1995*, contains a wealth of information about climate science, climate change impacts, response options, and economic/social dimensions. The IPCC Web site, www.ipcc.ch, provides information about this publication, subsequent technical reports on specific topics, and the ongoing (as of 1998) Third IPCC Assessment.

The RFF climate Web site, www.weather-vane.rff.org, contains a number of Issues Briefs and other short articles providing information on climate change economics and policy issues.

34

Adapting to Climate Change

Pierre R. Crosson and Norman J. Rosenberg

Adaptation and mitigation are sometimes treated as mutually exclusive approaches for dealing with global warming, but it is now acknowledged that they can be pursued jointly and that there are tradeoffs between them. A critical policy issue is the determination of which mix of adaptation and mitigation measures will maximize the benefits of efforts to reckon with climate change. Unfortunately, much less is known about how to adapt to that change than how to mitigate it. Despite this fact, the developed and developing countries have a mutual interest in devising adaptive responses, even if agreement on mitigation strategies remains elusive.

There is now reasonable scientific consensus that the continued loading of the atmosphere with radiatively active trace gases such as carbon dioxide, methane, nitrous oxide, chlorofluoro-carbons (CFCs), and other gases will cause the troposphere (the lower portion of the atmosphere) to warm. As a consequence of this warming, climatic conditions throughout the world would change; less certain is the nature of these changes and where they would occur. Still less certain are the rate at which the atmosphere might warm, the attendant rate of change in climatic conditions, whether transient climate changes will occur (for example, cooling in a locale before it warms), and whether climate might change so that the frequency and severity of extreme events—such as droughts, storms, floods, and freezes— might be altered. Despite these uncertainties, it is highly likely that greenhouse-forced warming could have significant impacts on water resources, unmanaged ecosystems, agriculture, forestry, and fisheries, and on the societies and economies dependent upon them. Although the impacts likely would be beneficial in some regions, there is a serious risk that the world as a whole might be considerably worse off. Prudence argues that we should mitigate or eliminate the risk of climate change, if we can.

However, atmospheric science also makes it clear that some amount of greenhouse warming is probable in the next century. The warming potential of the greenhouse gases that have already accumulated in the atmosphere has probably not been fully expressed because of the great capacity of the oceans to

Originally published in *Resources*, No. 103, Spring 1991.

absorb heat before they warm noticeably. Hence, even if emissions of greenhouse gases were reduced quickly enough to avoid further accumulation of them in the atmosphere, some additional warming would be likely. If even in the best of cases we cannot totally avoid greenhouse warming and consequent climate change, prudence suggests that we look for ways to adapt to whatever the change may be.

Adaptation and mitigation are sometimes treated as mutually exclusive approaches for dealing with global warming, and arguments in support of one approach may be treated as threats by advocates of the other. In a 1987 article, "Global Climate Change: Toward a Greenhouse Policy," in *Issues in Science and Technology*, Jessica Matthews described "adaptionists" as those who emphasize learning to live with greenhouse warming and "preventionists" as those who emphasize the need to slow and eventually halt warming. This schismatic classification is, we believe, a misreading of the relationship. Surely few adaptionists, if any, believe that permanently increasing warming would pose no threat to global society. And surely the most committed preventionist, if convinced that some warming is inevitable, would deem it incumbent upon government to undertake adaptive action to reduce the consequent threats to life and property.

That the adaptation/mitigation argument has progressed from the either/or stage is evidenced in a number of ways. Scientists attending the second World Climate Conference held in Geneva, Switzerland, in November 1990 recognized the need for research to strengthen our understanding of the potential impacts of climate change and of ways to adapt to it. Another manifestation is the creation within the executive branch of the U.S. government of the committee on Mitigation and Adaptation Research Strategies (MARS) to coordinate interagency activities on those strategies.

Finding the Right Mix

Adaptation and mitigation policies are simultaneously complements and substitutes. The policies are complements in the sense that they can be, and should be, pursued jointly. They are also substitutes, meaning that there are tradeoffs between them as policies for dealing with global warming. Many, if not most, of the resources that could be devoted to the development of adaptive responses to climate change could also be devoted to pursuit of mitigation. The more resources that are devoted to one course, the fewer available for the other.

As an economic issue, the critical question is: What are the costs and benefits of alternative levels of effort devoted to the two courses of action? The benefits are the social values of damages averted; costs are the social values of the resources devoted to the aversion effort that could have been turned to some other purpose. Many of the costs and benefits could be expressed in dollars—for example, the value of the labor and capital used to build barriers against a rise in sea level. Other costs and benefits, however, could not be adequately expressed in dollars—for example, the loss of ecological values in unmanaged forests or the community values preserved where successful adaptation permits continued farming in a region disadvantaged by climate change. The nonquantifiable costs and benefits likely would be of major importance. Despite the uncertainty about them, they must be taken into account in thinking about the relative merits of adaptive and mitigative responses to climate change.

Whatever the answers to the economic question, the critical policy issue is to find the mix of adaptation and mitigation measures that maximizes the net social benefits of efforts to deal with climate change. This policy mix defines the total amount of resources that should be devoted to dealing with climate change and also the socially optimum allocation of resources between adaptation and mitigation.

The outcome of this assessment of the relative merits of adaptation and mitigation strategies would not be as tidy as the foregoing statement might suggest. The great uncertainty about the costs and benefits of the two approaches and the political struggle among the various interests with a stake in the outcome assure that choices about the mix of strategies would be anything but clearcut. The point here is that however fuzzy the decision-making process, the choices should reflect recognition that, because there are tradeoffs between adaptation and mitigation, the concept of an optimum mix of the two approaches is meaningful.

Two Kinds of Adaptive Response

It is also important to recognize that there are two kinds of adaptive response to climate change, to which we now turn. One response includes all those things people would be induced to do within the existing institutional and policy regime. The other consists of institutional and policy changes that would be called for where and when the existing regime proved inadequate to deal with the impacts of climate change. The distinction is important because the resources available to undertake changes in institutions and in policies are always limited. These resources can be conserved to the extent that adaptations undertaken within the existing institutional and policy regime are successful.

Examples of the two kinds of adaptive response to climate change can be found in agriculture. Studies of the impacts of climate change on agriculture show that in many areas, including the U.S. Midwest, crop yields (output per acre) might fall with higher temperatures and less precipitation. The fall in yields would increase production costs to farmers, inducing them to investigate existing technologies and management practices for better ways to adapt to the changed climate. Farmers might turn to conservation tillage, a technique that conserves more soil moisture than the more commonly used tillage techniques. They might also adopt already available crop varieties that are better adapted to the hotter and drier climate, and invest in irrigation to counter the decline in precipitation. All of these adaptations are examples of measures that people would be induced to undertake within the existing institutional and policy regime.

However, in some circumstances these induced adaptations may be judged inadequate in the sense that after they have been made, society appraises the remaining costs of climate change as unacceptably high. In such a case, institutional or policy changes would be called for to develop additional adaptations that would bring the remaining costs within acceptable limits.

If farmers find that the alternatives available to them from among existing technologies and management practices are inadequate to compensate for the negative impacts of climate change, they may face the prospect of going out of farming, and perhaps leaving a region altogether. This prospect could stimulate agricultural research institutions and those charged with responsibility for agricultural policy to invest more in research to develop a new set of technologies and practices better adapted to the changed climatic regime. Institutional rules for allocating irrigation water might also be changed to give farmers greater flexibility in using water on their own farms and in transferring it among farms.

Prospects for Adaptation

The power of adaptation to offset negative consequences of climate change, or to permit exploitation of favorable consequences, has been little studied. It is likely that adaptive responses would be powerful in some circumstances and weak in others. A study conducted by Resources for the Future of the impacts of and responses to climate change in the four-state region of Missouri, Iowa, Nebraska, and Kansas showed that adaptations would significantly reduce the negative impacts of a hotter and drier climate on crop yields. RFF researchers projected climate conditions of the 1930s (the dust bowl years), which are consistent with predictions of hotter and drier weather produced by some climate models, on that region as it might be in 2030. Simulation models of plant growth indicated that, in the absence of adaptations, the production of corn, sorghum, and soybeans (three of the principal crops in the region) would be about 22 percent less in 2030 than if the climate did not change. When allowance was made for adaptations that farmers could make, including new technologies developed by research institutions, the simulation models showed a decline in production of corn (the most sensitive to climate change of the crops studied) of only 9 percent.

These results are, of course, speculative, but they are consistent with the history of the adaptability of farmers and of the ability of agricultural research institutions to respond to changing conditions of resource scarcity with which farmers must deal. The results are suggestive, therefore, of the power of adaptation in responding to climate change and of the importance of distinguishing between the two kinds of adaptive response.

In order to fully capture the benefits of the adaptive strategy to climate change, much more knowledge about the payoffs of various kinds of adaptation will be needed. The same could be said, of course, about mitigation strategies. But in at least one way adaptation is more complicated than mitigation. The physics of the greenhouse effect is understood, as are the ways to diminish the threat of global warming. Adaptation, however, raises a different set of problems stemming from the fact that we do not know how climate will change in any particular region and, hence, cannot know what the impacts of climate change will be. Regional climate changes are unpredictable as yet, and the prospects for improved predictability in the near term are poor. Thus the investment of great effort and resources now in developing specific adaptations to climate change for specific industries or infrastructures in specific regions would probably not pay off well.

One exception is adaptation to a rise in sea level, which will affect all of the world's shores, although not uniformly. As the atmosphere and (eventually) the seas warm, sea level will rise, threatening coastal areas around the world. But here, too, the possible rise in sea level is difficult to predict; estimates range from less than 0.5 meters to more than 1.5 meters during the course of the next century. In fact, it seems likely that the greatest impacts on land and people adjacent to the sea could be more the result of changes in wind force and direction than of a rise in sea level per se.

Finding a Sensible Adaptation Strategy

What, then, is a sensible strategy for adapting to future changes in climate? First, we must gain a better understanding of the sensitivity and vulnerability of specific regions, industries, ecosystems, and societies to the normal range of climatic variability, and what can be done to diminish this sensitivity and vulnerability. For example, the North American Great Plains and many other regions of the world are subject to recurrent droughts. What technical and institutional measures can be applied to diminish the impacts of drought so that these regions can be made more resilient than they are today? Knowledge gained from answering this question would be directly applicable in the event that droughts in these regions become more severe or more frequent.

Second, research establishments should be working now to develop better responses to climate variability. Such research would produce many of the techniques needed for adaptation to climatic change because the primary threat of that change lies in more severe and more frequent extreme events.

Third, as knowledge of the dynamics of climatic change improves or as signs of change are perceived, or both, scientific and engineering resources should be assigned to the development of the specific adaptations needed. This would require, of course, that the scientific establishment remain capable of effective reaction from now until the time at which adaptations must be put into action.

The adaptive strategy might have a high payoff, and research to identify opportunities for adaptation and to provide knowledge and techniques needed to adapt would be a central part of that strategy. The developed countries seem well positioned to follow this strategy because their reliance on markets promotes flexibility in reallocating resources and their research establishments are strong. But what of the developing countries, with their smaller endowments of means and resources? Will they have the capacity to adapt as easily as the countries with higher per capita income?

There is no reason to believe that the developing countries, as a group, will be exposed to worse climatic changes than will the developed countries. It is certain, however, that their margin of survival would be smaller and that their opportunities for adapting to climatic change might be severely limited where the institutional and technical infrastructure, including research capacity, is weaker. In a paper entitled "Potential Strategies for Adapting to Greenhouse Warming: Perspectives from the Developing World" in the RFF volume *Greenhouse Warming: Abatement and Adaptation* (1989), N. S. Jodha, an agricultural economist from India, argues that farmers in developing countries use age-old techniques to cope in times of stress, and that these provide an arsenal from which to draw when climate change imposes a need for adaptation. Jodha provides many examples of the use of these techniques in India. There are exceptions, of course. In areas where agriculture is already risky because of severe climate or poor soils, particularly in the semiarid tropics, any detrimental climate change, however small or slow, can accentuate the risks and have serious impacts.

Because the developing countries are preoccupied with raising their currently low standards of living, they have shown relatively little interest in mitigating global warming. As Jodha shows, however, these countries, without necessarily having any greater interest in an adaptive strategy, nonetheless have accumulated substantial experience in adaptation, particularly in agriculture. Of course, the developing countries, like the developed countries, will need more knowledge of the prospective impacts of climate change and of possibilities for effective adaptive responses. There is a mutuality of interest here between developed and developing countries that may foster cooperative efforts in devising strategies for adaptation, even if agreement on strategies for mitigation remains elusive.

35

A Framework for Climate Change Policy

Michael A. Toman

Differing interpretations of the evidence—and differing interests—complicate efforts to negotiate goals and actions regarding climate change. While no easy cookbook-style recipe can indicate what should guide thinking about risks and policies, several maxims are worth applying.

A great deal of controversy surrounds the issue of climate change. Some say that climate change is one of the greatest threats facing humankind, one that calls for immediate and strong controls on greenhouse gases from fossil fuel burning. Others say that the risks are weakly documented scientifically, that adaptation to a changing climate will substantially reduce human vulnerability, and that little action is warranted other than study and development of future technological options. The same kinds of divides arise in discussing policy options to reduce greenhouse gas emissions, with some predicting net benefits to the economy and others fearing the loss of several percentage points of national income.

These disagreements surface in the efforts of the international community to negotiate goals and actions under the 1992 Framework Convention on Climate Change. They reflect different interpretations of the evidence and different interests. To help sort through the tangle, I have summarized some ways to think about climate change risks and policies that may be useful in considering both international agreements and actions by the United States.

Decision Framework

While no easy recipe indicates what should go into a climate change decision framework, several maxims are worth applying.

Originally published in *Resources,* No. 127, Spring 1997.

Think Comprehensively about Climate Change Risks

Efforts to gauge climate change risks and the benefits of reducing emissions should be as broad as possible. Elements to consider include the impacts on market goods like agriculture; effects on human health; effects on nonmarket resources like wilderness areas and wetlands, which provide both recreational values and ecological functions; and the ancillary benefits of greenhouse gas reduction, such as improved air quality.

Given the current state of knowledge, it will be difficult to attach monetary values to a number of risk reductions and costs. This uncertainty is likely to persist for many risk categories (especially those related to ecological impacts) even if uncertainty about the physical manifestations of climate change declines. However, lack of information should not be confused with negligible risk. To be useful to decisionmakers, moreover, an assessment of climate change risks should go beyond a sequence of "best guess" or "worst case" estimates of atmospheric changes, biophysical impacts, and socioeconomic impacts to consider the variability of possible consequences.

Think Long-Term

The risks posed by climate change depend on the path of changes in the atmospheric concentration of greenhouse gases over many decades and centuries, not just on the emissions of these gases over a relatively short period of time. We are dealing with the cumulative effect of many smaller influences on the biosphere—an effect with a great deal of natural inertia.

Having to confront the distant future greatly complicates risk assessment and the development of consensus for policy actions. To be effective, at least some actions must anticipate long-term impacts before all of the scientific evidence is clear. Our political system arguably is less effective at taking such actions than responding to a single large and immediate concern. On the other hand, the long-term nature of climate change risks means we can hone our scientific understanding and policy responses over time; we need not do everything right away.

Address Adaptation

In areas such as agriculture, managed forestry, and human settlements, intuition and experience in other contexts suggest a medium-to-high degree of potential adaptability to natural changes, given enough lead time and investment. Adaptation possibilities include development of new plant varieties and crop patterns, changes in irrigation technology, relocation of coastal infrastructure, and expanded protection of wetlands to compensate for their potential future damage.

Adaptation may be difficult in some cases, for example, where damage occurs to natural ecosystems whose functions are not well understood. But even when adaptation capacity seems very limited, it should not automatically be treated as negligible. Improving the capacity to adapt where it is weak—as in many poor, developing countries—may be one of the most effective ways to respond to some climate change risks, at least until the cost of stabilizing atmospheric concentrations of greenhouse gases falls.

Think Internationally

Rich and poor countries argue over how the burden of greenhouse gas emissions reductions should be allocated. The ongoing tension can only be resolved by negotiation among the parties themselves. However, long-term global climate change risks will not diminish to any significant degree until total *global* emissions are reduced, and this will require global cooperation, not just action by today's rich countries. This point deserves to be underscored in light of the likely future decline in the share of total emissions from advanced industrial countries (currently about 50 percent) as economic growth proceeds in other areas. The efficacy of any policies the United States pursues to reduce climate change risks thus will depend on the actions taken by others.

Keep Distributional Issues in Mind

Climate change risks and response capacities vary with income level. There is also a fundamental asymmetry between the timing of response costs—which will largely be borne by the current generation—and the benefits of reduced climate change—which will largely accrue to future generations. This asymmetry means we cannot simply compare the costs of reducing the risk with the value of enjoying the ultimate benefits. Instead, we must assess both the costs members of the current generation would bear and the strength of our concerns for those who would be vulnerable in the future. These are economically and ethically complex questions about which we know little, and they require mature political debate.

Estimate Control Costs Comprehensively and Realistically

Just as it is important to think broadly about climate change risks, it is also important to have solid estimates of the direct and indirect economic impacts of greenhouse gas abatement. Some people argue that market inefficiencies are so rife, and opportunities for innovation so plentiful, that emissions abatement is a low-cost proposition or one that might even benefit the economy. This point of view is in sharp contrast to the outputs of economic models indicating that stabilizing emissions may cost as much as one percent (or even more) of a country's gross domestic product (implying that deeper cuts in emissions to reduce greenhouse gas concentrations in the atmosphere would be even more expensive).

Most people who have looked at the debate seem to agree that some low-cost improvements in energy efficiency exist, for example, by reducing subsidies and other market distortions. However, it is open to question whether these opportunities are substantial compared with, say, the amount of abatement needed to stabilize greenhouse gas emissions. Against the backdrop of future increases in global energy demand, the cost of longer-term reductions in greenhouse gas emissions cannot help but rise unless further progress occurs in the development of nonfossil energy alternatives. In assessing medium- to long-term costs, it is a mistake to assume either technical progress as a panacea for reducing abatement costs or no technical progress at all.

Another argument is that our tax system is so distorted that we can levy energy taxes to reduce

greenhouse gas emissions and use the proceeds to lower other taxes that hamper economic growth. However, recent analysis highlights the limits of this "double dividend." The basic conclusion of this analysis is that broader-based taxes like those on income generally create less overall economic distortion than narrower-based taxes like those on energy. Thus, adjusting other taxes might dull the economic pain of an added energy tax, but not to negligible levels. Moreover, any tinkering with the tax system is possible only if politicians take the difficult step of imposing higher energy taxes in the first place.

Most studies of greenhouse gas abatement costs assume the application of idealized least-cost policy measures like a comprehensive "emissions trading" program or a comprehensive "carbon tax" based on the carbon content of different fossil fuels. Abatement costs will be higher (perhaps considerably so) if less than ideal policies are used in practice. The debate about which greenhouse gas reduction targets are appropriate cannot be conducted independent of discussions about what concrete measures can and should be used to actually restrict emissions.

Implications for Policymaking

The decision framework I have described has several implications for formulating policy.

Incorporate Economic Incentives into Emissions-Reduction Policy

These incentives include carbon taxes on energy sources and various forms of tradable permit systems that would effectively establish quotas on emissions but allow their trade. Sources with higher control costs could (in effect) pay emitters with lower control costs to assume more of the reduction burden, thus lowering the total cost to society.

Allow Flexibility in the Timing of Cumulative Emissions Reductions To Reduce Overall Costs

The potential cost savings from intertemporal flexibility in meeting a particular long-term emissions-reduction goal depend on the assumptions made, but it appears that savings of 50 percent or more are possible. Taking this approach does not mean that all or even most policy actions are deferred to the future, or that flexibility should compromise credibility. It simply means that the emphasis is placed on sequential decisions—some of which are better taken sooner and others later. Unless we start with a longer-term perspective, it is impossible to consider such tradeoffs.

Provide Opportunities for Emissions Reductions Wherever Possible

One example of an abatement incentive program that takes place outside industrialized countries is the so-called "joint implementation" approach, whereby emitters in, say, the United States, can satisfy any emissions-reduction requirements they face through actions that reduce emissions in other countries. Formal emissions-trading programs among sources in countries with quantified emissions-reduction targets also are possible. Significant practical questions need to be answered to structure flexible yet verifiable programs for international (and intertemporal) emissions trading. However, the magnitude of the potential cost savings underscores the value of seeking to overcome these challenges. Depending on the assumptions made, savings of 50 percent or more seem possible.

Build Knowledge and Improve Technology

Even if we do all the best things possible to reduce emissions, given the current state of knowledge, economic growth—especially in developing countries—will continue to push up greenhouse gas emissions and atmospheric concentrations. Unlike limiting pollutant gases such as sulfur dioxide, for which a variety of technical control options is available, limiting carbon-dioxide emissions currently requires reduced energy use, greater energy efficiency, or substitution of energy sources with lower carbon content.

To avoid unacceptable climate change risks ultimately will require a fundamental change in our energy systems toward much greater reliance on other energy sources—solar, biomass, and possibly nuclear. To make the transition economically manageable will require continued or enhanced investments in basic and applied knowledge.

The government has an inescapable role to play not just in creating the incentives for private parties to seek better technologies but also in funding the development of basic knowledge about technology as well as climate change impacts. At the same time, we must recognize that our understanding of what policy can actually do to induce climate-friendly innovation is weak at best. We must also recognize that diverting resources from other areas to research on low-carbon energy systems may well reduce innovation elsewhere in the economy—technical progress is not a free good.

Increase Emphasis on Adaptation

Adaptation is part of an optimal response strategy in any event. Indeed, it is the means of transcending the narrower concern about our vulnerability to climate change to a broader concern with global-scale changes that place stress on natural systems and pose threats to human well-being. Furthering human capacity to adapt to climate change entails investment in improved understanding of the options and their international application. It also entails adjusting economic and other distortions that limit adaptation potential (such as assistance programs that subsidize coastal development or water use). In many cases, the best climate policy may have little to do with greenhouse gases or climate per se, and much more to do with developing better basic social infrastructures for natural resource conservation and use and for public health protection.

Suggested Reading

William D. Nordhaus, Ed. 1998. *Economics and Policy Issues in Climate Change*. Washington, DC: Resources for the Future,.

Toman, Michael. 1998. Research Frontiers in the Economics of Climate Change. *Environmental and Resource Economics* 11(April/June): 603–621.

The three-volume report on the Second Assessment by the Intergovernmental Panel on Climate Change, published in 1996 by Cambridge University Press under the heading *Climate Change 1995*, contains a wealth of information about climate science, climate change impacts, response options, and economic/social dimensions. The IPCC Web site, www.ipcc.ch, provides information about this pub-

lication, subsequent technical reports on specific topics, and the ongoing (as of 1998) Third IPCC Assessment.

The RFF climate Web site, www.weathervane. rff.org, contains a number of Issues Briefs and other short articles providing information on climate change economics and policy issues.

36 Global Trade in Greenhouse Gas Control
Market Merits and Critics' Concerns

Jonathan Baert Wiener

A world market for "greenhouse gas" emissions abatement services could lower the costs of preventing global climate change, widen the availability of climate-friendly technology, and engage more countries in emissions reduction efforts. So why is the United States having such a hard time getting other countries to like the idea?

Governments around the world are negotiating to reduce the amount of heat-trapping "greenhouse" gases (GHGs) we emit into the atmosphere. The challenge is to cut the emissions that may be changing the world's climate without hobbling the world's economies. One of the ways in which the international community could meet this challenge is to create a world market for emissions abatement. (See the box, "Tax versus Trade: The Pros and the Cons" for a comparison of the two.)

Market Options

Any international treaty intended to prevent global warming would need to impose and enforce limits on nations' emissions of GHGs. One approach would be to require that each nation stay within its limit on its own. The market alternative is to require the same global limit while allowing flexibility across nations in the locations where actual reductions are achieved.

Two kinds of international markets for GHG emissions abatement can be envisioned. One is a formal "cap and trade" system similar to the one adopted by the United States in 1990 to control the sulfur dioxide (SO_2) emissions that cause acid rain. An international treaty would establish a global cap on aggregate GHG emissions for some period of time and specify shares of emission allowances for each participating country. The governments of these countries would allocate their allowances to the private sector.

Originally published in *Resources*, No. 129, Fall 1997.

Worldwide allowance trading would reallocate abatement efforts to those who could do so most cost-effectively: emitters with high costs of abatement would seek to buy additional allowances, and emitters with low costs of abatement would undertake additional controls and seek to sell unneeded allowances. Organized exchanges would facilitate trades.

For each accounting period established by treaty, a country's report of its actual emissions (subject to monitoring and verification) would be compared with the allowances held by its emitters. If a country's emissions exceeded total allowances held, it would be out of compliance with the treaty.

The second kind of market envisioned is "informal." An international agreement would set national limits on emissions but not allocate formal allowances. Participating countries could meet their targets not only by investing in GHG emissions reductions at home but also by purchasing credits for emissions reductions in other countries, including in countries not subject to an overall emissions target.

This informal system is similar to the "pollution offsets" programs that the United States has employed for new emissions sources in certain areas. It is essentially the system of "joint implementation" (JI) outlined in the Framework Convention on Climate Change signed at Rio de Janeiro in 1992. In the sequel in Kyoto in December 1997, the United States is expected to press other parties to the framework convention both to institute a formal international market in tradable GHG emissions allowances among countries with caps and to recognize official credits for JI worldwide.

The Case for Emissions Trading

Why is the United States keen on establishing a market for international trade in emissions allowances? One of the key draws is cost-effectiveness. The cost of reducing GHG emissions varies significantly from place to place. Yet the global environmental benefits are essentially independent of where emissions are reduced. Numerous studies

Tax v. Trade: The Pros and the Cons

How does imposing a tax on greenhouse gas emissions compare with instituting a market for allowance trading? In economic theory, the two could achieve identical results. In practice they could be quite different.

A tax would offer more certainty about the costs involved in emissions reductions, since the tax rate would be fixed in advance, whereas the price of emissions allowances could vary. But a tax would offer less certainty about the amount of emissions control achieved, since it would not establish emissions caps.

A tax would not incur the transaction costs of allowance trading, but it would incur administrative costs to collect.

A tax system could be circumvented: national subsidies could be funneled to high-emitting industries to buffer the tax, distorting competition and increasing emissions.

And a tax would not create an automatic mechanism for transfers of resources and technology from richer to poorer countries. Such transfers are critical to getting developing countries engaged in GHG emissions reductions and thus to getting competitiveness-conscious industrialized countries to act as well.

indicate that flexibility as to where GHG emissions abatement can take place would cut the estimated total cost of compliance with emissions caps considerably—perhaps by 50 percent or more.

The United States has used allowance trading to achieve some of its greatest environmental successes, such as phasing out lead in gasoline and cutting emissions of SO_2. The cost savings in the lead and SO_2 cases have been substantial—as much as 50 percent or more compared with a control policy in which no trades were allowed. The SO_2 policy has also stimulated energy efficiency investments and the use of new abatement technologies. And the SO_2 experience suggests that a more cost-effective, market-based policy enabled Congress to sign on to more pollution control than it would have if control were more expensive. Similarly, reducing the cost of GHG abatement would likely lead countries to undertake more abatement than they otherwise would.

Depending on how international GHG abatement responsibilities are allocated, allowance trading could direct the flow of substantial resources from richer to poorer countries, where abatement costs appear to be relatively lower. These resources—perhaps exceeding all current official development aid—would help developing countries shift to a more prosperous but lower-emissions development path and would attract their participation in the GHG abatement regime at a time when their emissions will soon account for over half the world total. (And if developing countries do not participate, industrialized countries concerned about their economic competitiveness relative to developing countries are unlikely to sign on.)

Generic Concerns

Despite its advantages, many countries express concerns about creating an international emissions trading system. Most of the concerns they cite, however, would apply to any internationally based emissions control regime. Some are problems that trading could actually ease.

A fundamental challenge for any treaty is deterring "free riders"—nonparticipating countries that benefit from efforts to reduce emissions without adhering to limitations themselves. If free riding were not deterred, the entire collective regime might unravel. Adding allowance trading to a GHG treaty could make free riding less tempting. For industrialized countries, it would lower the treaty "price of admission" by allowing them to cut GHG emissions in the most cost-efficient way. For developing countries, allowance trading would raise the profits to be made from treaty participation, since industrialized nations would purchase allowances and credits from them.

Emissions "leakage" is another problem that would afflict any subglobal treaty, whether it employed trading or not. The problem occurs when reductions achieved in one place only encourage emissions to grow where caps do not apply. Such leakage could arise in the short term as emissions controls lowered world fossil fuel prices and in the long term as industries relocated to avoid emissions controls.

Informal JI projects probably could lead to some leakage on a local scale, if credits purchased from a project resulted in emissions growth elsewhere in a country not subject to a cap. Care would be needed in project design and the calculation of JI credits to account for such leakage. (Formal allowance trading among capped countries poses no such local leakage concern.) On a global scale, however, both JI and formal trading could reduce total emissions by inhibiting free riding and attracting more global participation in emissions control. By lowering abatement costs, GHG markets would give industries less reason to relocate to escape controls.

Another concern is the effect that a market would have on the ability to forecast "baselines"— what amount of emissions might occur in a given time and place and how much abatement actually was achieved. Under JI it might be difficult to gauge what emissions otherwise would be in a host country not subject to an emissions cap. (A formal trading market among capped countries does not raise this concern.) But prohibiting JI credits because of uncertainty would eliminate the opportunity to engage countries without national emissions caps in early GHG control efforts, as well as the opportunity to obtain low-cost abatement services in those countries. A better approach is to allow both cap-and-trade *and* JI, and to use benchmark rules to assign uncertainty-adjusted credit to JI projects. Investors could seek extra credits for providing more reliable emissions accounting, thus creating incentives to improve measurement capabilities in developing countries.

Critics also worry that it would be difficult to allocate emissions allowances among countries. But this problem is unavoidable in any climate agreement; emissions trading just makes allocations explicit. And without trading, dispositive national caps would be much harder to negotiate. If they had flexibility to reallocate emissions allowances through trading after a treaty had been signed, negotiators would face less pressure to devise ideal, permanent allocations in the treaty itself.

Specific Concerns

Some concerns do apply with special force to market-based emissions trading regimes. First, should the costs of arranging transactions in an emissions abatement market be high, they would impede trades and raise total costs. These "transaction costs" include searching for trading partners, negotiating deals, securing regulatory approval, monitoring and enforcing deals, and insuring against the risk of failure. Evidence from previous U.S. "environmental markets" such as the lead phasedown, the Los Angeles smog control program, and the Fox River water pollution control program suggests that such costs can determine the success or failure of a trading system.

The transaction costs of JI appear to be very high. Partners are hard to identify, each negotiation is novel, each project must be approved by the host and investor governments, and each investor must monitor its own projects. Moreover, JI typically involves investors supporting and bearing the risk of entire projects. JI transaction costs could be reduced, however, through brokers (many of which are emerging), information exchanges, streamlined approval processes, accredited monitoring agents (including environmental nongovernmental organizations), mutual funds and other means of risk diversification, and official credit.

The costs of transacting in a formal allowance trading market would be much lower, especially if fungible allowances were traded on organized exchanges. Indeed, reducing transaction costs would be a central goal of a formal system.

Second, a global allowance market could be impeded if national governments interfered in global trading. To be fully cost-effective, the entities actually responsible for GHG abatement must do the trading. Assigning allowances and credits to these entities will galvanize decentralized competition, creativity, and flexibility. But this approach might not be carried out well (or at all) in countries where the state is an active supervisor or owner of industry.

And national governments might try to influence the world abatement market to their advantage, obstruct allowance trades, or otherwise depart from the conditions of well-functioning markets assumed in the estimates of cost savings. Such meddling might be limited by international trade law, depending on how this law ends up applying to GHG allowances.

Third, concentrated power over allowance or credit prices could arise on the sellers' side, through a cartel or a large state-run energy company, or on the buyers' side, through a sole-purchasing agent for industrialized countries. Unlike domestic antitrust law, international law has no basic framework to combat such "market power." Climate treaty features such as less-than-unanimous voting rules for admitting new participants into the abatement market, or automatic phased inclusion of countries upon meeting pre-set criteria, could thus be crucial.

Understanding the Opposition

On balance, international GHG emissions trading appears to offer compelling advantages—lower emissions-reduction costs, valuable resource and technology flows, and greater participation in emissions reduction efforts. So what explains the opposition?

Clearly, some of it is due to misunderstanding and to genuine doubts that the system will work as envisioned. And some of it reflects a fear that such a market would lead to "carbon colonialism," if wealthy investors could depress allowance and credit prices, leading poorer countries to sell out their futures at a loss. This is a sincere concern about market power and must be addressed on its merits.

But other motivations appear to be at work as well. Some may feel it is unfair to include poor countries in a market-based control regime; but allowance trading would benefit (not harm) poor countries, and excluding developing economies would invite leakage and undermine a treaty's environmental effectiveness. Others may reject trading because their objective is not so much to protect the climate as it is to combat what they view as immorally extravagant lifestyles and excessive energy consumption. Some bureaucrats may disfavor private market transactions because they gain from their ability to manipulate official government aid more adroitly.

Wealthy countries with comparatively low abatement costs (say, in Europe) may prefer a less flexible control regime than emissions trading. Although less flexibility would cost them a little, it would cost their trade rivals (the United States and Japan) even more—a new global version of the "predation by regulation" phenomenon.

Opposition might also mask a desire to gain leverage over the greenhouse gas emissions reduction goal (target or cap). Advocates of aggressive climate protection may be withholding support for trading until it is paired with a stringent cap—risking a costly treaty or no agreement at all. Meanwhile, skeptics of aggressive climate policy may fear that cost-effective policy tools are an all-too-enticing "fast train to the wrong station," inducing premature

adoption of an overly stringent cap. Of course, the goal of climate policy should be chosen with great care. Yet these skeptics' gambit of urging a higher-cost "slow train" (in the hopes that it will derail any GHG limitations agreement) may just invite "Murder on the Orient Express"—a treaty that is both higher in cost and less environmentally effective—a "lose–lose" luxury train to the wrong station.

Suggested Reading

Cropper, Maureen L., and Wallace E. Oates. 1992. Environmental Economics: A Survey. *Journal of Economic Literature* 30(2, June): 675–740.

Stavins, Robert N. 1997. Policy Instruments for Climate Change: How Can National Governments Address a Global Problem. *University of Chicago Legal Forum* 1997: 293– 329.

Wiener, Jonathan Baert. 1999. Global Environmental Regulation: Instrument Choice in Legal Context. *Yale Law Journal* 108.

———. 1999. The Political Economy of Global Environmental Regulation. *Georgetown Law Journal* 87.

Part 9

Sustainable Development

37 The Difficulty in Defining Sustainability

Michael A. Toman

For ecologists "sustainability" connotes preservation of the status and function of ecological systems; for economists, the maintenance and improvement of human living standards. Disagreements about the salient elements of the concept hamper determination of appropriate responses for achieving sustainability. Key topics about which disagreement arises include inter-generational fairness, the substitutability of natural and other resources, and the carrying capacity of natural ecosystems.

Disparate perspectives on these topics might be bridged through the concept of the safe minimum standard, which posits a socially determined demarcation between moral imperatives to preserve and enhance natural resource systems and the free play of resource tradeoffs.

"Sustainability" has become a new watchword by which individuals, organizations, and nations are to assess human impacts on the natural environment and resource base. A concern that economic development, exploitation of natural resources, and infringement on environmental resources are not sustainable is expressed more and more frequently in analytical studies, conferences, and policy debates. This concern is a central theme in the international deliberations leading up to the United Nations Conference on Environment and Development in June 1992. To identify what may be required to achieve sustainability, it is necessary to have a clear understanding of what sustainability means.

Like many evocative terms, the word sustainability (or the phrase "sustainable development," which more strongly connotes concerns of particular importance to developing countries) means many things to different people and can be used in reference to a number of important issues. The term inherently evokes a concept of preservation and nurturing over time. The World Commission on Environment and Development (known popularly as the Brundtland Commission) labeled sustainable development in its 1987 report *Our Common Future* as "development that meets the needs of the present without compromising the ability of future generations to meet their own needs." Thus sustainability involves some notion of respect for the interests of our descendants. Beyond this point, however, uncertainty and disagreement are rife.

In scholarly usage, the term sustainability originally referred to a harvesting regimen for specific reproducible nat-

Originally published in *Resources*, No. 106, Winter 1992.

ural resources that could be maintained over time (for example, sustained-yield fishing). That meaning has been considerably broadened by ecologists in order to express concerns about preserving the status and function of entire ecological systems (the Chesapeake Bay, the biosphere as a whole). Economists, on the other hand, usually have emphasized the maintenance and improvement of human living standards, in which natural resources and the environment may be important but represent only part of the story. And other disciplines (notably geography and anthropology) bring in concerns about the condition of social and cultural systems (for example, preservation of aboriginal knowledge and skills).

Beyond ambiguity of meaning there also is disagreement about the prospects for achieving sustainability. The Brundtland Report foresees "the possibility for a new era of economic growth, one that must be based on policies that sustain and expand the environmental resource base." Some scholars, notably the economist Julian Simon, question whether sustainability is a significant issue, pointing out that humankind consistently has managed in the past to avoid the specter of Malthusian scarcity through resource substitution and technical ingenuity. Others, notably the ecologists Paul and Anne Ehrlich and the economist Herman Daly, believe that the scale of human pressure on natural systems already is well past a sustainable level. They point out that the world's human population likely will at least double before stabilizing, and that to achieve any semblance of a decent living standard for the majority of people the current level of world economic activity must grow, perhaps fivefold to tenfold. They cannot conceive of already stressed ecological systems tolerating the intense flows of materials use and waste discharge that presumably would be required to accomplish this growth.

Ascertaining more clearly where the facts lie in this debate and determining appropriate response strategies are difficult problems—perhaps among the most difficult faced by all who are concerned with human advance and sound natural resource management. Progress on these fronts is hampered by continued disagreements about basic concepts and terms of reference. To narrow the gaps, it may be helpful first to identify salient elements of the sustainability concept about which there are contrasts in view between economists and resource planners on the one hand, and ecologists and environmental ethicists on the other.

Key Conceptual Issues

As noted above, intergenerational fairness is a key component of sustainability. The standard approach to intergenerational tradeoffs in economics involves assigning benefits and costs according to some representative set of individual preferences, and discounting costs and benefits accruing to future generations just as future receipts and burdens experienced by members of the current generation are discounted. The justifications for discounting over time are first, that people prefer current benefits over future benefits (and weight current costs more heavily than future costs); and second, that receipts in the future are less valuable than current receipts from the standpoint of the current decisionmaker, because current receipts can be invested to increase capital and future income.

Critics of the standard approach take issue with both rationales for unfettered application of discounting in an intergenerational context. They maintain that invoking impatience entails the exercise of the current generation's influence over future generations in ways that are ethically questionable. The capital growth argument for intergenerational discounting also is suspect, critics argue, because in many cases the environmental resources at issue— for example, the capacity of the atmosphere to absorb greenhouse gases or the extent of biological diversity—are seen to be inherently limited in supply.

These criticisms do not imply that discounting should be abolished (especially since this could increase current exploitation of natural and environmental capital), but they do suggest that discounting might best be applied in tandem with safeguards on the integrity of key resources like ecological life-support systems. Critics also question whether the preferences of an "average" member of the current generation should be the sole or even primary guide to intergenerational resource tradeoffs, particularly if some resource uses threaten the future wellbeing of the entire species but are only dimly experienced by current individuals. Adherents of "deep ecology" even take issue with putting human values at the center of the debate, arguing instead that other elements of the global ecological system have equal moral claims to be sustained. Even if one accepts that human values should occupy center stage, it is difficult to gauge what the values held by future generations might be.

A second key component of sustainability involves the specification of what is to be sustained. If one accepts that there is some collective responsibility of stewardship owed to future generations, what kind of "social capital" needs to be intergenerationally transferred to meet that obligation? One view, to which many economists would be inclined, is that all resources—the natural endowment, physical capital, human knowledge and abilities—are relatively fungible sources of well-being. Thus large-scale damages to ecosystems such as degradation of environmental quality, loss of species diversity, widespread deforestation, or global warming are not intrinsically unacceptable from this point of view; the question is whether compensatory investments for future generations in other forms of capital are possible and are undertaken. Investments in human knowledge, technique, and social organization are especially pertinent in evaluating these issues.

An alternative view, embraced by many ecologists and some economists, is that such compensatory investments often are infeasible as well as ethically indefensible. Physical laws are seen as limiting the extent to which other resources can be substituted for ecological degradation. Healthy ecosystems, including those that provide genetic diversity in relatively unmanaged environments, are seen as offering resilience against unexpected changes and preserving options for future generations. For natural life-support systems, no practical substitutes are

possible, and degradation may be irreversible. In such cases (and perhaps in others as well), compensation cannot be meaningfully specified. In addition, in this view environmental quality may complement capital growth as a source of economic progress, particularly for poorer countries. Such complementarity also would limit the substitution of capital accumulation for natural degradation.

In considering resource substitutability, economists and ecologists often also differ on the appropriate level of geographical scale. On the one hand, opportunities for resource tradeoffs generally are greater at the level of the nation or the globe than at the level of the individual community or regional ecosystem. On the other hand, a concern only with aggregates overlooks unique attributes of particular ecosystems or local constraints on resource substitution and systemic adaptation.

A third key component of sustainability is the scale of human impact relative to global carrying capacity. As already noted, there is sharp disagreement on this issue. As a crude caricature, it is generally true that economists are less inclined than ecologists to see this as a serious problem, putting more faith in the capacities of resource substitution (including substitution of knowledge for materials) and technical innovation to ameliorate scarcity. Rather than viewing it as an immutable constraint, economists regard carrying capacity as endogenous and dynamic.

The Safe Minimum Standard

Concerns over intergenerational fairness, resource constraints, and human scale provide a rationale for some form of intergenerational social contract (though such a device can function only as a "thought experiment" for developing our own moral precepts, since members of future and preceding generations cannot actually be parties to a contract). One way to give shape to such a contract is to apply the concept of a safe minimum standard, an idea that has been advanced (sometimes with another nomenclature) by a number of economists, ecologists, philosophers, and other scholars.

To simplify somewhat, suppose that damages to some natural system or systems can be entirely characterized by the size of their cost and degree of irreversibility. Since ecologists do not view all the effects of irreversibility as readily monetizable, these two attributes of damages are treated separately (see Figure 1). The magnitude of cost can be interpreted in terms of opportunity cost by economists or as a physical measure of ecosystem performance by ecologists.

Irreversibility reflects uncertainty about system performance and the resulting human consequences. At one extreme, very large and irreversible effects may threaten the function of an entire ecosystem. At a global level, the threat could be to the cultural if not the physical survival of the human species. In Figure 1, this extreme is represented at the upper lefthand corner. At the other extreme (the lower righthand corner), small and readily reversible effects are relatively easily mediated by private market transactions or by corrective government policies based on comparisons of benefits and costs.

There is uncertainty about how rapidly the threat to current and future human welfare grows as damages become costlier and irreversibility becomes more likely. The safe minimum standard posits a socially determined dividing line between moral imperatives to preserve and enhance natural resource systems and the free play of resource tradeoffs. To satisfy the intergenerational social contract, the current generation would rule out in advance actions that could result in natural impacts beyond a certain threshold of cost and irreversibility. Rather than depending on a comparison of expected benefits and costs from increased pressure on the natural system, such proscriptions would reflect society's value judgment that the cost of risking these impacts is too large. Possible resources for which society would not risk damages beyond a certain cost and degree of irreversibility include wetlands, other sources of genetic diversity, the climate, wilderness areas, Antarctica, and other ecosystems with unique functional or aesthetic values (like the Grand Canyon).

There is a distinct difference between the safe minimum standard approach and the standard prescriptions of environmental economics, which involve obtaining accurate valuations of resources in benefit-cost assessments and using economic

Figure 1. Diagram of the safe minimum standard for balancing natural resource tradeoffs and imperatives for preservation.

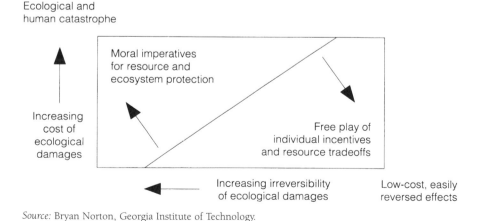

Source: Bryan Norton, Georgia Institute of Technology.

Update

Since I wrote this piece in 1992, the term "sustainability" has continued to be embraced by advocates of a variety of positions, threatening to render the term not just ambiguous but meaningless in policy debates. Within the economics literature, the term continues to connote concerns with intergenerational equity and irreversibility that seem not to be adequately addressed through the usual discounted-present-value approach of benefit-cost analysis. Along with refinement of the concept and its relationship to standard economic decision criteria, recent research has begun to address empirical assessment of sustainability, especially through the definition and calculation of a variety of *indicators*. The citations of my work in Suggested Readings are included as much for their bibliographies as their own content. The citation to the 1996 report of the President's Council on Sustainable Development illustrates the diffuse application of the concept in the formation of policy recommendations.

incentives to achieve efficient resource allocation given these valuations. Whether a resource-protection criterion is established by imperatives through an application of the safe minimum standard concept or by tradeoffs through cost-benefit analyses, that criterion can be cost-effectively achieved by using economic incentives. However, for impacts on the natural environment that are uncertain but may be large and irreversible, the safe minimum standard posits an alternative to comparisons of economic benefits and costs for developing resource-protection criteria. It places greater emphasis on potential damages to the natural system than on the sacrifices experienced from curbing ecological impacts. The latter are seen as likely to be smaller and more readily reversible. In addition, the

safe minimum standard invokes a wider, possibly less individualistic set of values in assessing impacts. Since societal value judgments determine the level of safeguards, public decisionmaking and the formation of social values are explicit parts of the safe minimum standard approach.

This illustrative discussion of course provides no actual guidance on where and how (if at all) such a dividing line between imperatives and trade-offs should be drawn. The location of the line will depend on the range of individual beliefs in society and available knowledge about human impacts on ecosystems. For example, ecologists who are concerned mainly about irreversibility and believe that ecological systems are fragile might draw an essentially vertical line, with a large area covered by moral imperatives for ecosystem protection; economists who are concerned mainly about expected cost and believe that the well-being of future generations should be highly discounted might draw an essentially horizontal line, with little (or no) scope for moral imperatives. Acquisition of additional knowledge also will alter the relative weight given to imperatives and tradeoffs for specific ecosystems or the environment as a whole. In addition, how the delineation would be made depends on complex social decision processes, some of which probably have not yet been constructed.

The safe minimum standard thus does not provide an instant common rallying point for resolving the disagreements discussed here. However, this concept does seem to provide a frame of reference and a vocabulary for productive discussion of such disagreements. Such discussion would refine understanding of what sustainability means and the steps that should be taken to enhance prospects for achieving it.

Research Needs

There is a need for much additional interdisciplinary work to refine the concept of sustainability. Along with basic concept definitions, extensions of economic and ecological theory to more fully account for the objectives and constraints of sustainability would be useful. To clarify some of the points of disagreement already outlined, substantial interdisciplinary data-gathering and analysis also would be required. This empirical work should address issues in developing countries and in developed countries, and those relevant to the entire world.

The tension between ecological and economic perspectives on sustainability suggests several ways in which both economists and ecologists could adapt their research emphases and methodologies to make the best use of interdisciplinary contributions. Economists could usefully expand analyses of resource values to consider the function and value of ecological systems as a whole, making greater use of ecological information in the process. Economic theory and practice also could be extended to consider more fully the implications of physical resource limits that often are not reflected in more stylized economic constructs. In addition, research by economists and other social scientists (psychologists and anthropologists) could help to improve understanding of how future generations might value different attributes of natural environments. Finally, the sustainability debate should remind economists to carefully distinguish between efficient allocations of resources—the standard focus of economic theory—and socially optimal allocations, which may include intergenerational (as well as intragenerational) equity concerns.

For ecologists, the challenges include providing information on ecological conditions in a form that could be used in economic valuation. Ecologists also must recognize the importance of human behavior, particularly behavior in response to economic incentives—a factor often given short

shrift in ecological impact analyses. Finally, it must be recognized that human behavior and social decision processes are complex, just as ecological processes are. What may appear as self-evident to the student of natural environment need not seem so for the student of human society, and vice versa.

Suggested Reading

Lile, Ron, Michael Toman, and Dennis King. 1998. "Assessing Sustainability: Some Conceptual and Empirical Challenges," RFF Discussion Paper 98-42; forthcoming in *International Journal of Environment and Pollution*, 1998, vol. 10.

Norton, Bryan, and Michael Toman. 1997. Sustainability: Ecological and Economic Perspectives. *Land Economics* 73(November): 553–568. (This article appears in a special issue of *Land Economics* on sustainability that contains a number of other excellent papers on the subject.)

President's Council on Sustainable Development. 1996. *Sustainable America: A New Consensus.* Washington, DC: Government Printing Office. (See also the critique of this report in *Resources,* Issue 123, Spring 1996).

Toman, Michael. 1994. Economics and "Sustainability": Balancing Tradeoffs and Imperatives. *Land Economics* 70(November): 399–413.

Toman, Michael. 1998. "Sustainable Decision-making: The State of the Art from an Economics Perspective," RFF Discussion Paper 98-39; forthcoming in M. O'Connor and C. Spash (eds.), *Valuation and Environment: Principles and Practices;* Cheltenham, UK: Edward Elgar Publishing.

38 Sustainable Agriculture

Pierre R. Crosson

Most future increases in global demand for food are expected to arise by 2050. By that time, demand could increase by 2.5 to 3.0 times the present level. The global agricultural system will fail to increase food production that much over the next 60 years if policies to achieve agricultural sustainability focus primarily on increasing the supplies of energy, land, water, climate, and genetic resources in the present state of knowledge. The potential supplies of these resources simply are inadequate. The only hope of sustainably meeting the future increase in demand for food is to invest in expanding the supply of knowledge about agricultural production.

Concern about the world's ability to feed itself dates at least from the time of the English economist Thomas Robert Malthus in the early nineteenth century. The concern has waxed and waned since then, but the adequacy of global agricultural capacity still figures prominently on the policy agendas of many countries and international organizations concerned about economic development. It surely will be prominent in the deliberations of the United Nations Conference on Environment and Development in June 1992.

A sustainable agricultural system is one that can indefinitely meet demands for food and fiber at socially acceptable economic and environmental costs. There is unavoidable ambiguity in the meaning of socially acceptable costs. No consensus has emerged about what standards should be used to judge acceptability. Yet concern about costs drives the current discussion about sustainability in agriculture and development generally. If we are to think fruitfully about the concept of sustainability in agriculture we cannot avoid thinking about costs.

Concern about sustainability reflects a sense of intergenerational obligation. With respect to agriculture, this means that each generation is obliged to manage its affairs so as to provide subsequent generations with the opportunity to engage in agricultural production at acceptable economic and environmental costs.

Sustainability cannot be discussed usefully without specifying the spatial scale of production units and the possibilities for movement of goods and people among units. In the absence of

Originally published in *Resources*, No. 106, Winter 1992.

such possibilities, the agricultural system of a region may be unsustainable because it cannot meet the demands on it at costs the people of the region find acceptable. Where trade and emigration are possible, the relevant spatial scale is greater, a region can substitute lower-cost food and fiber for its own high-cost production, and people can move from one region to other regions where costs are lower. Thus the agricultural system for a group of regions (or countries) linked by trade and migration of people may be quite sustainable even though the systems for each separate region (or country), without the linkages, would be unsustainable. Most farmers are connected through trade to markets for their output in their immediate region and often to more distant regional, national, and international markets. Thus the spatial scale appropriate for discussions of sustainable agriculture is global.

A discussion of sustainable agriculture must also specify the scale of the demands for production imposed on the system; in general, the problems of achieving sustainability become more difficult as demand for the system's output increases. The quantitative dimension of sustainability thus is crucially important.

Taken together, the above concepts create a workable meaning of sustainable agriculture. That meaning has a temporal dimension—the indefinite future; a spatial dimension—the world as a whole; a quantitative dimension—the demands placed on the system now and in the future; and a normative dimension—the need to meet those demands over time at economic and environmental costs that society deems to be acceptable. In considering the sustainability of the present agricultural system in these respects, it is useful to begin with prospective future demands on the system.

The Global Demand Scenario

If current population projections by the United Nations are accurate, most of the future increase in global demand for food will occur by about 2050. By then global population will be close to the expected ultimate total of 10 billion to 12 billion

(the present global population is 5.2 billion). In addition, if the global system as a whole proves to be sustainable, per capita income in the less developed countries (LDCs) will have risen to the point at which additional income would stimulate little additional spending on food because at that income level most people would be adequately nourished. In more developed countries (MDCs), per capita income already is at that point. Thus the critical period for the global agricultural system is roughly the next 60 years. If the system can sustainably meet the increase in demand over that period, it probably will be indefinitely sustainable.

Research at Resources for the Future (RFF) indicates that the projected increase in global population, combined with a plausible increase in per capita income in the LDCs, could increase global food demand 2.5 to 3.0 times the present level by the middle of the next century. The sustainability question is whether the global agricultural system will be able to increase food production that much over that period at acceptable costs. The answer to the question will depend on the ability of the system to mobilize the resources—the social capital—necessary to sustain the production increase.

The Concept of Social Capital

The question of sustainability can be put in terms of the kinds and amounts of social capital that intergenerational equity requires to be passed from one generation to the next. Social capital consists of all the natural and human-made resources used to produce goods and services valued by people. For agricultural sustainability, social capital includes supplies of energy, land, irrigation water, plant genetic material, climate, and knowledge embedded in people, technology, and institutions.

Energy
Over the next several decades global energy supplies are likely to be increasingly constrained by both rising real prices and concerns about the environmental costs of fossil fuels—among them the costs of the greenhouse effect on the global climate.

Experience since the run-up in energy prices in the 1970s suggests that farmers should be able to adjust reasonably well to future increases in energy costs, should they occur. There is little doubt, however, that eventually the costs of fossil fuels will rise high enough to pose a threat to sustainability, not only in agriculture but also in the economy as a whole. Avoidance of the threat will require development of renewable and other nonfossil sources of energy. When this must occur is uncertain; but that in time it must occur is not.

Land

The supply of land has both quantitative and qualitative dimensions. The United Nations Food and Agriculture Organization estimates that worldwide some 1.5 billion hectares currently are in crops of all kinds. Sketchy estimates indicate some 1.8 billion additional hectares have the soil and climate conditions suitable for crop production. However, for several reasons this estimate surely overstates the amount of land that could be converted to crop production over coming decades at acceptable economic and environmental costs. Much of the potential cropland is of inferior quality in comparison with current cropland. Moreover, most of it is in Africa and Latin America, but much of the future increase in demand for food will be in already land-scarce Asia. Asian countries will be able to draw on imports to offset some of their land constraints, but concern about food self-sufficiency probably would limit this response. Asian countries are not likely to view a hectare of potential land in Africa and Latin America as equivalent to a hectare within their own borders.

Estimates of potential cropland are also overstated because they do not take account of the opportunity costs of converting the land to agriculture. Yet these costs could be significant. Much land around urban areas in LDCs will be priced out of the agricultural market by demands to accommodate rising urban populations. And the clearing of forests in order to graze animals and raise crops already is seen by many as having high opportunity costs because of the losses of plant and animal

genetic diversity that clearing is believed to entail. Governments in the tropics are under increasing pressure from governments of MDCs and the international environmental community to reduce these losses by curbing forest clearing, and the pressure likely will continue to grow.

As noted, the average quality of most potential cropland is less than that of land presently in crops. In addition, the quality of agricultural land can be and is degraded by soil erosion, salinity buildup in irrigated areas, compaction from overuse of heavy tractors or trampling by animals, loss of nutrient supply through overgrazing, and other kinds of damage. Global land degradation through these various processes is widely believed to be severe. However, work done at the World Bank and elsewhere indicates that the evidence of land degradation is too sparse to warrant firm conclusions about the extent of the problem. Research at RFF and at the U.S. Department of Agriculture indicates that soil erosion in the United States, widely believed to be a major threat to the sustainability of the nation's agriculture, is not in fact a serious problem. Comparable studies have not been conducted for other countries. It is worth noting, however, that global crop yields (output per hectare) continue to increase, as they have for the last 40 years, indicating that on a global scale soil erosion has not so far seriously impaired land quality.

Water

About 17 percent of the world's cropland, producing about one-third of global crop output, is irrigated. Almost 75 percent of this land is in the less developed countries, 62 percent of it in Asia—mostly in India, China, and Pakistan. Africa has a little more than 4 percent of the global total of irrigated agricultural land, and Latin America about 6 percent.

World Bank estimates indicate that, based solely on soil and climate factors, the present area of irrigated land worldwide could be increased about 50 percent. However, these estimates, like those for potential cropland, almost surely overstate the real potential for additional irrigation. The estimates give too little weight to the economic and environmental costs of additional irrigation. World Bank studies of India's experience show that the real economic costs of recent irrigation projects were substantially higher than the costs of earlier ones, in large part because the best sites were developed first. Nor do the estimates of potential irrigation take proper account of sharply rising demands for nonagricultural uses of water in urban areas and for instream flows to protect aquatic habitat.

Much irrigation water is inefficiently used, not only because it is typically priced well below its true social value but also because much of it is managed by large, unwieldy public bureaucracies. Even if these inefficiencies were removed—a formidable undertaking—the potential for expanding global irrigation at socially acceptable economic and environmental costs surely is well below that suggested by the World Bank estimates.

Climate

Although there now is a strong scientific consensus that the global climate will change over the next 50 to 100 years because of the greenhouse effect, there is no consensus about the consequences of this for global agricultural capacity. Studies conducted for the Intergovernmental Panel on Climate Change and by the U.S. Department of Agriculture suggest that climate change might reduce global agricultural capacity by 15 to 25 percent. However, these estimates make no allowance for the ability of farmers to adjust to the changed climate or for agricultural research institutions to develop new technologies better adapted to the changed climate. Research at RFF on the impacts of climate change on agriculture in the midwestern United States indicates that these various adjustment processes could virtually eliminate the negative effects of a hotter and drier climate in the Midwest.

Steps taken to limit climate change would reduce the damage to the social capital represented by the climate. In the best of circumstances, however, the climate will contribute little if anything to meeting the prospective increase in global demand for food and fiber.

Genetic Materials

Crops and animals are under continuing assault from a host of pests and diseases and from climatic vicissitudes. Maintenance of present levels of crop and animal production requires a sustained effort by plant and animal breeders to develop new varieties better able to resist this assault. Expanding agricultural production on the needed scale will require an even more intensive effort by breeders. To succeed in this, breeders must have access to a broad range of genetic material for developing more resistant and productive varieties of plants and animals. The plant and animal gene pool, therefore, is a critical resource for achievement of sustainable agriculture.

Most of the research on the supply of genetic resources for agriculture has dealt with plants. "Banks" to protect plant genetic materials have been set up by private firms and governments—most prominently, by the U.S. government—and by the Consultative Group on International Agricultural Research (CGIAR). These gene banks serve not only as repositories for plant genetic materials but also as distributors of the materials to plant breeders worldwide.

A study for the World Bank of the CGIAR system criticized some details of the system's performance but overall gave it high marks. Studies by World Bank researchers of the gene bank system as a whole pointed to some potentially serious weak spots in LDCs, particularly in Africa, but also concluded that in general the system is robust. The key question is whether the global gene bank system will continue to receive the support from national governments and international institutions that it will need to maintain that state of health. If it does, the plant genetic resource should be adequately protected. However, as the resource already is reasonably well managed, improvements in its management are unlikely to add much to its supply.

Knowledge

Given the present state of knowledge, the above discussion points to the conclusion that the potential supplies of energy, land, water, climate, and genetic resources would be quite inadequate to meet the prospective increase in global demand for food and fiber at acceptable economic and environmental costs. The implication is that most of the burden of sustainably meeting future demand must be carried by increasing the productivity of these combined resources. Achieving the necessary increases in productivity will require a substantial increase in the social capital represented by knowledge of agricultural production embedded in people, technology, and institutions.

Thus the critical question for agricultural sustainability is whether the global supply of knowledge can be expanded on the requisite scale. Although the answer must be uncertain, there are grounds for optimism. Compared with the other resources, the supply of knowledge about agricultural production is subject to few physical constraints. Knowledge accumulates; it is never used up and, in today's world, it is quickly and cheaply transmitted to the remotest regions of the globe. Reflecting these characteristics, agricultural knowledge has grown enormously over the last several decades and has accounted for most of the 2.5- to 3.0-fold increase in global agricultural production since the end of World War II. The international agricultural research system and the national agricultural research systems in more developed coun-

tries appear up to the future task if they continue to get adequate financial support. Private firms in those countries also are promising sources of new knowledge—for example, in biotechnology. Capacity to expand knowledge also is well developed in Asia, but is less satisfactory in Latin America, and least satisfactory in Africa. This capacity must be increased. In addition, agricultural research institutions will have to focus more on technologies and practices less dependent on irrigation and on fossil fuels, and more friendly to the environment than those now in common use.

Governments all around the world are moving toward greater use of agricultural markets, and this will strengthen farmers' incentives to use the new knowledge as it becomes available. The governments of many LDCs, however, have consistently underinvested in the education of rural people. This potentially serious obstacle to the needed expansion in knowledge must be overcome.

Expanding knowledge on the scale needed to achieve a sustainable agricultural system 2.5 to 3.0 times as large as the present one poses a formidable challenge to the global community. The historical record suggests that the challenge can be met. The potential consequences of failure provide perhaps the strongest assurance that it will be.

An Almost Practical Step toward Sustainability

Robert Solow

You may be relieved to know that this talk will not be a harangue about the intrinsic incompatibility of economic growth and concern for the natural environment. Nor will it be a plea for the strict conservation of nonrenewable resources, even if that were to mean dramatic reductions in production and consumption. On the other hand, neither will you hear mindless wish fulfillment about how ingenuity and enterprise can be counted on to save us from the consequences of consuming too much and preserving too little, as they have always done in the past.

Actually, the argument I want to make seems to be particularly appropriate on the occasion of the fortieth anniversary of Resources for the Future; it is precisely about resources for the future. And it is even more appropriate for a research organization: I hope to show how some fairly interesting pure economic theory can offer a hint—though only a hint—about a possible improvement in the way we talk about and think about our economy in relation to its endowment of natural resources. The theoretical insight that I will present suggests a potentially important line of empirical research and a possible guideline for long-term economic policy. Then I will make a naive leap and suggest that, if we talked about the economy in a more sensible and accurate way, we might actually be better able to conduct a rational policy in practice with respect to natural and environmental resources. That is probably foolishness, but I hope you will find it a disarming sort of foolishness.

An invited lecture on the occasion of the fortieth anniversary of Resources for the Future; originally published as a pamphlet in 1992.

Previewing the Arguments

It will be useful if I tell you in advance where the argument is leading. It is a commonplace thought that the national income and product accounts, as currently laid out, give a misleading picture of the value of a nation's economic activity to the people concerned. The conventional totals, gross domestic product (GDP) or gross national product (GNP) or national income, are not so bad for studying fluctuations in employment or analyzing the demand for goods and services. When it comes to measuring the economy's contribution to the well-being of the country's inhabitants, however, the conventional measures are incomplete. The most obvious omission is the depreciation of fixed capital assets. If two economies produce the same real GDP but one of them does so wastefully by wearing out half of its stock of plant and equipment while the other does so thriftily and holds depreciation to 10 percent of its stock of capital, it is pretty obvious which one is doing a better job for its citizens. Of course the national income accounts have always recognized this point, and they construct net aggregates, like net national product (NNP), to give an appropriate answer. Depreciation of fixed capital may be badly measured, and the error affects net product, but the effort is made.

The same principle should hold for stocks of nonrenewable resources and for environmental assets like clean air and water. Suppose two economies produce the same real net national product, with due allowance for depreciation of fixed capital, but one of them is wasteful of natural resources and casually allows its environment to deteriorate, while the other conserves resources and preserves the natural environment. In such a case we have no trouble seeing that the first is providing less amply for its citizens than the second. So far, however, the proper adjustments needed to measure the stocks and flows of our natural resources and environmental assets are not being made in the published national accounts. (The United Nations has been working in this direction for some years, so the situation may change, although only with respect to environmental accounting.) The nature of this problem has been understood for some time, and individual scholars, beginning with William D. Nordhaus and James Tobin in 1972, have made occasional passes at estimating the required corrections.

That is hardly news. The additional insight that I want to explain is that there is a "right" way to make that correction—not perhaps the easiest or most direct way, but the way that properly charges the economy for the consumption of its resource endowment. The same principle can be extended to define the right adjustment that must be made to allow for the degradation or improvement of environmental assets in the course of a year's economic activity. The properly adjusted net national product would give a more meaningful indicator of the annual contribution to economic well-being.

The corrections are more easily defined than performed. The necessary calculations would undoubtedly be more error-prone than those the U.S. Department of Commerce already does with respect to the depreciation of fixed capital. Nevertheless, I would suggest that talk without measurement is cheap. If we—the country, the government, the research community—are serious about doing the right thing for the resource endowment and the environment, then the proper measurement of stocks and flows ought to be high on the list of steps toward intelligent and foresighted decisions.

The second and last step in my argument is more abstract. It turns out that the measurements I have just been discussing play a central role in the only logically sound approach to the issue of sustainability that I know. If "sustainability" is anything more than a slogan or expression of emotion, it must amount to an injunction to preserve productive capacity for the indefinite future. That is compatible with the use of nonrenewable resources only if society as a whole replaces used-up resources with something else. As you will see when I return to this point for a full exposition, the very same calculation that is required to construct an adjusted net national product for current evaluation of economic benefit is also essential for the construction of a strategy aimed at sustainability. This conclusion confirms the importance of a serious effort to dig out the relevant facts.

That is a brief preview of what I intend to say, but before going on to say it, I would like to mention the names of the economists who have contributed most to this line of thought. They include Professors John Hartwick of Queen's University in Canada, Partha Dasgupta of the University of Cambridge, England, and Karl-Göran Mäler of the Stockholm School of Economics; my sometime colleague Martin L. Weitzman, now of Harvard University; and, more on the practical side, Robert Repetto of the World Resources Institute. I have already mentioned the early work of Nordhaus and Tobin; Nordhaus has continued to contribute common sense, realism, and rigorous economic analysis. Finally, I should confess that I have contributed to this literature myself. My idea of heaven is an occasion when a piece of pretty economic theory turns out to suggest a program of empirical research and to have implications for the formulation of public policy.

Finding the True Net Product of Our Economy

Now I go back to the beginning and make my case in more detail. Suppose we adopt a simplified picture of an economy living in some kind of long run. What I mean by that awkward phrase is that we are going to ignore all those business-cycle problems connected with unemployment and excess capacity or overheating and inflation. From quarter to quarter and year to year this economy fully exploits the resources of labor, plant, and equipment that are available to it.

To take the easiest case—that of natural resources—first, imagine that this economy starts with a fixed stock of nonrenewable resources that are essential for further production. This is an over-simplification, of course. Even apart from the possibility of exploration and discovery, the stock of nonrenewable resources is not a pre-existing lump of given size, but a vast quantity of raw materials of varying grade, location, and ease of extraction.

Those complications are not of the essence, so I ignore them.

It is of the essence that production cannot take place without some use of natural resources. But I shall also assume that it is always possible to substitute greater inputs of labor, reproducible capital, and renewable resources for smaller direct inputs of the fixed resource. Substitution can take place on reasonable terms, although we can agree that it gets more and more costly as the process of substitution goes on. Without this minimal degree of optimism, the conclusion might be that this economy is like a watch that can be wound only once: it has only a finite number of ticks, after which it stops. In that case there is no point in talking about sustainability, because it is ruled out by assumption; the only choice is between a short happy life and a longer unhappy one.

Life for this economy consists of using all of its labor and capital and depleting some of its remaining stock of resources in the production of a year's output (GDP approximately). Part of each year's output is consumed, and that gives pleasure to current consumers; the rest is invested in reproducible capital to be used for production in the future. There are various assumptions one could make about the evolution of the population and employment. I will assume them to have stabilized, since I want to talk about the very long run anyway. Next year is a lot like this year, except that there will be more plant and equipment, if net investment was positive this year, and there will be less of the stock of resources left.

Each year there are two new decisions: how much to save and invest, and how much of the remaining stock of nonrenewable resources to use up. There is a sense in which we can say that this year's consumers have made a trade with posterity. They have used up some of the stock of irreplaceable natural resources; in exchange they have saved and invested, so that posterity will inherit a larger stock of reproducible capital.

This intergenerational trade-off can be managed well or badly, equitably or inequitably. I want to suppose that it is done well and equitably. That means two things. First, nothing is simply wasted; production is carried on efficiently. Second, although the notion of intergenerational equity is much more complicated and I cannot hope to explain it fully here, the idea is that each generation is allowed to favor itself over the future, but not too much. Each generation can, in turn, discount the welfare of all future generations, and each successive generation applies the same discount rate to the welfare of its successors. To make conservation an interesting proposition at all, the common discount rate should not be too large.

You may wonder why I allow discounting at all. I wonder, too: no generation "should" be favored over any other. The usual scholarly excuse—which relies on the idea that there is a small fixed probability that civilization will end during any little interval of time—sounds farfetched. We can think of intergenerational discounting as a concession to human weakness or as a technical assumption of convenience (which it is). Luckily, very little of what I want to say depends on the rate of discount, which we can just imagine to be very small.

Given this discounting of future consumption, we have to imagine that our toy economy makes its investment and resource-depletion decisions so as to generate the largest possible sum of satisfactions over all future time. The limits to this optimization process are imposed by the pre-existing stock of resources, the initial stock of reproducible capital, the size of the labor force, and the technology of production.

This assumption of optimality is an embarrassing load to carry around. Its function is primarily to allow the semi-fiction that market prices accurately reflect scarcities. A similar assumption is implicit whenever we use ordinary GDP as a measure of economic well-being. In practice, no doubt, prices reflect all sorts of distortions arising from monopoly, taxation, poor information, and other market imperfections. In practice one can try to make adjustments to market prices to correct for the worst distortions. The conceptual points I want to make would survive. They are not to be taken literally in any case, but more as indicators of the sort of measurements we should be aiming at in principle.

Properly Charging the Economy for the Consumption of Its Resource Endowment

Now I come to the first major analytical step in my argument. If you look carefully at the solution to the problem of intergenerational resource allocation I have just sketched, you see that an excellent approximation of each single period's contribution to social welfare emerges quite naturally from the calculations. It is, in fact, a corrected version of net domestic product. The new feature is precisely a deduction for the net depletion of exhaustible resources. (I use the phrase "net depletion" because it is possible to extend this reasoning to allow for some discovery and development of new resources. In the pure case, where all discovery and development have already taken place, net and gross depletion coincide.)

The correct charge for depletion should value each unit of resource extracted at its net price, namely, its real value as input to production minus the *marginal* cost of extraction. As Hartwick has pointed out, if the marginal cost of mining exceeds average cost, which is what one would expect in an extractive industry, then the simple procedure of deducting the gross margin in mining (that is, the value of sales less the cost of extraction) will overstate the proper deduction and thus understate net product in the economy. If I may use the jargon of resource economics for a moment, the correct measure of depletion for social accounting prices is just the aggregate of Hotelling rents in the mining industry. That is the appropriate way to put a figure on what is taken from the ground in any given year, that year's withdrawal from the original endowment of nonrenewable resources.

This proposal presents two practical difficulties for national income accounting. The first is that observed market prices have to be corrected for the worst of the distortions I have just listed (that is, the distortion that would result from deducting the gross margin in mining—overstatement of the proper deduction and understatement of the net product in the economy). Making adjustments to market prices to correct for distortions is attempted routinely by the World Bank and other agencies in making project evaluations in developing countries. We seem to ignore the problem of such distortions when we use our own national income accounts to study and judge the economies of advanced countries. If we are justified in that practice, the same casual treatment may be satisfactory in this context. (Not always, however: the large observed fluctuations in the price of oil cannot be accepted as indicating "true" values.) Either way, this is a surmountable problem.

I am not sure whether it is safe to be so casual about the second practical difficulty that my proposal for deducting net depletion of exhaustible resources presents for national income accounting. In principle, the proper measurement of resource rents requires the use of a numerical approximation to the marginal cost of mining. As I said, if marginal cost exceeds average cost by a lot, then taking the easy way out (just deducting the gross margin in mining) would entail a large error by overstating the depreciation of the resource stock. It seems to me that this is exactly where the fund of knowledge embodied in an organization like RFF can find its application. Tentative calculations for the main extractive industries would tell us something important about the true net product of our own economy. That would be important not merely because it would allow a more accurate evaluation of the path the economy has been following, but also, as you will see, because the measurement of resource rents should be an input into policy decisions with a view to sustainability.

Correcting National Accounts to Reflect Environmental Amenities

Pretty clearly, similar ideas should apply to a program of correcting the conventional national accounts to reflect environmental amenities. Much more attention has been lavished on environmental accounting than on resource accounting, and I have very little to add. Henry M. Peskin's work (much of which was done here at RFF) goes back

to the early 1970s, and the Organisation for Economic Co-operation and Development, the World Bank, and the U.S. Department of Commerce are preparing a framework for integrating national income and environmental accounts. The sooner it happens the better. My only comment is a theoretical one. Without too much strain, it may be possible to treat environmental quality as a stock, a kind of capital that is "depreciated" by the addition of pollutants and "invested in" by abatement activities. In such cases the same general principles apply as to other forms of capital. The same intellectual framework will cover reproducible capital, renewable and nonrenewable resources, and environmental "capital."

The data problems may be altogether different, of course, especially when it comes to the measurement of benefits, a nicety that does not arise in the case of resource depletion. But the underlying treatment will follow the same rules. This counts for more than fastidiousness, I think. It would be a real achievement if it were to become a commonplace that capital assets, natural assets, and environmental assets were equally "real" and subject to the same scale of values, indeed the same bookkeeping conventions. Deeper ways of thinking might be affected.

That completes the first phase of my argument, so I will summarize briefly. The very logic of the economic theory of capital tells us how to construct a net national product concept that allows properly for the depletion of nonrenewable resources, and also for other forms of natural capital. Carrying out those instructions is far from easy, but that only makes the process more interesting. The importance of doing the work and doing it right is that theory underlines the basic similarity among all forms of capital, and that is a lesson worth learning. It will be reinforced by routine embodiment in the national accounts. Perhaps RFF could take the lead, as it has done with respect to environmental costs and benefits.

Analyzing Sustainable Paths for a Modern Industrial Society

Now I want to start down an apparently quite different path, but I promise that it will eventually link up with the unromantic measurement issues I have discussed so far, and will even reinforce the argument I have made.

I do not have to remind you that "sustainability" has become a hot topic in the last few years, beginning, I suppose, with the publication of the Brundtland Commission's report, *Our Common Future,* in 1987. As far as I can tell, however, discussion of sustainability has been mainly an occasion for the expression of emotions and attitudes. There has been very little analysis of sustainable paths for a modern industrial economy, so that we have little idea of what would be required in the way of policy and what sorts of outcomes could be expected. As things stand, if I express a commitment to sustainability, all that tells you is that I am unhappy with the modern consumerist life-style. If I pooh-pooh the whole thing, on the other hand, all you can deduce is that I am for business as usual. It is not a very satisfactory state of affairs.

Understanding What It Is That Must Be Conserved

If sustainability means anything more than a vague emotional commitment, it must require that something be conserved for the very long run. It is very important to understand what that something is: I think it has to be a generalized capacity to produce economic well-being.

It makes perfectly good sense to insist that certain unique and irreplaceable assets should be preserved for their own sake; nearly everyone would feel that way about Yosemite or, for that matter, about the Lincoln Memorial, I imagine. But that sort of situation cannot be universalized: it would be neither possible nor desirable to "leave the world as we found it" in every particular.

Most routine natural resources are desirable for what they do, not for what they are. It is their capacity to provide usable goods and services that we value. Once that principle is accepted,

we are in the everyday world of substitutions and trade-offs.

For the rest of this talk, I will assume that a sustainable path for the national economy is one that allows every future generation the option of being as well off as its predecessors. The duty imposed by sustainability is to bequeath to posterity not any particular thing—with the sort of rare exception I have mentioned—but rather to endow them with whatever it takes to achieve a standard of living at least as good as our own and to look after their next generation similarly. We are not to consume humanity's capital, in the broadest sense. Sustainability is not always compatible with discounting the well-being of future generations if there is no continuing technological progress. But I will slide over this potential contradiction because discount rates should be small and, after all, there is technological progress.

All that sounds bland, but it has some content. The standard of living achievable in the future depends on a bundle of endowments, in principle on everything that could limit the economy's capacity to produce economic well-being. That includes nonrenewable resources, of course, but it also includes the stock of plant and equipment, the inventory of technological knowledge, and even the general level of education and supply of skills. A sustainable path for the economy is thus not necessarily one that conserves every single thing or any single thing. It is one that replaces whatever it takes from its inherited natural and produced endowment, its material and intellectual endowment. What matters is not the particular form that the replacement takes, but only its capacity to produce the things that posterity will enjoy. Those depletion and investment decisions are the proper focus.

Outlining Two Key Propositions

Now it is time to go back to the toy economy I described earlier and to bring some serious economic theory to bear. There are two closely related logical propositions that can be shown to hold for such an economy. The first tells us something about the properly defined net national product, calculated with the aid of the right prices. At each instant, net national product indicates the largest consumption level that can be allowed this year if future consumption is never to be allowed to decrease.

To put it a little more precisely: net national product measures the maximum current level of consumer satisfaction that can be sustained forever. It is, therefore, a measure of sustainable income given the state of the economy—capital, resources, and so on—at that very instant.

This is important enough and strange enough to be worth a little explanation. How can this year's NNP "know" about anything that will or can happen in the future? The theorist's answer goes something like this. The economy's net product in any year consists of public and private consumption and public and private investment. (I am ignoring foreign trade altogether. Think of the economy as representing the world.) The components of investment, including the depletion of natural resources, have to be valued. That is where the "rightness" of the prices comes in. If the economy or its participants are forward-looking and far-seeing, the prices of investment goods will reflect the market's evaluation of their future productivity, including the productivity of the future investments they will make possible. The right prices will make full allowance even for the distant future, and will even take account of how each future generation will look at its future.

This story makes it obvious that everyday market prices can make no claim to embody that kind of foreknowledge. Least of all could the prices of natural resource products, which are famous for their volatility, have this property; but one could entertain legitimate doubts about other prices, too. The hope has to be that a careful attempt to average out speculative movements and to correct for the other imperfections I listed earlier would yield adjusted prices that might serve as a rough approximation to the theoretically correct ones. We act as if that were true in other contexts. The important hedge is not to claim too much.

While it is closely related to the proposition that NNP measures the maximum current level of consumer satisfaction that can be sustained forever, the second theoretical proposition I need is considerably more intuitive, although it may sound a little mysterious, too. Properly defined and properly calculated, this year's net national product can always be regarded as this year's interest on society's total stock of capital. It is absolutely vital that "capital" be interpreted in the broadest sense to include everything, tangible and intangible, in which the economy can invest or disinvest, including knowledge. Of course this stock of capital must be evaluated at the right prices. And the interest rate that capitalizes the net national product will generally be the real discount rate implicit in the whole story. Investment and depletion decisions determine the real wealth of the economy, and each instant's NNP appears as the return to society on the wealth it has accumulated in all forms. There are some tricky questions about wage incomes, but they are off the main track and I shall leave them unanswered.

Maintaining the Broad Stock of Society's Capital Intact

Something interesting happens when these two propositions are put together. One of them tells us that NNP at any instant is a measure of the highest sustainable income achievable, given the total stock of capital available at that instant. The other proposition tells us that NNP at any instant can be represented as that same stock of capital multiplied by an unchanging discount rate. Suppose that one goal of economic policy is to make investment and depletion decisions this year in a way that does not erode sustainable income. Then those same decisions must not allow the aggregate capital stock to fall. To use a Victorian phrase, preserving sustainability amounts to maintaining society's capital intact.

Let me say that in a slightly different way, speaking more picturesquely of generations rather than of instants or years. Each generation inherits a capital stock in the very broad and inclusive sense that matters. In turn, each generation makes consumption, investment, and depletion decisions. It enjoys its own consumption and leaves a stock of capital for the next generation. Of course, generations do not make decisions; families, firms, and governments do. Still, if all those decisions eventuate in a very large amount of current consumption, clearly the next generation might be forced to start with a lower stock of capital than its parents did. We now know that this is equivalent to saying that the new sustainable level of income is lower than the old one. The high-consumption generation has not lived up to the ethic of sustainability.

In the opposite case, consider a generation that consumes very little and leaves behind it a larger stock of capital than it inherited. That generation will have increased the sustainable level of income, and done so at the expense of its own consumption. Obviously that is what most past generations in the United States have done. Equally obviously, they were helped by ongoing technological progress. I have left that factor out of account, because it makes things too easy. It could probably be accommodated in the theoretical picture by imagining that there is a stock of technological knowledge that is built up by scientific and engineering research and depreciates through obsolescence. We know so little about that process that the formalization seems almost misleading. But the fact is very important.

A concern for sustainability implies a bias toward investment. That does not mean investment *über alles;* it means just enough investment to maintain the broad stock of capital intact. It does not mean maintaining intact the stock of every single thing; trade-offs and substitutions are not only permissible, they are essential. Unfortunately I have to make the limp statement that the terms on which one form of capital should be traded off against another are given by those adjusted prices— "shadow prices" we call them—and they involve a certain amount of guesswork. The guesswork has to be done; it cannot be avoided by defining the problem away. It is better that the guesswork be based on careful research than that the decision be fudged.

Connecting Up the Arguments

Knowing What and How Much Should Be Replaced

Now I can connect up the two halves of my argument. Every generation uses up some part of the earth's original endowment of nonrenewable resources. There is no alternative. Not now anyway. Maybe eventually our economy will be based entirely on renewables. (The theory I have been using can be applied then too, with routine modifications.) Even so, there will be a long meanwhile. What should each generation give back in exchange for depleted resources if it wishes to abide by the ethic of sustainability? We now have an answer in principle. It should add to the social capital in other forms, enough to maintain the aggregate social capital intact. In other words, it should replace the used-up resources with other assets of equal value, or equal shadow value. How much is that? The shadow value of resource depletion is exactly the aggregate of Hotelling rents. It is exactly the quantity that should be deducted from conventional net national product to give a truer NNP that takes account of the depletion of resources. A research project aimed at estimating that deduction would also be estimating the amount of investment in other forms that would just replace the productive capacity dissipated in resource depletion. This is sometimes known as Hartwick's rule: a society that invests aggregate resource rents in reproducible capital is preserving its capacity to sustain a constant level of consumption.

Once again, I should mention that the same approach can be applied to environmental assets— the most complete treatment is by Karl-Göran Mäler—and to renewable resources—as in the work of John Hartwick. The environmental case is more complex, because even a stylized model of environmental degradation and rehabilitation is more complex than a model of resource depletion. The principle is the same, but the execution is even more difficult. Remember that even the simplest case offers daunting measurement problems.

Translating Sustainability into Policy

It is possible that the clarity brought to the idea of sustainability by this approach could lift the policy debate to a more pragmatic, less emotional level. But I am inclined to think that a few numbers, even approximate numbers, would be much more effective in turning discussion toward concrete proposals and away from pronunciamentos.

Suppose that the Department of Commerce published routinely a reasonable approximation to the "true" value of each year's depletion of nonrenewable resources. We could then say to ourselves: we owe to the future a volume of investment that will compensate for this year's withdrawal from the inherited stock. We know the rough magnitude of this requirement. The appropriate policy is to generate an economically equivalent amount of net investment, enough to maintain society's broadly defined stock of capital intact. Of course, there may be other reasons for adding to (or subtracting from) this level of investment. The point is only that a commitment to sustainability is translated into a commitment to a specifiable amount of productive investment.

By the way, the same sort of calculation should have a very high priority in primary producing countries, the ones that supply the advanced industrial world with mineral products. They should also be directing their—rather large—Hotelling rents into productive investment. They will presumably want to invest more than that, because sustainability is hardly an adequate goal in poor countries. In this perspective, the cardinal sin is not mining; it is consuming the rents from mining.

It goes without saying that this concrete translation of sustainability into policy leaves a lot of questions unanswered. The split between private and public investment has to be made in essentially political ways, like the split between private and public saving. There are other reasons for public policy to encourage or discourage investment, because there are social goals other than sustainability. One could hope for more focused debate as trade-offs are made more explicit.

I want to remind you again that environmental preservation can be handled in much the same way. It is a more difficult context, however, for several reasons. Many, though not all, environmental assets have a claim to intrinsic value. That is the case of the Grand Canyon or Yosemite National Park, as noted earlier. The claim that a feature of the environment is irreplaceable, that is, not open to substitution by something equivalent but different, can be contested in any particular case, but no doubt it is sometimes true. Then the calculus of trade-offs does not apply. Useful minerals are in a more utilitarian category, and that is why I dealt with them explicitly.

Yet another difficulty is the deeper uncertainty about environmental benefits and costs. Marketed commodities, like minerals or renewable natural resources, are much simpler. I have admitted, fairly and squarely, how much of my argument depends on getting the shadow prices approximately right. Ordinary transaction prices are clearly not the whole answer; but they are a place to start. With environmental assets, not even that benchmark is available. I do not need to convince this audience that the difficulty of doing better does not make zero a defensible approximation for the shadow price of environmental amenity. I think the correct conclusion is the one stated by Karl-Göran Mäler: that we are going to have to keep depending on physical and other special indicators in order to judge the economy's performance with regard to the use of environmental resources. Even so, the conceptual framework should be an aid to clear thinking in the environmental field as well.

Maybe this way of thinking about environmental matters offers a way out of a dilemma facing less developed countries. The dilemma arises because they sometimes find that the adoption of developed-country environmental standards makes local industries uncompetitive in world markets. The poor countries then seem to have a choice between cooperating in the degradation of their own environment or acquiescing in their own poverty. At least when pollution is localized, the resolution of the dilemma appears to be a controlled trade-off between an immediate loss of environmental amenity and a gain in future economic well-being. Temporary acceptance of less-than-the-best environmental conditions can be made more palatable if the "rents" from doing so are translated into productive investment. Higher incomes in the future could be spent in part on environmental repair, of course, but it is general well-being that counts ultimately.

Notice that I have limited this suggestion to the case of localized pollution. When poor countries in search of their own economic goals contribute to global environmental damage, much more difficult policy questions arise. Their solution is not so hard to see in principle, but the practical obstacles are enormous. In any case, I leave those problems aside.

Concluding Comments

That brings me to the end of my story. I have suggested that an innovation in social accounting practice could contribute to more rational debate and possibly to more rational action in the economics of nonrenewable resources and the approach to a sustainable economy. There is a trick involved here, and I guess I should confess what it is. In a complex world, populated by people with diverse interests and tastes, and enmeshed in uncertainty about the future (not to mention the past), there is a lot to be gained by transforming questions of yes-or-no into questions of more-or-less. Yes-or-no lends itself to stalemate and confrontation; more-or-less lends itself to trade-offs. The trick is to understand more of what and less of what. This lecture was intended to make a step in that direction.

Part 10

Environmental Problems in Developing and Transitional Countries

40 Environmental Policies, Economic Restructuring, and Institutional Development in the Former Soviet Union

Michael A. Toman and R. David Simpson

Foreign aid in the form of technical assistance might be useful in helping the newly independent states of the former Soviet Union deal with past and current pollution. But such aid probably will not have a lasting, positive impact in the absence of reforms in the countries' basic social institutions. Without development of the institutions of a market economy, environmental measures are unlikely to be successful. Obstacles to investments that promote economic and environmental improvements must be removed if these improvements are to be achieved.

Technical and regulatory efforts to improve the management of environmental quality in the former Soviet Union are a focus of programs to provide foreign assistance to the newly independent states that once made up that country. To assess the prospects for the success of these assistance programs, policymakers in the United States and other Western countries must address many basic questions about the new states' environmental policies and their transitions from planned economies to market economies. These questions fall into three categories.

First, what can we expect regarding the investment incentives of firms in these countries? How many low-cost investments that improve both the environment and the economy will the firms undertake, and why are these investments not already being undertaken?

Second, how will the restructuring of enterprises and institutions alter the responses of polluters to environmental policy instruments? Conversely, how do the challenges of restructuring that face enterprises affect the design of environmental policy?

Third, how will environmental policies and fiscal policies interact? How will environmental policies interact with industrial and overall social security policies? For example, where enterprises that are not viable in the long run are being temporarily maintained on social grounds, what investments should be pursued to reduce enterprise losses and environmental damages?

Based on our observations in Russia and Ukraine, and on extensive discussions with experts in those countries, we

Originally published in *Resources,* No. 116, Summer 1994.

believe that real progress on environmental problems in the countries of the former Soviet Union will lag until there are substantial and far-ranging reforms in basic economic, legal, and social institutions. We do not deny that some targeted technical assistance could produce substantial improvements in environmental quality and quality of life for individuals affected by the assistance. Without basic institutional reforms, however, it is doubtful that these countries will have the capacity to continue the progress made possible by foreign assistance and to generate substantial environmental improvement on their own.

Our justification for this conclusion goes beyond the observation that the states of the former Soviet Union remain poor and that their resources available for environmental investment remain limited. It also goes beyond the observation that, in the absence of development assistance, excessively strict environmental regulation likely will be politically unpalatable.

Given these countries' current social institutions, it will be costly, if not impossible, to succeed in translating a public demand for environmental improvement into concrete action. The necessary political, legal, and economic accountability needed to do this effectively does not yet exist. Moreover, even if there were agreement on the need for change, institutional failures in the economy would likely raise the cost of enforcing environmental standards well above even the levels experienced under inefficient command-and-control programs in otherwise functional market economies.

These observations in turn raise doubts about the cost-effectiveness of major environmental assistance programs in the former Soviet Union without significant institutional reform there. Evidence is growing that improvement of environmental quality is a highly valued objective in the countries in question. However, environmental quality, as well as economic performance, might be better served first by assistance that helps the countries of the former Soviet Union to develop the institutions of a market economy, including the associated legal institutions of property, liability, and contract law.

Environmental Policies in Russia and Ukraine

Environmental policies in Russia and Ukraine illustrate both the disarray in the environmental policies of the countries of the former Soviet Union and the difficulties in improving these policies without progress toward the development of market economies. Environmental policies in these countries consist of a hybrid of standards for emissions and fees on emissions in excess of the standards. Environmental regulators tax pollution at two rates: all emissions are subject to a low tax rate, but emissions in excess of standards set for each source are subject to a higher rate. In principle, environmental regulators also have the power to order polluters to reduce emissions or to cease operations if the emissions pose a serious threat to public health or ecological integrity.

Emissions standards are based on essentially arbitrary distinctions among hundreds of pollutants. Far more standards exist than regulators can monitor or enforce. Standards also are set rigidly for individual sources of emissions, without regard for differences among the emissions reduction costs for each source or for differences among the impacts of each polluter on actual pollution concentrations. Regulators express concern that flexibility in the ways polluters are allowed to comply with emission standards, as with emission permit trading, would expand the opportunities of firms to exceed their emissions allowances—although current rules already require that emissions sources be monitored. This concern is ironic (even surrealistic), given the current scale of pollution violations.

In principle, these problems could be lessened by overhauling environmental regulations. However, other shortcomings in environmental regulation reflect economic and social concerns, as well as environmental concerns, and thus are harder to address.

The administration of pollution fees in Russia and Ukraine is problematic in several respects. Because expenditures for environmental protection are financed mostly by fees on polluters rather than from general revenues, environmental regulators are faced with a fundamental contradiction: to address environmental concerns arising from one set of activities, they must tax pollution from other, quite possibly unrelated, activities. If regulators were to charge pollution fees high enough to encourage substantial reductions of pollution, they would risk undercutting the tax base.

Revenue raising largely motivates the setting of emissions standards. To maintain tax revenues, the government often sets emissions standards a few percentage points below prevailing emissions levels. If the system worked to reduce emissions, it would require a ratcheting upward of standards to raise revenues, compromising firms' incentives to make long-term investments in environmental improvement. However, pollution fees simply are too low to achieve much environmental improvement, especially for state enterprises that do not face the normal budget constraints of a market economy. Moreover, at current rates of inflation, increases in pollution fees are rendered negligible shortly after they are announced.

With fees having little impact on pollution, the only other line of defense is legal sanctions against egregious violators of pollution standards. In practice, however, the problems of economic and political transition render this option largely ineffectual as well. Because so much of the economy in the former Soviet Union remains under state control, attempts to enforce environmental sanctions become intramural conflicts among government ministries. In this situation, the rule of law with regard to the environment often is quite weak, especially in light of the strong vested interests in maintaining enterprise operations that we discuss below.

Aside from problems related to the administration of emissions standards and pollution fees, the allocation of funds for pollution cleanup and reduction projects leaves something to be desired. The determination of priorities for environmental expenditures is not necessarily linked to environmental benefits. Some effort to identify such benefits is made when different expenditure proposals are considered. However, an important criterion for allocating funds appears to be the financial need of the local government or enterprise proposing a pollution cleanup or reduction project. Thus funds are often made available for projects that local governments or firms cannot finance on their own, with little regard for the benefits the projects generate by reducing serious health or environmental risks arising from pollution.

Obstacles to Making Win-Win Investments

In the long run, the industrial pollution problems of the former Soviet Union can only be overcome by major investments in more efficient and cleaner production processes and equipment. Many such investments probably could be undertaken at very low cost and result in both substantial environmental benefits and lower production costs. The existence of such "win-win" investments begs an important question, however: If such options are available, why have they not been pursued? Some of the reasons that these possibilities continue to be unexploited may be traced to the Soviet legacy, others to the difficulties of the transitional period, and still others to problems of information and oversight common, to greater or lesser degrees, in all economies.

One impediment to win-win investments is the morass of regulation and licensing requirements left over from central planning. These requirements make starting new businesses and instituting substantial reforms in existing ones extremely difficult. As a result, they discourage the establishment or retooling of firms that are both more profitable and less polluting.

A second part of the Soviet legacy that impedes win-win investments is the tradition of propping up faltering firms with public funds. This tradition undermines incentives for both increased efficiency and pollution reduction. If enterprise managers know that they will be bailed out with public funds, regardless of the performance of their firms, they have little incentive to seek cost-saving production innovations. Moreover, they may have little incentive to adopt even low-cost solutions to their environmental problems if they believe that these solutions will sooner or later be financed out of public funds.

One ongoing impediment to win-win investments is obstacles to both foreign and domestic private investment. Such investment is limited by several factors. First, the process of privatization has just begun in Russia and is even less advanced in many other states of the former Soviet Union. Second, the institutions that characterize capital markets and the banking system in Western economies are just now coming into being in these countries. For example, corporate law is very incomplete, and accounting procedures that would enable outside investors to determine the value of potential investments have not yet been adopted. Third, taxes on the profits of firms in the countries of the former Soviet Union are substantial. These taxes, along with exchange controls and high inflation (which triggers high interest rates), limit the attractiveness of new investment.

A second ongoing obstacle to win-win investments is the limited capacity of the labor market to adjust to the transition from planned to market economies. Labor mobility remains limited because workers have traditionally obtained all social services (including housing) through the enterprises that employ them and because internal migration is subject to state control. Another circumstance that has made it difficult for the labor market to adapt to the transition is the financial stake employees have in some newly privatized enterprises. Share ownership in such enterprises is largely concentrated among workers and managers, increasing the financial losses that employees would face if these enterprises fail. Because employees realize that they face risks of financial loss and unemployment in emerging market economies and because they distrust the prospects for success with new investment, they are exerting political pressure for their governments to subsidize firms or to take other measures to prevent firms from failing.

A seemingly simple solution for overcoming the obstacles to win-win investments would be to allow firms to sink or swim on their own merits. However, this solution may not be feasible under the current circumstances of the transition from planned to market economies. These circumstances may combine to deprive even deserving enterprises of the financing they will need to survive. Thus, in deciding which otherwise failing enterprises receive financial aid and regulatory leniency, decision makers in the countries of the former Soviet Union must distinguish between firms that make obsolete products using archaic production methods and those that are trying to make needed products using newer, cleaner production methods. Moreover, given restrictions on the social "safety net," decision makers need to take into account social factors that may outweigh considerations of narrowly defined economic efficiency in making these decisions.

Thus the limitations on private activity to improve the environment and the economy at the same time also hinder governmental policies for pollution control. While the possibility of publicly funded bailouts exists, polluting firms will be unresponsive to economic sanctions. Moreover, legal or economic sanctions that threaten employment, the viability of enterprises, and the social fabric will be vigorously opposed by enterprise managers, employees, and the branches of government that still oversee polluting industries. As long as public and private decision makers remain unaccountable for their decisions, firms' capacity to change their environmental behavior is very limited.

Institutional Reforms

Environmental policies per se probably will have relatively little effect until there is progress toward greater general strengthening of economic and legal institutions. The development and maturation of the institutions of capitalism in the former Soviet Union may be facilitated by increased macroeconomic stability, the establishment of legislation to govern corporate conduct and reduce regulatory barriers to the creation of new businesses, the reform of financial markets, and the revamping of the provision of public goods and social security.

While the transition from planned to market economies proceeds, it is important that environmental policy move in tandem with the general development of economic, legal, and social institutions. For example, while some flexibility in approaches to the enforcement of environmental regulations is called for, it is also important that decision makers be able to predict the effects of these approaches. Actions that would further decrease the confidence of potential investors could be counterproductive, even if they achieved some short-term environmental improvement.

It is also important that the institutions of environmental policy reflect the changing technical capabilities of regulators, the evolution of judicial and other institutions, and the increased stability of firms. Case-by-case reviews of compliance strategies should be replaced by general regulations that incorporate flexibility in compliance. Incentive-based measures, such as limited emissions-permit trading programs, should be established and expanded as opportunities arise. A more concerted effort to set priorities for environmental expenditures and to limit soft enterprise budgets (budgets that are based on the expectation of publicly provided funds to make up losses) would probably improve the efficiency of environmental expenditures while the economy as a whole makes the tran-

sition toward greater private financing of environmental improvements. Finally, some simplification of the environmental standards themselves would be beneficial.

Building better economic and political institutions is time consuming and does not offer the immediate and tangible rewards that technical support systems may afford. However, institutional reforms are crucial if the technical support programs for environmental improvement that are now being championed in foreign assistance debates are to be successful.

Suggested Reading

Bluffstone, Randall, and Bruce A. Larson, Eds. 1997. *Controlling Pollution in Transition Economies: Theories and Methods.* Cheltenham, U.K.: Edward Elgar.

Kaderják, Péter, and John Powell, Eds. 1997. *Economics for Environmental Policy in Transition Economies: An Analysis of the Hungarian Experience.* Cheltenham, U.K.: Edward Elgar.

Peszko, Grzegorz, and Tomasz Zylicz. 1998. Environmental Financing in European Economies in Transition. *Environmental and Resource Economics* 11(3–4): 521–38. (Special issue, *Frontiers of Environmental and Resource Economics: Testing the Theories,* Thomas Sterner and Jeroen C. J. M. van den Bergh, Eds.)

41

The Allocation of Environmental Liabilities in Central and Eastern Europe

James Boyd

Existing soil and groundwater contamination is likely to affect future industrial development in Central and Eastern Europe, as large-scale pollution cleanup costs are potentially tied to industrial property transactions and the division of liability for these costs is uncertain. Determining how pollution cleanup costs should be allocated between governments and current or future property owners will not be easy.

Retroactive liability is unlikely to be a desirable or a feasible means of assigning such costs. Publicly financed liability funds, by widely distributing cleanup costs, create a better climate for foreign and domestic investment than does a U.S.-style system of retroactive liability. However, pooled fund programs should be operated on a short-term basis only, as they may reduce private incentives to invest in pollution reduction.

Privatization and market reform in the economies of Central and Eastern Europe are occurring against a backdrop of severe environmental degradation left by decades of inadequate government attention to environmental conditions. The legacy of soil and groundwater pollution inherited by the new governments of Central and Eastern Europe not only creates direct health and ecological costs but is also likely to affect future industrial development and investment. The contentious development and expensive implementation of legal and regulatory approaches to mitigating environmental degradation in the United States and the European Economic Community (EEC) suggest that environmental problems in Central and Eastern Europe—where economies are much weaker, environmental problems much greater, and legal and regulatory institutions much less developed than in the West—will only be resolved at great economic and political cost.

In theory, an effective environmental liability system in Central and Eastern Europe will serve to deter the future generation of pollution by threatening polluters with liability costs arising from improper waste generation or disposal. However, a more immediate, practical consequence of new liability rules is the assignment of responsibility for existing pollution. This assignment raises an important question—namely, how should the costs of removing or reducing existing pollution be allocated between governments and current or future property owners? The answer is complicated by the financial weakness of both governments and property owners in Central and Eastern

Originally published in *Resources,* No. 112, Summer 1993.

Europe, the costs and uncertainty involved in the quantification of environmental risks, the political nature of liability reform, and the need to promote domestic and foreign investment.

While the costs of remediating existing pollution in the former Soviet bloc cannot be precisely estimated, they are clearly huge. The estimated cost of meeting EEC or U.S. environmental standards in Poland, for example, is as high as $300 billion. There is great uncertainty regarding the costs of remediating existing pollution due to the lack of lending and insurance institutions familiar with risk assessment and to the virtual nonexistence of accounting, zoning, or regulatory requirements for documenting risk-generating processes or technologies. Large-scale pollution costs are thus potentially tied to transactions involving industrial property in Central and Eastern Europe.

Uncertainty regarding potential liabilities is exacerbated by the lack of established liability concepts, legal precedent, and consistent enforcement principles in Central and Eastern Europe. The formerly communist countries have no common-law traditions—such as those in the United States—that allow environmental claims based on concepts such as nuisance, trespass, negligence, or strict liability. Instead, they use a civil law approach that imposes damages almost exclusively in cases where there has been a violation of a government standard or regulation. A civil law, rather than a common law, definition and enforcement of liabilities presents an opportunity for governments to coordinate and achieve cost-effective resolutions to the cleanup of existing pollution. However, it also creates uncertainty for potential investors. Because such a system is defined neither by precedent nor by a consistent application of judicial principles, the scope and division of potential liabilities is unclear.

The unique environmental and institutional conditions in the countries of Central and Eastern Europe argue for liability approaches that may differ from those advocated in countries with more advanced legal systems, less pollution, and greater economic vitality. Given these unique conditions, the environmental liability systems established in

Central and Eastern Europe should be influenced by two goals. First, in light of the need for economic growth, liability rules consistent with the promotion of privatization and foreign investment should be favored. Second, because Central and Eastern European governments lack the funds to pay the entire cost of cleaning up existing pollution, legal and regulatory policies should be designed to target public revenues toward the environmental hazards that pose the greatest threat.

To pursue these goals, liability initiatives in Central and Eastern Europe should distinguish between the timely and effective implementation of liability rules governing the creation of future environmental risks, and the efficient cleanup of pollution generated in the past. These are entirely different issues. The first concerns the question of how to create incentives for future pollution reductions, while the second concerns the question of how to efficiently achieve a distributional goal—that is, how the costs arising from pollution created in the past should be borne. With respect to the latter, it can be easily argued that both moral and legal responsibility for existing pollution lies primarily with the former Soviet bloc governments themselves. The question of who should bear the costs of remediating existing site contamination is particularly important since it is likely to affect patterns of foreign and domestic investment, and clearly affects the value of initial asset endowments distributed in the process of privatization.

The Argument against Strict and Retroactive Liability

One liability approach that might be instituted in Central and Eastern Europe is strict and retroactive environmental liability. This type of liability holds the current owner of a property fully liable for pollution cleanup and compensation costs, even when the pollution was generated by past owners or users of the property. While strict and retroactive liability strongly deters the future generation of pollution, its application in the United States has prompted debate over its inequitable allocation of responsibility for cleanup costs and its potentially adverse impact on property development.

Independent of judgments about its effects on pollution reduction and economic activity in the United States, strict and retroactive liability is unlikely to be desirable or even feasible in Central and Eastern Europe for several reasons. First, given the weak condition of the economies of Central and Eastern Europe—which is due in large part to capital scarcities—such liability could impose costs high enough to force many domestic producers to declare bankruptcy or liquidate their assets. Given the magnitude of existing environmental hazards, the full internalization of costs based on a strict and retroactive application of liability might yield negative real asset values for a significant fraction of industrial properties. These consequences are inefficient, since bankruptcy and asset dissolution involve costs in the form of abandoned capital, lost firm-specific human capital, and reduced competition. In any event, shortages of capital and the tenuous financial position of newly privatized firms suggest that liability rules dependent on firms' ability to liquidate or otherwise free capital to compensate for environmental damages will be ineffective.

Second, a strict and retroactive liability system is not likely to lead to an effective prioritization of cleanups. Under such a system, only the most unpolluted properties would be sought for development, and resources would therefore be devoted to the cleanup of relatively unpolluted properties.

Third, a strict and retroactive liability system will likely stifle foreign investment, which is critical to the acquisition of skills, technology, and capital by Central and Eastern Europe. Foreign investors' concerns about retroactive liability derive from their experience with huge retroactive liability costs in domestic markets, the fact that their firms' capital is relatively available to be tapped in the event of liability actions, and the lack of political stability—and hence investor influence—in Central and Eastern Europe.

Fourth, the political and ethical "polluter pays" motivation for strict and retroactive liability does not in general apply in Central and Eastern Europe.

Because former governments and managers of cooperatives are most to blame for existing pollution, there is little ethical justification for the new owners of privatized properties to be liable for the past sins of others.

Fifth, the distributional impact of strict and retroactive liability poses a threat not only to the success and timeliness of cleanups of existing pollution but to the success of liability rules aimed at future pollution reduction. As in the West, environmental policies in Central and Eastern Europe will be derived and enforced in a political context, and their distributional impacts will largely determine their legislative and political success. The fact that liability rules have large distributional, and hence political, consequences can influence the evolution and enforcement of environmental pollution law. Because the profitability (or existence) of new enterprises is potentially threatened by strict and retroactive liability, resources will be directed at the political system to redistribute costs. One natural way to do this is to seek changes in the liability rules themselves. Given the political context in which liability laws are formulated, it follows that rules dealing with future liabilities should be separated from rules dealing with retroactive liabilities in order to enhance political acceptance of a tough prospective liability system. A more equitable distribution of the costs arising from existing pollution makes laws aimed at future pollution reduction more politically and economically sustainable.

From a practical standpoint, however, separating the costs of existing pollution from the costs of pollution being generated by new property owners is difficult, since precise definitions and divisions of responsibility for pollution require costly risk assessment efforts. Even in the United States and EEC, an initial, noncomprehensive environmental audit can cost hundreds of thousands of dollars for a major industrial site. Because the risk assessment capabilities of Central and Eastern European governments and industries are greatly inferior to those of their counterparts in the West, it is unlikely that an accurate division of responsibility for pollution is possible at a reasonable social cost. When existing pollution is widespread but difficult to detect with conventional site assessment methods at the point when ownership of a property is transferred, and when advanced risk assessment technologies and expertise are in short supply, a precise technical—let alone legal—separation of responsibility for pollution cleanup may be unrealistic.

Contractual Mechanisms for Allocating Retroactive Liability

Given uncertainties regarding the scale of and the liabilities implied by existing pollution, different contractual mechanisms to allocate liability may be needed to improve the efficiency of privatization and foreign investment decisions. The desirability of alternative contract forms is largely a function of the type of information available and the point in time at which information is acquired. With respect to the latter, assessment of liabilities may occur either ex ante, at the point of transaction, or ex post, following the transaction.

Knowledge of either responsibility for or the extent of pollution may be available to only one of the parties to a property transaction. For example, a government may have knowledge of existing pollution risks but choose not to reveal them prior to a transaction. On the other hand, if the government is unable to ascertain when pollution was generated and grants amnesty for retroactive liability, an investor might inflate the value of risks he or she claims to have inherited at the point of sale.

In the unlikely event that both the buyer and the seller have complete knowledge of all the risks posed by pollution on a property, there are two primary contractual possibilities. One is for the seller (the government) to guarantee that no liability will be assessed for existing hazards. The other is for the government to impose strict retroactive liability on the buyer but discount the price of a property to account for the costs of such liability. The virtue of the latter approach is that the property transaction would be immune to "renegotiation" by the government. Therefore, subsequent disputes over which hazards did or did not exist at the point of sale would be avoided.

When the buyer cannot observe contamination of a property ex ante, he or she is purchasing an asset of unknown quality. Given this, an optimal contract requires insurance against levels of risk that differ from those revealed by the seller. Should the seller know that the property is clean, that person can simply guarantee to compensate the buyer for expenses resulting from any subsequently revealed contamination; alternatively, he or she can absolve the buyer of liability. If contamination created before the sale can be separated from that created after the sale, an optimal contract would release the buyer from retroactive liability costs. Having done so, the asset price would reflect the property's value net of retroactive liabilities. When the buyer cannot observe contamination ex ante, retroactive liability for the buyer is clearly not desirable, since the costs of liability are not known at the point of sale and so cannot be accurately discounted from the asset price.

If pollution generated by the buyer cannot be separated from pollution existing at the point of sale, however, a liability amnesty would give the buyer a loophole to escape the costs of the pollution he or she generated. In this case, government assumption of liability may be inefficient. The conflict between buyer uncertainty over liability and the creation of loopholes by liability amnesties underscores the importance of environmental audits, which allow for an accurate separation of responsibility for pollution.

If a clear separation of responsibility is not practical, the question that remains is how to distribute the costs of existing pollution while creating incentives for the reduction of future pollution generation. An imperfect but potentially desirable approach is to provide relief from retroactive liability through the provision of government funds earmarked for cleanups. Two distinct forms of liability funds exist. One form is pooled funds, which provide public money for cleanups and compensation—money that would be provided by property owners in a Superfund-type liability system. The other form is the liability escrow account, in which a fraction of a property's purchase price is set aside and earmarked for cleanup costs defined at a later date. A crucial difference between the two funds is that escrow funds provide money to clean up pollution at one specific property, while pooled funds provide money for the cleanup of any number of properties.

Pooled Funds

Pooled funds, which are conceptually related to "no-fault" pollution insurance, have been instituted or proposed in several countries to deal with large-scale environmental risks. They can differ in terms of duration, limits on the nature and scale of costs covered, and criteria—such as compliance with regulatory standards—that must be met in order for property owners to be eligible for reimbursement of cleanup costs. In all cases, however, only a fraction of liabilities are borne by property owners, with the balance being borne by the pooled fund.

Pooled funds are contrary to the notion that the polluter should pay cleanup costs. However, the use of public moneys for cleanups in Central and Eastern European countries is more easily justified than in western countries, since decades of state ownership and central planning in the former imply that responsibility for existing pollution lies largely with the governments of Central and Eastern Europe.

Because they widely distribute the costs of environmental cleanups, pooled funds may represent the least economically disruptive mechanism for dealing with large retroactive liabilities—liabilities that could otherwise force the abandonment of properties or the bankruptcy of property owners. The administration of such funds allows for the coordination and rationalization of a nation's risk reduction activities. With centralized control of the system, pollution mitigation measures at sites presenting the greatest social risks could, in principle, receive priority. The caveat is that an effective, centralized system of risk identification and ranking does not currently exist and is difficult and costly to implement.

Firms that expect large retroactive liabilities to ultimately force the closure of their enterprises pose a particularly serious pollution threat. If the enforcement of liability is delayed due to an overburdened legal system or the slow pace of regulators in identifying pollution sources, such firms have no incentive to reduce pollution in the period before enforcement occurs. Faced with the likely prospect of closure due to existing property contamination, such firms will find it profitable to pollute at will until the government forces them to cease their operations. A benefit of publicly provided cleanup funds, then, is that they increase the expected value of the firms, reduce the likelihood that firms will close as a result of retroactive liability costs, and thus lead firms to make investments in pollution control based on the now realistic ability to continue profitable operation well into the future. Since legal and regulatory enforcement of liability claims is likely to take some time in Central and Eastern Europe, this benefit of publicly provided cleanup funds is particularly important.

There are potentially significant problems associated with the use of pooled funds, however. Because the current owners of properties where cleanups are to be conducted will be reimbursed for the costs of cleanup, one problem is that price competition in the market for risk remediation services may be lessened unless the government polices a bidding process for such services. The market for environmental cleanup services, while sure to become increasingly competitive, is currently not competitive. Another problem is that property owners have an incentive to engage in costly cleanup activities that might have little social benefit because the owners are, in effect, insured by the government. Limited fund levels combined with inflated remediation costs could swiftly deplete a pooled fund's reserves.

A more fundamental problem associated with pooled funds derives from the inability to adequately separate risks existing before a fund is set up from those created during the fund's lifetime. If pooled fund programs cannot effectively distinguish between retroactive and prospective sources of pollution, amnesty for retroactive liability would carry a significant danger—namely, that newly created hazards will be claimed as hazards created in the past, thus allowing property owners to escape liability for pollution they generated. Realistically, a complete assessment of retroactive liabilities in Central and Eastern Europe will take years. Thus, while pooled funds may have desirable short-term benefits, it is unequivocally undesirable for them to become a permanent fixture in a government's environmental policy portfolio. If firms believe that all or even a fraction of their potential future liabilities will be borne by a subsidized government fund, the private incentive to invest in pollution reduction would be reduced. Thus an important political question is how a pooled liability fund program can be effectively operated on a short-term basis and then phased out to create liability assignments that effectively internalize environmental costs.

Escrow Funds

Several governments in Central and Eastern Europe allow fractions of up to 100 percent of the purchase price of a property to be set aside in escrow and used for the cleanup of pollution on the property. In practice, escrow funds guard against the incentive problems created by a pooled, no-fault liability fund by placing clear limits on the amount of public funding that will be provided for cleanups. In addition, because they typically expire after a period of months, escrow funds limit the ability of new property owners to escape the costs of pollution that they generate in the future.

Like pooled funds, escrow funds provide some insurance against existing pollution costs for property purchasers. Compared with the alternative of strict and retroactive liability, such insurance discourages a government from setting a costly standard for cleanups once it sells a property. The reason is that costly cleanups deplete the escrow funds—which would revert to government coffers if not drawn down.

However, new property owners may have an incentive to deplete escrow funds as much as possible, since for them the funds represent a source of costless pollution remediation financing. The result is that government funds that could be used to address pressing environmental problems might instead be used to address relatively unimportant environmental problems. Investors will clearly seek to purchase the least polluted properties. If funds are dedicated to the further cleanup of these relatively clean properties, government revenue that could be used to reduce environmental risks at relatively polluted sites would be reduced. As compared with escrow funds, pooled funds provide the government greater authority to determine the allocation of cleanup funds.

The Creation of an Effective Liability System

Both pooled fund and escrow fund systems will create a beneficial form of insurance against pollution liabilities, and thus will stimulate foreign and domestic investment and potentially smooth the transition to a tough future liability system. However, public funding of pollution remediation should be viewed as only a short-term means of addressing existing soil and groundwater cleanup issues in Central and Eastern Europe. Any permanent government subsidy of soil and groundwater cleanups will only continue to distort private property owners' incentives to reduce pollution.

Pooled funds present the best opportunity for targeting public funds to the cleanup of pollution posing the greatest health and ecological threats. However, they also represent a form of subsidy that might be politically difficult to dismantle. The challenge for the governments of Central and Eastern Europe, then, is to provide public funding for pollution remediation, but in a way that leads private property owners to believe that in the future they will be responsible for the social costs of polluting activities. While it may be tempting to give generous liability amnesties to foreign firms in order to encourage investment, doing so may lead foreign firms to export environmental risks to

Central and Eastern Europe. Limitations on existing liabilities are desirable, but there is nothing to recommend investment incentives created by weakened liability rules aimed at reducing future pollution generation.

The strict enforcement of private property rights (and the assignment of liabilities) is not a particularly effective way—in the short term—to guarantee a rational social approach to pollution reduction. The reason is that private interests pursued through a liability system need not coincide with the social interest when resources are severely limited. The economies of Central and Eastern Europe are not currently robust enough to support large resource expenditures aimed at the resolution of legal disputes.

It remains the case that the costs of existing pollution must be distributed in some fashion. Moreover, a system of incentives for future risk reduction is desperately needed if current environmental conditions are to improve. Pooled liability funds are likely to be the most politically and economically effective mechanism for distributing costs and reducing risks. However, they should be subject to safeguards—specifically, a limit on the duration of coverage, and requirements for eligibility to claim reimbursement of cleanup costs (such as the installation of pollution reduction technologies). Pooled funds are a necessary compromise between strict and retroactive liability and unrealistic attempts to perfectly and quickly separate responsibility for existing and future pollution.

Suggested Reading

Boyd, James. 1996. Environmental Liability Reform and Privatization in Central and Eastern Europe. *European Journal of Law and Economics* 3(1).

Goldenman, Gretta, and others. 1994. *Environmental Liability and Privatization in Central and Eastern Europe: A Report for the Environmental Action Program for Central and Eastern Europe.* Nowell, MA: World Bank and Kluwer Publishers.

Randall, Thomas. 1994. The Impact of Environmental Liabilities on Privatization in Central and Eastern Europe: A Case Study of Poland. *University of California–Davis Law Review* 28: 165–217.

42

Cost-Effective Control of Water Pollution in Central and Eastern Europe

Charles M. Paulsen

Lack of controls on point sources of pollutant discharges—primarily sewage treatment plants—has contributed to the degradation of surface water quality in Central and Eastern Europe. Neither relying on existing pollution control nor adopting the West's best-available pollution control technology and minimum pollutant discharge policies is likely to be a feasible course of action for the region. However, a recent case study involving the Nitra River basin in the Slovak Republic suggests that the region can realize substantial improvements in water quality at a fraction of the cost of command-and-control policies used in the West by taking into account the relative contributions to pollution and pollution control costs of individual point sources and basing pollution control efforts on those contributions and costs.

Since political transformations there in 1989, Central and Eastern Europe has increasingly come to realize the severity of the degradation of its surface water quality. Most major rivers and lakes in the region have pollutant concentrations far above international standards. In addition to posing health threats, contamination of the region's surface water has economic consequences. For example, pollutant discharges into the Baltic and the Black seas have already seriously reduced the output of once-productive fisheries.

Policies designed to improve the region's water quality will have to grapple with the declining industrial and agricultural output, concomitant decreases in material living standards, and shortages of investment capital faced by all the region's national governments. Given these conditions, the countries of Central and Eastern Europe could simply choose to delay adoption of the best-available pollution control technology and minimum pollutant discharge policies of Western Europe and North America until their economies can afford them. In the meantime, this decision would mean relying on existing pollution control facilities to deal with water quality problems caused by so-called point sources of water pollution—primarily industrial and municipal sewage treatment plants. As the region's economies improve, presumably more money would become available for the capital investments that are required for construction of sewage treatment plants with state-of-the-art pollution control. The region's governments would meanwhile stand to gain an advantage from delaying investment in water quality

Originally published in *Resources*, No. 113, Fall 1993.

improvement: the longer they wait to undertake such investment, the greater the likelihood that noncompetitive industries will fail, obviating the need to invest in new or improved plants to treat the industries' sewage.

Delaying efforts to improve water quality is problematic, however. Although pollutant discharges into the region's waters can be expected to decrease as industries close, change their product mix, or update their production processes, it is likely that municipal sewage loads will increase as more and more households and newly formed businesses are connected to public water and sewer networks. In addition, the public may demand that water quality issues be addressed in the present rather than in the future. The downfall of many of the formerly Communist governments was brought about in part by environmental movements, and anecdotal evidence suggests that a substantial demand for improved environmental quality still exists in many Central and Eastern European countries.

One alternative to delaying water quality improvement efforts would be an immediate attempt to implement a minimum discharge policy, whereby sewage treatment plants would be required to reduce pollutant discharges into surface water in line with European Community (EC) standards for wastewater treatment. However, the cost of such a policy might well be more than governments in the region are willing (or able) to pay, given that the per capita cost of meeting such standards exceeds per capita gross domestic product (GDP) in three of five countries in Central and Eastern Europe (see Table 1). Although countries in the region might be able to borrow a portion of the capital investment required to construct new or improve existing sewage treatment plants in order to meet EC wastewater treatment standards, it might not be wise for them to do so. Debt as a percentage of GDP is already high in many Central and Eastern European countries. Moreover, it is likely to increase as investment in industrial modernization and communications and transportation infrastructure proceeds.

Together, three factors—poor surface water quality, a demand for improvements in such quality, and scarce financial resources—suggest that neither long delays in wastewater treatment nor immediate implementation of a minimum discharge policy is appropriate. If the desire to improve surface water

Table 1. Resources of and Potential Costs in U.S. Dollars to Improve Water Quality in Central and Eastern Europe

County	Population (millions), 1992[1]	GDP (millions of dollars), 1992[1]	Per capita GDP, 1992[1]	Per capita cost to meet European Community water quality standards, 1992[2]	Total debt as percentage of GDP, 1991[1]	Percentage change in industrial production, 1990–1992[1]
Bulgaria	8.47	6,903	815	3,755	not available	−54
Former Czech and Slovak Federal Republic	15.66	36,093	2,305	4,927	27	−40
Hungary	10.30	35,494	3,446	2,116	78	−32
Poland	38.30	72,579	1,895	1,230	61	−32
Romania	23.20	14,152	610	1,422	not available	−54

[1]Figures are from *The Economist* (March 13, 1993).
[2]Figures are from *Der Standard* (June 17, 1993).

Figure 1. Typical river basin.

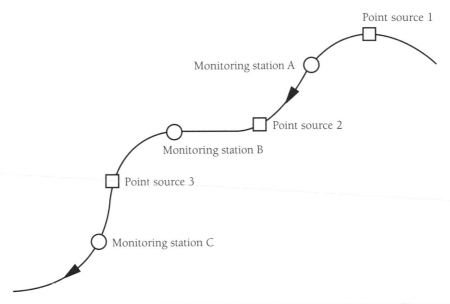

quality and the necessity of minimizing pollution control costs are important factors in decisions made by the governments of Central and Eastern Europe, a policy that attempts to improve water quality cost-effectively would seem to offer a means of realizing the most improvement per dollar invested.

Behavior of Pollutants in River Basins

Since most of Central and Eastern Europe's water supply is drawn from rivers, these bodies of water can be expected to be the primary focus of efforts to improve water quality. In order to understand which such efforts are likely to be cost-effective, it is necessary to take into account two behavioral patterns of pollutants in a typical river basin. To illustrate these patterns, suppose that our typical river basin has three point sources of pollutant discharges and three monitoring stations where water quality is measured, and that point source 1 is located highest upstream, followed further downstream by monitoring station A, point source 2, monitoring station B,

point source 3, and monitoring point C (see Figure 1). The first behavior pattern to consider is that pollutants from each source of discharges into the basin move only in a downstream direction, resulting in higher quality of water upstream and lower quality of water downstream. Thus the quality of water passing by monitoring stations A, B, and C will be affected by pollutants discharged from point source 1, while the quality of water passing by monitoring station A will be affected only by pollutants discharged from point source 1. The second behavioral pattern to consider is that most conventional pollutants—nitrogen and phosphorus, for example—either decay naturally and so are effectively removed from the river as they move downstream or settle out of the water column and become entrained in the sediment of the river bed.

The downstream movement and the natural decay or settling out of conventional pollutants in rivers have several implications for management strategies to enhance water quality. First, even if all point sources of a pollutant discharge the same quantity of the pollutant into our typical river basin and cost the same amount to control (an extremely unlikely circumstance), the relative importance of each point source with respect to improving water quality at the critical monitoring station will differ. If the worst water quality is found at monitoring station A, only the control of discharges from point source 1 would make any contribution to improving water quality. If, on the other hand, the worst quality water is found at monitoring point C, control of discharges from point sources 1, 2, and 3 would contribute to water quality improvements. In the latter case, it is likely that discharges from point source 3 will have far greater effects on water quality at monitoring station C than will discharges from point source 1. Thus the location of point sources makes a difference in the effects of the point sources on water quality at various places in the river basin.

It is particularly important to consider differential effects on water quality due to the location of point sources when the financial resources needed to reduce pollutant discharges are scarce. When this is the case, an analysis of potential pollution control policies that accounts for the location of pollution sources along our typical river basin may be needed to identify the policy that will meet ambient water quality targets in the most cost-efficient way. If water sampled at monitoring stations B and C meets such targets, while water sampled at monitoring station A does not, a policy that attempts to achieve the requisite pollution control at least cost would focus on controlling pollutant discharges from point source 1. In the more likely case that water sampled at monitoring station C has the worst ambient quality, environmental authorities would need information on the relative contributions of all three point sources to water quality degradation, as well as on the relative costs of controlling discharges from each of the sources, in order to construct a policy that meets ambient quality standards at least cost. The basic idea is that the more a point source contributes to environmental degradation, the more it should control its pollutant discharges. Similarly, the less it costs a source to control its discharges, the more the source should control discharges relative to other sources.

Nitra River Basin Case Study

A study I conducted with László Somlyódy of the International Institute for Applied Systems Analysis in Laxenburg, Austria, suggests that Central and Eastern Europe might be able to improve its ambient water quality substantially by considering the relative effects and pollution control costs of point sources of pollutant discharges into river basins, and to do so in a way that would be cheaper than adopting the minimum discharge and best-available technology policies of Western Europe and North America. The study of alternative water quality enhancement policies accounts for the location of each major point source of discharges into the Nitra River basin, which is located in a heavily industrialized area of the Slovak Republic; the pollution control costs of each of these sources; and the effects of each source's discharges on the basin's ambient water quality. It focuses on concentrations of dissolved oxygen, which are often used as a broad measure of the quality of water and the health of aquatic ecosystems, and it considers the effects of three types of policies to increase such concentrations. The first policy is to require point sources to increase the current concentration of dissolved oxygen in the basin by reducing pollutant discharges to the lowest possible level using the best-available pollution control technologies. The second policy is for the region in which the Nitra River basin is located to implement what for it would be the least-cost strategy for increasing the concentration of dissolved oxygen in the basin to 4.0 milligrams per liter (mg/l), a concentration high enough to sustain fish and other forms of aquatic life. The third policy is for the region to implement what would be the least-cost strategy for increasing this concentration to 6.0 mg/l.

A comparison of the costs of each of these three policies reveals that the minimum discharge/best-available technologies (MD/BAT) policy is the most expensive (see Table 2). While this policy would increase the concentration of dissolved oxygen in the Nitra River basin to 6.9 mg/l, it would do so at an annual cost of approximately $14.4 million (U.S.

Table 2. Comparison of Base Case and Alternative Policies to Increase the Concentration of Dissolved Oxygen in the Nitra River Basin

Policy	Minimum concentration of dissolved oxygen (mg/l)	Annual cost (millions of U.S. dollars)	Percentage of cost of MD/BAT policy
Maintain status quo (base case)	0.7	0	not applicable
Minimum discharge/best-available technologies (MD/BAT)	6.9	14.4	100
Regional least-cost (4 mg/l)	4.0	2.8	19
Regional least-cost (6 mg/l)	6.0	6.6	46

dollars). In contrast, the annual cost of each of the least-cost policies is less than half this figure. The least-cost policy to increase the concentration of dissolved oxygen to 6.0 mg/l would entail an annual cost of $6.6 million; the least-cost policy to increase this concentration to 4.0 mg/l would entail an annual cost of only $2.8 million. Both least-cost policies represent a substantial improvement over maintenance of the status quo (the base case), even though the cost of the latter is zero. This is because the currently low concentration of dissolved oxygen in the basin—0.7 mg/l—is likely to be detrimental to many forms of aquatic life.

The above cost comparisons illustrate the likely ratio of cost savings that could be achieved through the use of a least-cost policy to increase concentrations of dissolved oxygen. The 4.0 mg/l concentration could be achieved at less than 20 percent of the cost of the MD/BAT policy, while the 6.0 mg/l concentration could be achieved at less than 50 percent of the cost of this policy. The question that arises is whether similar cost savings would be realized if least-cost pollution control policies were applied to river basins larger than the Nitra River basin. Given the magnitude of potential cleanup costs relative to GDP in Central and Eastern Europe, the answer to this question is doubtless of considerable interest to the region's governments.

Adoption of Least-Cost Pollution Control Policies

Despite the fact that resource economists have been advocating their use for more than two decades, least-cost pollution control policies are the exception rather than the rule in practice. Although the United States has recently adopted one such policy—trading among electric power plants of permits to emit sulfur dioxide—it and many other countries in the West have traditionally made little attempt to design and implement pollution control policies that are efficient in the sense that they will lead to ambient standards being met at the lowest possible cost. There are many reasons why such policies are not promulgated more often. They include technical difficulties in projecting the economic and environmental effects of alternative policies, concerns about whether pollution control costs will be evenly distributed among pollution sources, and the lack of institutions to coordinate management of environmental resources.

Given that cost-effective pollution control policies are not the norm in the West, it might be expected that Central and Eastern European countries would be hesitant to adopt them. However, these countries' severe resource constraints and their institutional flexibility—the result of recent political transformations in the former Soviet bloc—tend to make such policies particularly attractive and potentially easier to implement than in the West. This combination of conditions suggests that Central and Eastern European governments may be more attuned to the arguments of resource economics than Western governments have been to date.

Suggested Reading

IIASA (International Institute for Applied Systems Analysis). 1996. Water Resources: Good to the Last Drop? *Options* (Summer). Available online at http://www.iiasa.ac.at/docs/Admin/INF/OPT/Summer96/water.html.

Novotny, Vladimir, and Laszlo Somlyody, Eds. 1995. *Remediation and Management of Degraded River Basins with Emphasis on Central and Eastern Europe.* NATO ASI Series. Berlin: Springer-Verlag.

Toman, Michael. 1994. *Pollution Abatement Strategies in Central and Eastern Europe.* Washington, DC: Resources for the Future.

IIASA also carried out a research study, Water Quality Management of Degraded River Basins in Central and Eastern Europe, from 1993–96. A description of this study and its publications are available from IIASA's Web site, http://www.iiasa.ac.at; click on Research, and then on Water Resources.

43

Chongqing
A Case Study of Environmental Management during a Period of Rapid Industrial Development

Walter O. Spofford Jr.

An RFF assessment of Chongqing's environmental regulatory framework identified the two national programs most effective in controlling the municipality's air and water pollution—and the flaws that mar them. Although the RFF team identified ways to improve implementation of the two programs in urban and industrial regions, China may be unable to take corrective action immediately, given current economic realities. Breaking the cycle in which environmental degradation follows on industrial development will require foreign investments and loans from multinational development banks.

China began to develop its regulatory framework for environmental management and pollution control just after the first United Nations Conference on Environment in Stockholm in 1972. Today, China's framework is comprehensive and well developed, comprising a vast set of environmental laws, programs, and standards promulgated by the state, provincial, and local governments, all underpinned by the "Environmental Protection Law of the People's Republic of China" (1989), first adopted on a trial basis in 1979.

Despite the framework's comprehensiveness, analyses of environmental programs at the local level reveal several weaknesses, which China's National Environmental Protection Agency in Beijing is beginning to address. These include gaps in program coverage and conflicts between programs, as well as the fact that no current laws control air pollution from domestic sources or water pollution from municipal sewage. A lack of systematic integration of economic development and environmental protection goals is another problem, subtle and pervasive.

Members of an RFF research team found sufficient evidence of these general weaknesses to explain why Chongqing's own environmental regulatory framework is not working as intended. Trends in ambient levels of air and water quality and the city's high levels of air and water pollution make the point as well. (See the box, "Chongqing Profile.")

Originally published in *Resources*, No. 123, Spring 1996.

Environmental Management in Chongqing

In conducting their case study, the RFF team discovered that two of the eight programs established by the state for urban and industrial pollution control were far more effective than the others in controlling air and water pollution in Chongqing. These are environmental management for construction projects—or *Three Synchronizations*, as it is called in China—and the pollution levy system. In addition to singling them out for deserved recognition, the team also discovered weaknesses that prevent even these programs from living up to their promise.

Three Synchronizations

Introduced in Chongqing in 1977, this program's purpose is to ensure that new construction projects include pollution abatement facilities to meet state emission and effluent standards. Under the program, a new industrial enterprise or one that wishes to expand or change its production process must register its plans with the local environmental protection bureau and design (first synchronization), construct (second synchronization), and begin to operate (third synchronization) pollution control facilities simultaneously with the principal part of the enterprise's production activities.

The program has had a significant impact on controlling pollution from new sources in Chongqing. According to the Chongqing Environmental Protection Bureau, 70 percent of all investments in pollution control systems and equipment made by firms in the municipality are the direct result of Three Synchronizations regulations.

Despite these encouraging results, however, Chongqing Municipality has tended recently to let economic development take precedence over environmental protection, eager as the municipality is to catch up to the per capita incomes of coastal cities in China that have experienced dramatic economic growth. Waning enforcement of Three Synchronizations regulations is reflected in the program's low fines for noncompliance. From 1990 to 1993, the amounts charged for violating the third

synchronization requirements at seventy-three industrial enterprises in Chongqing averaged less than the annual salary of a typical worker.

Also problematic is the rapid rise in the number of township and village industrial enterprises (TVIEs) in Chongqing, most of which are engaged in highly polluting activities. TVIEs tend to be small, and because there are so many of them—90,000 in the rural area—it is impossible for Chongqing's environmental protection bureaus to ensure universal monitoring and inspection.

Indeed, a different approach to enforcement is needed. Monitoring and enforcement procedures to control pollution at industrial enterprises in Chongqing's urban area cannot be practically applied to control pollution from TVIEs dispersed over thousands of square kilometers of rural countryside. The compliance rate of these enterprises with Three Synchronizations regulations is no more than 22 percent and perhaps much lower.

Pollution Levy System

Second only to Three Synchronizations in catalyzing investments in industrial pollution controls in Chongqing, the pollution levy system is nevertheless afflicted with problems, too. Introduced in 1980, it consists of a combination of fees levied on industrial enterprises whose pollutants exceed state emission and effluent standards and a series of fines and other charges levied on those who violate system regulations.

In effect, the system offers a carrot and wields a stick to control emissions and effluents. The carrot consists of grants and low-interest loans that industrial enterprises may receive for the construction of industrial pollution control facilities. The stick consists of fees levied for exceeding state emission and effluent standards.

Thus far, the combination of carrot and stick has not been sufficient to induce a high level of compliance with the standards. As under Three Synchronizations, the penalties for noncompliance levied by the system are too low. Set by the central government, the amounts of such fees are far below the marginal costs of operating and maintaining

Chongqing Profile

Chongqing is a severely polluted city of 15 million people located in China's most populous province—Sichuan Province in Southwest China. The central government moved dirty heavy industries there in the 1960s to prepare for an anticipated Soviet invasion.

As China's fifth largest city in terms of industrial output, Chongqing has heavy industry that accounts for 60 percent of the value added in the industrial sector; high-sulfur coal is the principal source of energy. The municipality has approximately 8,000 industrial enterprises in the urban area of which about 1,000 are state-owned and 7,000 are owned collectively.

Based on 1988 data, Chongqing ranked first among twenty-three large cities in China for levels of sulfur dioxide and eighth for levels of total suspended particulate matter. Acid rain, with a pH ranging between 3.5 and 4.5, also is a serious problem. For the past ten to fifteen years, the municipality's growth in real terms has averaged just under 10 percent annually. In terms of income per capita, however, Chongqing lags far behind other major Chinese cities, with less than 30 percent the per capita income of Shanghai and less than 40 percent that of Beijing.

waste treatment facilities and cannot be raised or lowered by local governments.

Soft budget constraints have also had their impact. State-owned industrial enterprises in poor economic health are often allowed to pass pollution levy fees on to the state as deficits to be covered by government subsidies or to escape from payment altogether.

The pollution levy system's carrot has encouraged initiative on the part of industrial enterprises in Chongqing that are not in compliance with emission and effluent standards to the extent that, in order to qualify for a grant or loan from the pollution levy fund, they themselves must finance at least 50 percent of the cost of controlling the pollutants they emit.

The success of the system has its limits, however. The grants and loans available through the fund do not necessarily go to those projects that promise the most cost-effective regional pollution control. This is because the grants and loans are earmarked for industrial enterprises that have paid pollution levy fees and because the amount of a grant or loan to a particular enterprise cannot exceed 80 percent of the total fees the company has paid. Moreover, the pollution levy system funds cannot be used for investments in sewerage systems to collect and treat industrial wastewater, even though such systems are often more cost-effective than "end-of-pipe" treatment.

Intended to give industrial enterprises an economic incentive to comply with emission and effluent standards, China's pollution levy system was designed for a second purpose as well—to raise revenues for the use of the environmental protection bureaus. In Chongqing, this includes the city's own environmental protection bureau as well as some twenty-one other bureaus operating at the municipality's district and county levels.

As required by State Council regulations, Chongqing must set aside 80 percent of the pollution levy fees collected to pay for industrial pollution control systems and equipment. The local environmental protection bureaus may use the remaining 20 percent and all of the fines and charges for regulatory violations to cover the costs of their equipment, supplies, staff training, and public education programs.

Increasingly, however, local environmental protection bureaus have come to rely on these fees to cover far more—up to 90 percent in some cases—of their operating budgets. This distortion of the system reflects the inadequate means of public finance for environmental regulatory agencies in Chongqing, which goes far beyond the environmental regulatory framework per se. Responsibility for raising income for operating budgets has placed fundraising objectives far higher on the priority list of the environmental protection bureaus than their main regulatory responsibility, which is to enforce environmental laws, regulations, and standards.

As it stands, the need to raise revenues for operating budgets has virtually institutionalized the pollution levy system in China—not because the system is efficient or even effective in controlling emissions and effluents, but because it is a principal source of financial support for local environmental protection bureaus.

Study Findings

As the findings of the study suggest, Chongqing might explore ways to make its environmental protection bureaus less dependent on pollution levy fees as their principal source of financial support. The municipal government could provide additional public finance, or the environmental protection bureaus could charge a service fee for monitoring industrial emissions and effluents. Currently, the bureaus perform this service at no charge. Because there will be a shortage of capital for pollution control in Chongqing for several years to come, the Chongqing Environmental Protection Bureau, together with the Chongqing Planning Commission, should address the most critical environmental problems first.

Delegating more power to local governments also would help Chongqing integrate decisions on environment and development. For example, the central government might consider giving local governments the authority to increase pollution levy fees, as well as the authority to decide the most effective way to use the pollution levy fund, given local conditions.

In addition, the Chongqing Environmental Protection Bureau might consider indirect regulatory approaches for controlling emissions and effluents at township and village industrial enterprises, such as taxes on the sulfur and ash contents of fuels, as substitutes for pollution levy fees on emissions of sulfur dioxide and particulate matter. Finally, the Chongqing Municipal Government could increase the size of the staffs of the county environmental protection bureaus to better enforce Three Synchronizations regulations and emission and effluent standards with regard to TVIEs.

The Future

Given current economic and institutional realities, China may be unable to act on all of the findings of the study in the near term. However, as the economy of the region grows, the capital necessary to address some of Chongqing's knottier problems will accumulate and improvements in environmental programs could be implemented in pace with economic and other institutional reforms.

As is the case throughout China, economic development is a precondition for long-term environmental protection in Chongqing. Without the creation of wealth, it will be impossible to restructure and modernize the industrial sector, develop cleaner sources of energy, build needed urban and environmental infrastructure, and clean up existing contaminated sites within the city.

At least in the short run, then, meaningful environmental progress will require a willingness on the part of the international community to support and encourage China's investments in clean technologies and environmental infrastructure before the proceeds of economic development arrive. Such a response could make a decided difference, however. For despite the dire nature of some of China's environmental problems, they are not that much worse overall than what the United States, Europe, and Japan experienced only a few years ago.

Index

Other Materials of Interest from Resources for the Future

On the Web: http://www.rff.org

Our site contains essential information such as original research, press releases, discussion papers, issue briefs, project summaries, downloadable reports, full-text articles from our periodicals, and descriptive material on RFF programs and products.

Books

Economics and Policy Issues in Climate Change
Edited by William D. Nordhaus, Yale University

An extensive IPCC Working Group report published in 1995 examined the economic and social aspects of climate change. In this volume, eminent economists assess that IPCC report and address the questions that emerge from it. The result is an instructive and cogent look at the realities of climate change and some methods (and difficulties) of dealing with them.

1998 • 336 pages
ISBN 0–915707–95–0 hardback, $45.00

Sustainability of Temperate Forests
Roger A. Sedjo, Alberto Goetzl, and Steverson O. Moffat

"This book provides a good overview of sustainable forestry related to temperate zones. It also presents a good, practical discussion of the costs and benefits of moving towards application of sustainability practices in temperate forests. A good supplemental text for courses on forest management and administration, and on forest policy."

Hans M. Gregersen, *University of Minnesota*

1998 • 102 pages
ISBN 0–915707–98–5 paper, $14.95

Pollution Control in the United States: Evaluating the System
J. Clarence Davies and Jan Mazurek

"This is the most thorough and balanced analysis to date of the successes and failures of our pollution control system. Anyone concerned about the environment should read this book to understand why our regulatory system often fails, and how it must be changed if we want better results."

Norman J. Vig, *Carleton College*

1998 • 336 pages
ISBN 0–915707–87–X hardback, $48.00
ISBN 0–915707–88–8 paper, $29.95

Environmental and Resource Economics in the World of the Poor
Partha Dasgupta

An eminent economist and social thinker explains his views on why traditional environmental and resource economics have not met the needs of the developing world. The gaps in wealth are so great,

Dasgupta says, that many basic premises of our analyses are mistaken or irrelevant in other international contexts. A thoughtful contribution to our understanding of poverty.

1998 • 40 pages
ISBN 0–915707–91–8 paper, $9.95

Economic Analyses at EPA: Assessing Regulatory Impact
Edited by Richard D. Morgenstern

"The core of the book is a fascinating set of 12 retrospective case studies by economists who were actually involved in specific analyses... Highly recommended for upper-division undergraduates through faculty and professionals."

Choice

1997 • 496 pages
ISBN 0–915707–83–7 paper, $49.95

Regulating Pollution: Does the U.S. System Work?
J. Clarence Davies and Jan Mazurek

This report concisely describes America's pollution control system. "This report is exactly on target in its criticism of current environmental policies, in the many thoughtful questions posed, and in its well-considered recommendations for policy change."

Michael Kraft, *University of Wisconsin–Green Bay*

1997 • 50 pages
ISBN 0–915707–85–3 paper • $9.95

A Shock to the System: Restructuring America's Electricity Industry
Timothy J. Brennan, Karen L. Palmer, Raymond J. Kopp, Alan J. Krupnick, Vito Stagliano, and Dallas Burtraw

"RFF has succeeded admirably in providing a 138-page discussion of electric power restructuring that clearly and concisely covers the key issues. Every major issue is treated; all the main arguments are presented fairly."

Regulation

1996 • 160 pages
ISBN 0–915707–80–2 paper • $18.95

Readings in Risk
Edited by Theodore S. Glickman and Michael Gough

"Its organization, selection of papers, and concise but provocative introductory essays make it an understandable and desirable resource for a non-technical audience... Has its greatest value as a classroom tool."

Environmental Science and Technology

1990 • 262 pages
ISBN 0–915707–55–1 paper • $24.95